Bacterial Pili: Structure, Synthesis and Role in Disease

————————————

Advances in Molecular and Cellular Microbiology 27

Bacterial Pili: Structure, Synthesis and Role in Disease

Edited by

Michèle A. Barocchi

and

John L. Telford

Novartis Vaccines and Diagnostics, Siena – Italy

www.cabi.org

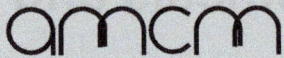 **Advances in Molecular and Cellular Microbiology**

Through the application of molecular and cellular microbiology, we now recognize the diversity and dominance of microbial life forms on our planet, which exist in all environments. These microbes have many important planetary roles, but for us humans a major problem is their ability to colonize our tissues and cause disease. The same techniques of molecular and cellular microbiology have been applied to the problems of human and animal infection during the past two decades and have proved to be immensely powerful tools in elucidating how microorganisms cause human pathology. This series has the aim of providing information on the advances that have been made in the application of molecular and cellular microbiology to specific organisms and the diseases that they cause. The series is edited by researchers active in the application of molecular and cellular microbiology to human disease states. Each volume focuses on a particular aspect of infectious disease and will enable graduate students and researchers to keep up with the rapidly diversifying literature in current microbiological research.

Series Editor

Professor Michael Wilson
University College London

Titles Available from CABI

17. *Helicobacter pylori* in the 21st Century
Edited by Philip Sutton and Hazel M. Mitchell

18. Antimicrobial Peptides: Discovery, Design and Novel Therapeutic Strategies
Edited by Guangshun Wang

19. Stress Response in Pathogenic Bacteria
Edited by Stephen P. Kidd

20. Lyme Disease: an Evidence-based Approach
Edited by John J. Halperin

22. Antimicrobial Drug Discovery: Emerging Strategies
Edited by George Tegos and Eleftherios Mylonakis

24. Bacteriophages in Health and Disease
Edited by Paul Hyman and Stephen T. Abedon

25. The Human Microbiota and Microbiome
Edited by Julian Marchesi

26. Meningitis: Cellular and Molecular Basis
Edited by Myron Christodoulides

27. Bacterial Pili: Structure, Synthesis and Role in Disease
Edited by Michèle A. Barocchi and John L. Telford

Titles Forthcoming from CABI

Microbial Metabolomics
Edited by Silas Villas-Bôas and Katya Ruggiero

Earlier titles in the series are available from Cambridge University Press (www.cup.cam.ac.uk).

CABI is a trading name of CAB International

CABI	CABI
Nosworthy Way	745 Atlantic Avenue
Wallingford	8th Floor
Oxfordshire, OX10 8DE	Boston, MA 02111
UK	USA
Tel: +44 (0)1491 832111	Tel: +1 (617)682-9015
Fax: +44 (0)1491 833508	
E-mail: info@cabi.org	E-mail: cabi-nao@cabi.org
Website: www.cabi.org	

A catalogue record for this book is available from the British Library, London, UK.

The Library of Congress has cataloged the hardcover edition as follows:

Bacterial pili : structure, synthesis, and role in disease / edited by Michèle A. Barocchi and John L. Telford.
        p. ; cm. -- (Advances in molecular and cellular microbiology ; no. 27)
    Includes bibliographical references and index.
    ISBN 978-1-78064-255-0 (hbk)
    I. Barocchi, Michèle A., editor of compilation. II. Telford, John L., editor of compilation. III. C.A.B. International, issuing body. IV. Series: Advances in molecular and cellular microbiology ; 27.
    [DNLM: 1. Fimbriae, Bacterial--classification. 2. Fimbriae, Bacterial--physiology. QW 51]

    QR342
    579.2'6--dc23

2013025141

ISBN-13: 978 1 78639 528 3 (PB)

Commissioning editor: Rachel Cutts
Editorial assistant: Emma McCann
Production editor: Simon Hill

Typeset by Columns XML Ltd.
Printed and bound in the UK by CPI Group (UK) Ltd, Croydon, CR0 4YY.
First printed in hardback in 2014. Transferred to POD paperback in 2019.

Contents

Contributors

Edward N. Baker, School of Biological Sciences, University of Auckland, 3A Symonds Street, Auckland 1010, New Zealand. E-mail: ted.baker@auckland.ac.nz

Michèle A. Barocchi, Novartis Vaccines and Diagnostics, Via Fiorentina 1, Siena, Italy 53100. E-mail: Michele.barocchi@novartis.com

Lori L. Burrows, Department of Biochemistry and Biomedical Sciences, McMaster University, Hamilton, Ontario, Canada L8S 4K1. E-mail: burrowl@mcmaster.ca

Peter J. Christie, Department of Microbiology and Molecular Genetics, University of Texas Medical School at Houston, Houston, TX 77030, USA. E-mail: Peter.J.Christie@uth.tmc.edu

Roberta Cozzi, Novartis Vaccines and Diagnostics, Via Fiorentina 1, Siena, Italy 53100. E-mail: roberta.cozzi@novartis.com

Lisa Craig, Department of Molecular Biology and Biochemistry, Simon Fraser University, Burnaby, British Columbia, Canada V5A 1S6. E-mail: licraig@sfu.ca

Haley Echlin, Departments of Pediatric Dentistry and Microbiology, University of Alabama at Birmingham, Schools of Dentistry and Medicine, Birmingham, AL 35294, USA. E-mail: hechlin@olemiss.edu

Michaella Georgiadou, MRC Centre for Molecular Bacteriology and Infection, Section of Microbiology, Imperial College London, London, UK. E-mail: m.georgiadou09@imperial.ac.uk

P. Lynne Howell, Program in Molecular Structure and Function, Hospital for Sick Children, Toronto, Ontario, Canada M5G 1X8. E-mail: howell@sickkids.ca

I-Hsiu Huang, Department of Microbiology and Molecular Genetics, University of Texas Health Science Center at Houston, Houston, TX 77030, USA; Institute of Basic Medical Sciences and Department of Microbiology and Immunology, College of Medicine, National Cheng Kung University, Tainan, Taiwan. E-mail: ihsiuhuang@mail.ncku.edu.tw

Domenico Maione, Novartis Vaccines and Diagnostics, Via Fiorentina 1, Siena, Italy 53100. E-mail: domenico.maione@novartis.com

Andrea G.O. Manetti, Novartis Vaccines and Diagnostics, Via Fiorentina 1, Siena, Italy 53100. E-mail: andrea.manetti@novartis.com

Monica Moschioni, Novartis Vaccines and Diagnostics, Via Fiorentina 1, Siena, Italy 53100. E-mail: daniela.rinaudo@novartis.com

Matthew Mulvey, Division of Microbiology and Immunology, Pathology Department, University of Utah, Salt Lake City, UT 84112, USA. E-mail: Matthew.Mulvey@path.utah.edu

Neil G. Paterson, Diamond Light Source Ltd., Didcot OX11 0DE, UK. E-mail: Neil.paterson@diamond.ac.uk

Vladimir Pelicic, MRC Centre for Molecular Bacteriology and Infection, Section of Microbiology, Imperial College London, London, UK. E-mail: v.pelicic@imperial.ac.uk

Gilles Phan, Institute of Structural and Molecular Biology, University College London and Birkbeck College, Malet Street, London WC1E 7HX, UK. E-mail: g.phan@ucl.ac.uk

Daniela Rinaudo, Novartis Vaccines and Diagnostics, Via Fiorentina 1, Siena, Italy 53100. E-mail: daniela.rinaudo@novartis.com

Colin Russell, Division of Microbiology and Immunology, Pathology Department, University of Utah, Salt Lake City, UT 84112, USA. E-mail: Colin.Russell@path.utah.edu

June R. Scott, Microbiology and Immunology, Emory University School of Medicine, Atlanta, GA 30329, USA. E-mail: scott@microbio.emory.edu

Tiziana Spadafina, E-mail: tspadafina@gmail.com

Stephanie Tammam, Program in Molecular Structure and Function, Hospital for Sick Children, Toronto, Ontario, Canada M5G 1X8. E-mail: stephanie.tammam@utoronto.ca

Ronald K. Taylor, Department of Microbiology and Immunology, Geisel School of Medicine at Dartmouth, Hanover, NH 03755, USA. E-mail: Ronald.K.Taylor@dartmouth.edu

John L. Telford, Novartis Vaccines and Diagnostics, Via Fiorentina 1, Siena, Italy 53100. E-mail: john.telford@novartis.com

Hung Ton-That, The University of Texas Medical School at Houston, Department of Microbiology and Molecular Genetics, 6431 Fannin Street, MSE R213, Houston, TX 77030, USA. E-mail: Ton-That.Hung@uth.tmc.edu

Gabriel Waksman, Institute of Structural and Molecular Biology, University College London and Birkbeck College, Malet Street, London WC1E 7HX, UK. E-mail: g.waksman@ucl.ac.uk or g.waksman@bbk.ac.uk

Hui Wu, Departments of Pediatric Dentistry and Microbiology, University of Alabama at Birmingham, Schools of Dentistry and Medicine, Birmingham, AL 35294, USA. E-mail: hwu@olemiss.edu

Foreword

Long filamentous structures on the surface of Gram-negative bacteria were first described more than 50 years ago. These structures are called fimbriae (from the Latin word for fringe) or pili (from the Latin for hair) and were recognized as clearly distinguishable from the filamentous flagellae that power bacterial motility. Both terms, fimbriae and pili, are still in use and are essentially interchangeable. In subsequent decades, the structure, biosynthesis and function of these surface organelles have been extensively studied. More recently, pili on the surface of Gram-positive bacteria have been described. In contrast to the pili commonly found on Gram-negative bacteria that are formed by protein–protein interactions between the pilin subunits, pili present in Gram-positive bacteria are generally formed by enzyme-catalysed covalent isopeptide bond formation between the subunits, leading to long, rigid hair-like structures.

Despite the various architectures found in different pili, all share a similar role in adhesion to host tissues and/or other bacteria. In fact, a common feature of pili is the presence of an adhesive tip protein attached to the end of the backbone formed by the structural pilins. Hence these structures are important in host colonization and biofilm formation. Furthermore, some pili have additional functions such as the transfer of genetic material and effector proteins or a type of twitching motility caused by the dynamic polymerization and de-polymerization leading to extension and contraction of the structures. The importance of these functions for bacterial pathogenesis has led to attempts to develop vaccines that induce immune responses targeting the structures that may interfere with the infectious process.

Over the past five decades, several distinct pilus types have been identified, most of which were described and characterized in Gram-negative bacteria. The best characterized of these cell-surface organelles are type I pili (expressed by enteropathogenic *Escherichia coli*), type IV pili (expressed by *E. coli*, and *Pseudomonas* and *Neisseria* species) and curli pili (expressed by some strains of *E. coli*). The first six chapters are about pili in Gram-negative bacteria and review the contribution of many scientists, beginning with the first experiments by Lederberg and Tatum in the late 1940s with their discovery of bacterial conjugation (even though at the time they were still unaware of the existence of pili). Chapters 7–13, focus on the discovery of pili in Gram-positive organisms, starting with *Corynebacterium diphtheria* and then focusing on Streptococci. Areas covered include the biogenesis of pili, their unusual structural characteristics that include inter- and intra-molecular isopeptide bonds and the role of pili in disease processes.

This book is written for anyone in the field of microbiology or those with an interest in bacterial pathogenesis and infectious diseases, and their application to therapeutics. The authors of the book are experts and cover a wide range of microorganisms, although it would be impossible to cover them all. Each chapter is a complete synthesis of current research and central ideas that are affecting the field of pilus research today. As with all scientific research, each discovery is made by the contribution of all those who came before that discovery. In this collection of expertise on the subject, we would like to acknowledge all of the pioneers of bacterial genetics, classical microbiologists and those involved in bacterial pathogenesis who have contributed to our understanding of these complex and intricate machines that support so many vital bacterial processes, such as DNA exchange, evolution, secretion, and adherence and colonization.

Michèle A. Barocchi
John L. Telford
June 2013

1 The *Vibrio cholerae* Toxin Coregulated Pilus: Structure, Assembly and Function with Implications for Vaccine Design

Lisa Craig[1] and Ronald K. Taylor[2]

[1]*Department of Molecular Biology and Biochemistry, Simon Fraser University, Burnaby, Canada* [2]*Department of Microbiology and Immunology, Geisel School of Medicine at Dartmouth, Hanover, USA*

1.1 Introduction

The aquatic Gram-negative bacterium *Vibrio cholerae* causes the deadly gastrointestinal disease cholera, which has wreaked havoc on civilizations throughout history. Seven cholera pandemics have been recorded since the 1800s, the most recent of which affects millions of people per year, with more than 100,000 deaths (Harris *et al.*, 2012). Cholera devastated parts of London the 18th century and was a driving force in the introduction of public health programmes and policy after John Snow, widely considered the father of modern epidemiology, applied statistical mapping methods to track deaths from cholera, ultimately linking them to specific drinking water supplies. *V. cholerae* most commonly enters the body through ingestion of contaminated water. It colonizes the small intestine without invading the intestinal epithelial cells, establishes microcolonies, reproduces and releases cholera toxin, an ADP-ribosylating enzyme that enters the epithelial cells and stimulates constitutive activation of adenylate cyclase. This causes water and ion channels in the epithelial cells and surrounding tissues to open, resulting in a massive efflux of fluids and electrolytes in the form of voluminous 'rice-water stools' that are characteristic of cholera disease (Kaper *et al.*, 1995). An afflicted person can lose as much as 20 litres of water in a single day, which rapidly leads to death from dehydration and organ failure. Not all *V. cholerae* cause cholera disease – there are hundreds of harmless *V. cholerae* serogroups living in salt and fresh water estuaries and in the ocean. However, two serogroups, O1 and O139, classified by their lipopolysaccharide O antigens, acquired a set of genetic elements, endowing them with the ability to colonize the human intestine and produce cholera toxin. These features allow *V. cholerae* O1 and O139 to exploit the human host, attaching in protective niches in the gut and reproducing in the tens of millions, and then inducing the host to expel their massive numbers into the environment to prey upon new human hosts.

The first genetic element pathogenic *V. cholerae* acquired was the Vibrio pathogenicity island 1 (VPI-1), which includes a large *tcp* operon encoding all of the proteins necessary to assemble a type IV pilus called the toxin coregulated pilus (TCP) (Taylor *et al.*, 1987).

These hair-like filaments are several microns in length but less than 10 nm in diameter (Li et al., 2012). They are displayed on the V. cholerae surface where they self-associate to hold the bacteria in microcolonies – tight bacterial aggregates that protect them from host defences (Kirn et al., 2000; Lim et al., 2010; Krebs and Taylor, 2011). This feature is essential for V. cholerae to survive in the human gut. But TCP serve another purpose that has been integral to the evolution of toxigenic V. cholerae: they are the primary receptors for a filamentous bacteriophage, CTXφ (Waldor and Mekalanos, 1996). This phage attaches to the pili and is somehow brought into the cell where its single-stranded DNA is duplicated. The CTXφ phage genome represents the second genetic element, which integrates into and is replicated with the large chromosome, or sometimes both chromosomes, of V. cholerae O1 and O139. Phage proteins are synthesized along with the V. cholerae proteins. Two of these phage proteins are not required for phage assembly. They are the cholera toxin A and B subunits, which assemble into an AB_5 toxin that binds to intestinal epithelial cells via the B subunit pentamer and ADP-ribosylates the cellular adenlyate cyclase via the catalytic A subunit. Thus, V. cholerae became an important human pathogen by horizontally acquiring the virulence factors TCP and cholera toxin. Importantly, expression of both virulence factors is controlled by a single transcriptional activator, ToxT, which is why TCP are called toxin coregulated pili. A third, less well-understood virulence factor is TcpF, a soluble protein encoded on the tcp operon and secreted by the TCP biogenesis apparatus. TcpF is critical for V. cholerae colonization in the infant mouse infection model but its mode of action is not known (Kirn et al., 2003; Megli et al., 2011). Here we present our current understanding of V. cholerae TCP with respect to its gene arrangement, subunit structure and filament architecture. We will present a theoretical mechanism by which these pili are assembled from thousands of copies of a single TcpA pilin subunit into a multifunctional helical filament that is critical for V. cholerae pathogenesis, and explain how dynamic assembly and disassembly may be intricately linked with pilus functions in

microcolony formation, TcpF secretion and CTXφ uptake. Finally, we will discuss progress towards TCP- and TcpF-based multicomponent cholera vaccines.

1.2 Evolution of TCP-positive V. cholerae

The expression of the genes that encode components for TCP biogenesis and function as well as those that encode the cholera toxin are coordinately controlled in response to many environmental signals that act on regulators encoded in both the ancestral genome and VPI-1 to orchestrate a regulatory cascade that exquisitely optimizes virulence factor expression in the small intestine (for a comprehensive review see Skorupski and Taylor, 2013). The genes encoding all of the components necessary for TCP assembly are clustered in a single operon, the tcp operon (Fig. 1.1A). These genes encode proteins that are directly involved in pilus biogenesis (the major pilin protein TcpA, outer membrane proteins TcpQ and TcpC, inner membrane accessory proteins TcpR and TcpD, periplasmic protein TcpS, the assembly ATPase TcpT, the inner membrane core protein TcpE and the prepilin signal peptidase TcpJ) as well as proteins of unknown function, which are not required for TCP assembly (the minor pilin TcpB and the secreted colonization factor TcpF). The master regulator of pilus and toxin gene expression, ToxT, is also encoded within the operon. The tcp operon is very similar in nucleotide sequence and gene synteny to two type IV pilus operons in enterotoxigenic E. coli (ETEC), cof and lng, which encode the CFA/III and longus pilus, respectively, and less similar to the bfp operon encoding the bundle forming pilus of enteropathogenic E. coli (EPEC) (Fig. 1.1A). All four pili belong to a subclass of type IV pili, the type IVb pili, which are present on enteric pathogens. Type IVb pili are classified based on the primary sequence of their major pilin. These pilins have a long signal sequence (in some cases 25 or more residues) and a variable amino acid at position 1 of the mature protein. In contrast, the type IVa pilins have a short signal peptide (6–8 residues) and an invariant phenylalanine at position 1. Type IVa pili are found on a diverse range of bacteria, including the pathogenic Neisseria, *Pseudomonas aeruginosa* and *Francisella tularensis*.

Genes encoding type IV pili are also present in Gram-positive bacteria, with the *Clostridium perfringens* pili being the most well-characterized (Varga *et al.*, 2006; Rodgers *et al.*, 2011). Type IV pilus genes are ubiquitous among Clostridia (Melville and Craig, 2013), one of the most ancient types of Eubacteria, suggesting they may have been acquired by Gram-negative bacteria by horizontal gene transfer. Although the gene synteny differs between the clostridial and *V. cholerae* type IV pilus operons, these systems are among the few examples in which all of the genes necessary for assembling type IV pili are encoded within a single cluster. In contrast, genes encoding type IVa pili are distributed in small clusters on distant regions of the bacterial chromosome (Pelicic, 2008). The relative simplicity of the *tcp* operon, with its small number of assembly proteins (nine, compared with 40 or more for some type IVa pili; Ayers *et al.*, 2010) suggests that it may be one of the most primitive type IV pilus systems in Gram-negative bacteria. The type IV pilus systems of *V. cholerae* and ETEC are unique in having only a single minor pilin and no retraction ATPase. In contrast, most other systems, including Clostridia and the closely related type IVb bundle forming pilus of EPEC, have multiple minor pilins and possess a retraction ATPase (Fig. 1.1A, B). Thus, the *V. cholerae* and ETEC systems represent the simplest and possibly most tractable systems in which to study type IV pilus assembly and functions.

1.3 TcpA, the Major Pilin of the Toxin Coregulated Pilus

The TCP building block is the pilin subunit, TcpA, a small protein of 199 amino acids with a molecular mass of ~20 kDa. TcpA are synthesized in the cytoplasm but presumably adjacent to the inner membrane as they possess no chaperone protein to escort them to the membrane. They are synthesized with a positively charged 26-residue signal peptide

that looks nothing like the short hydrophobic sequences recognized by the Sec apparatus. Indeed, this N-terminal peptide is cleaved after a glycine at the −1 position by a dedicated inner membrane signal peptidase, TcpJ, also encoded in the *tcp* operon (Fig. 1.1). The N-terminal Met1 of the mature protein is part of a 25-amino acid segment that is hydrophobic, with the exception of a glutamate at position 5. These N-terminal features – a positively charged signal peptide with Gly(−1), a 25-residue hydrophobic segment with a Glu5 for the mature protein – are characteristic of all type IV pilin proteins. In addition, type IV pilins are N-methylated on the N-terminal residue of the mature protein, a post-translational modification that is catalysed by the dual-function prepilin peptidase (Strom *et al.*, 1993). These characteristics are also shared with the pseudopilins of the type II secretion system, which is responsible for exporting a variety of hydrolases and toxins, including cholera toxin, from the periplasm across the outer membrane. Pseudopilins do not form surface-displayed filaments under normal expression conditions, but instead are thought to form short 'pseudopili' that act as pistons to extrude substrate (Hobbs and Mattick, 1993). The type IV pili and type II secretion pseudopili utilize a homologous set of biogenesis components and appear to have similar filament architectures. As further evidence of their relatedness, some type IV pilus systems have a secretory function (Kirn *et al.*, 2003; Kennan *et al.*, 2001), and T2S systems can be induced to form surface-displayed pseudopili when their major pseudopilin is overexpressed (Sauvonnet *et al.*, 2000; Vignon *et al.*, 2003).

The *V. cholerae* prepilin peptidase, TcpJ, is an aspartic acid protease with a predicted 8-transmembrane topology and an active site near the cytoplasmic side of the inner membrane (LaPointe and Taylor, 2000). Removal of the signal peptide from TcpA leaves it anchored in the membrane via its N-terminal hydrophobic segment, with its C-terminal ~170 amino acids exposed to the periplasm. The x-ray crystal structure of the periplasmic portion of TcpA (residues 29–199) reveals a globular domain comprised of

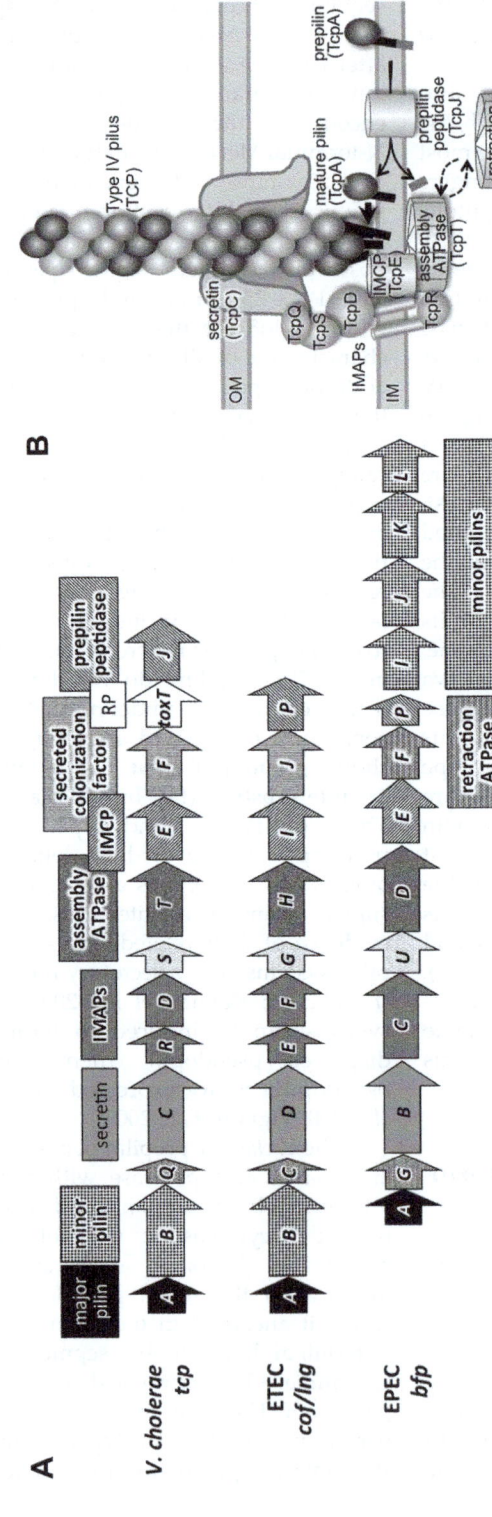

Fig. 1.1. Type IVb pilus operons and schematic of the Type IV pilus assembly apparatus. (A) Organization of the Type IVb pilus operons for *V. cholerae* TCP, ETEC CFA/III and longus, and EPEC bfp. Genes with the same pattern are homologous in sequence and their protein products have similar cellular localization and functions. The exception is *V. cholerae* TcpF and ETEC CofJ, which have no sequence or structural similarity but are both secreted by their respective Type IV pilus systems (Kern *et al.*, 2003, Yenn *et al.*, 2013). EPEC has a gene encoding a retraction ATPase syntenic to *tcpF* and *cofJ/lngJ* genes. IMAPs, inner membrane accessory proteins; IMCP, inner membrane core protein; RP, regulatory protein. The major pilin forms the pilus filament. The minor pilins have the conserved hydrophobic N-terminus seen in the major pilin but their functions have yet to be established. (B) Schematic of the core proteins conserved in Type IV pilus systems. TCP and the ETEC Type IVb pili do not possess a retraction ATPase. Protein names are provide for the TCP system.

an α-helical spine (α1C) and a twisted anti-parallel β-sheet that packs against this helix for three of its five strands (Craig *et al.*, 2003; Fig. 1.2A). This α-helix/β-sheet core is seen in all type IV pilin structures solved thus far. The segment between α1C and the β-sheet, called the αβ-loop, is an extended loop with a central 4-turn α-helix, α2, that crosses over α1C at right angles on one edge of the globular domain. The first two strands of the β-sheet are anti-parallel, after which the polypeptide chain exits the β-sheet to form another 4-turn α-helix, α3, that packs against both the β-sheet and α1C and then feeds into β3 which is followed by a meandering loop with a 1.5-turn α-helix at the edge of the globular domain opposite the αβ-loop. This loop is stabilized by a disulfide bond between Cys186 near its C-terminus and Cys120 in the first turn of α3. Finally, the chain re-enters the β-sheet as its central strand, β5, which is the most C-terminal segment of the protein. The topology of the TcpA globular domain is typical of the type IVb pilins and differs from the type IVa pilins, which have nearest-neighbour connectivity for the globular domain β-sheet, followed by a conserved C-terminal loop that is stabilized by a disulfide bond. The disulfide bond is a conserved feature of most type IV pilins, but its intervening segment, called the D-region, differs substantially in length and amino acid composition between the two pilin subtypes. Crystal structures of full-length type IVa pilins show that the N-terminal segment, which is conserved among all type IV pilin is an extended α-helix, α1N, that is continuous with α1C but protrudes from the globular domain. This segment has been modelled onto the TcpA globular domain crystal structure in Fig. 1.2A using the x-ray coordinates for Va1N in the *P. aeruginosa* PAK pilin structure (Craig *et al.*, 2003). This hydrophobic N-terminal 'stalk' anchors the pilin subunit in the inner membrane prior to pilus assembly, but also serves as the polymerization domain, holding subunits together in the helical pilus filament (Fig. 1.2A, B).

TCP is only one of two type IV pilus structures that have been determined by electron microscopy, which provides a medium-resolution molecular envelope in which to dock atomic resolution crystal structures of the pilin subunits, giving a 'pseudoatomic resolution' filament structure. The 20 Å resolution TCP reconstruction shows a helical filament in which the N-terminal α-helices of each of the pilin subunits taper to form a solid filament core and the globular domains are loosely packed on the filament surface (Fig. 1.2B, C) (Li *et al.*, 2012). Pilin subunits are related by an axial rise of 8.5 Å and an azimuthal rotation of 96.8°, which places the conserved Glu5 of each subunit in a position to neutralize the positively charged N-terminal amine (N1+) of the neighbouring subunit (Fig. 1.2D). Although TcpA has a larger globular domain and different topology than that of PilE, the type IVa pilin subunit from *Neisseria gonorrhoeae*, their pilus architectures are very similar. PilE is also arranged with its N-terminus in the gonococcus (GC) pilus filament core, positioned such that Glu5 can neutralize N1+ of its adjacent subunit, and the globular domains form the outer shell of the filament (Craig *et al.*, 2006). The helical symmetry of the VGC pilus filament is similar to that of TCP, with a subunit rise of 10.5 Å and a rotation of 100.5°. This common architecture for the type IVa and IVb pili buries the conserved N-terminal poly-merization domain while exposing variable regions that are directly involved in pilus functions. Subunits in both filament types are held together primarily by hydrophobic inter-actions among their N-terminal α-helices, forming remarkably stable filaments that are resistant to proteolysis, heat and chemical denaturation, yet a single substitution at Glu5 prevents pilus assembly (Aas *et al.*, 2007; Li *et al.*, 2012), illustrating the key role played by this residue.

1.4 Assembly of Pilus Filaments

The TCP biogenesis apparatus is among the simplest of all type IV pilus assembly machineries, with only nine proteins that are directly involved in building the pilus filament (Heidelberg *et al.*, 2000). Apart from the major pilin, TcpA, there are several key

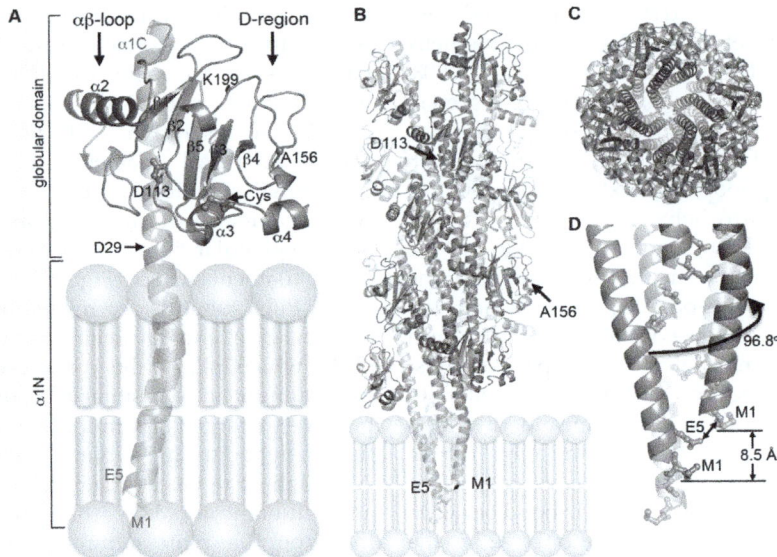

Fig. 1.2. Structure of the TcpA pilin subunit and pilus filament and their predicted orientation in the inner membrane. (A) Crystal structure of the TcpA globular domain (residues 29–199, PDB ID 1OQV) with the N-terminal 28 residues (α1N) modelled based on the x-ray coordinates of PAK pilin (1OQW). The side chains of residues Met1, Glu5, Asp113, Ala156 and the disulfide-bonded cysteines Cys120 and Cys186 are shown in ball-and-stick representation. The position of the phospholipid bilayer is estimated. (B) TCP pseudoatomic resolution structure based on the EM reconstruction (EMDB ID 1955), shown as if growing from the inner membrane, with the bottom-most subunit being the last subunit added. (C) Top view of the TCP filament to show the packing of N-terminal α-helices in the filament core. (D) Close-up of the N-terminal α-helices in the TCP filament to show the putative salt bridge between the Glu5 side chain and the amino nitrogen on Met1(N1+) of the next subunit up in the one-start helix of the filament. The symmetry parameters (96° rotation, 8.5 Å axial rise) that relate each subunit to the next in the one-start helix are shown.

assembly components that are conserved across both type IV pilus and type II secretion systems. The inner membrane assembly platform is comprised of an assembly ATPase, TcpT, a polytopic integral inner membrane core protein (IMCP), TcpE, and two inner membrane accessory proteins (IMAPs), TcpD and TcpR, each with a single membrane-spanning segment (Fig. 1.1B). The assembly ATPase, TcpT, belongs to a superfamily of secretion NTPases (Planet *et al.*, 2001). These homohexameric complexes are located on the cytoplasmic side of the inner membrane, recruited by an IMAP, and undergo a conformational change upon binding to and hydrolysing ATP (Robien *et al.*, 2003; Abendroth *et al.*, 2005; Satyshur *et al.*, 2007; Yamagata and Tainer, 2007; Misic *et al.*, 2010), that somehow facilitates type IV pilus

assembly on the periplasmic side of the membrane. TcpT associates with the inner membrane platform by binding via its N-terminal segment to the N-terminal domain of the IMAP TcpR, which is anchored to the membrane via its C-terminal segment (Tripathi and Taylor, 2007). The role of the IMCP, TcpE, is not known. Its central position in the inner membrane platform suggests that it may act as a relay, transmitting the conformational change of the ATPase upon ATP hydrolysis to extrude the growing pilus incrementally out of the inner membrane with each subunit addition (Craig *et al.*, 2006). The second IMAP, TcpD, has an N-terminal transmembrane segment and a C-terminal cytoplasmic domain. TcpD may interact with periplasmic proteins TcpS and the outer membrane associated protein TcpQ, acting as a linker between the inner membrane platform and the outer membrane secretin channel, TcpC, similar to the PilM/PilN/PilO and PilP complexes of the Neisseria, and Thermus thermophilus Pseudomonas, T4P systems (Tammam *et al.*, 2011; Berry *et al.*, 2012; Karuppiah *et al.*, 2013) and GspC/GspM/GspL of the T2S system (Korotkov *et al.*, 2006). Secretin channels are homomultimeric gated channels with a periplasmic vestibule and an outer membrane β-barrel through which the type IV pilus filaments grow and, in the case of the secretion systems, substrates pass (Reichow *et al.*, 2010; Berry *et al.*, 2012). *V. cholerae* TcpQ is required for localization and assembly of the secretin, TcpC (Bose and Taylor, 2005).

While many of the components of the type IV pilus and type II secretion systems have been characterized, the assembly mechanism remains poorly understood. Any assembly model must explain rapid filament assembly as well as disassembly, a process that occurs in many type IV pilus systems, in which surface-displayed pili are retracted via a second, 'retraction ATPase'. Pilus retraction occurs rapidly at a rate of ~1000 subunits per second, and facilitates such functions as DNA uptake and twitching motility, a specialized form of bacterial motility on semi-solid surfaces (Merz *et al.*, 2000; Graupner *et al.*, 2001; Skerker and Berg, 2001; Aas *et al.*, 2002; Maier *et al.*, 2002). Neither *V. cholerae* TCP nor the type II secretion systems appear to possess a retraction ATPase, although the type II pseudopilus must cycle rapidly between assembly and retraction if the piston model is correct. Thus, filament assembly appears to be a dynamic and reversible process. We propose a model whereby pilus assembly is driven in part by electrostatic complementarity between the negatively-charged Glu5, which is exposed to the acyl phase of the lipid bilayer in the membrane-embedded pilins (Fig. 1.2), and the positively charged amine on the N-terminal residue (N1+) of the terminal subunit in the growing pilus filament (Fig. 1.3). This charge complementarity would induce a pilin subunit, diffusing in the inner membrane, to dock into a gap at the base of the growing pilus filament. Next, ATP hydrolysis and release of

ADP by the assembly ATPase would induce a conformational change that is transmitted to the IMCP, driving the nascent filament, together with its newly docked pilin subunit, out of the membrane a short distance, equivalent to the rise between subunits of 8–10 Å. This extrusion step would prevent the newly added pilin subunit from diffusing back into the membrane and would open a new gap ~120° around the base of the filament for a new subunit to dock, stabilizing the extruded state and allowing the IMCP to collapse back to its resting state, ready for the next cycle. Thus, we envision pilus assembly as a processive process that can only proceed provided the system is continually supplied with pilin subunits and ATP. Should either of these supplied be interrupted, the filament would collapse one subunit at a time into the membrane, resulting in pilus retraction. So, according to our assembly model, filament retraction could occur even in the absence of a retraction ATPase.

1.5 TCP Functions

The most well-characterized function of TCP in *V. cholerae* colonization is microcolony formation. Microcolonies are clusters of many thousands of bacteria that form at an infection site. They differ from biofilms in that the bacteria within the colony are undifferentiated – they are essentially all viable, but those that are deep within the aggregates are protected from host defences. *V. cholerae* aggregation into microcolonies is mediated by TCP self-association. TCP are critical for *V. cholerae* colonization of human volunteers and the infant mouse, which is used as an infection model, and are required for 'auto-agglutination', an *in vitro* activity in which *V. cholerae* grown under TCP-expressing conditions form macroscopic aggregates that sink to the bottom of a tube in overnight cultures (Taylor *et al.*, 1987; Herrington *et al.*, 1988; Kirn *et al.*, 2000). TCP-mediated *V. cholerae* aggregation occurs due to stereo-chemical complementarity between pilus surfaces. Single amino acid changes can disrupt pilus:pilus interactions without affecting other pilus functions (Taylor *et al.*,

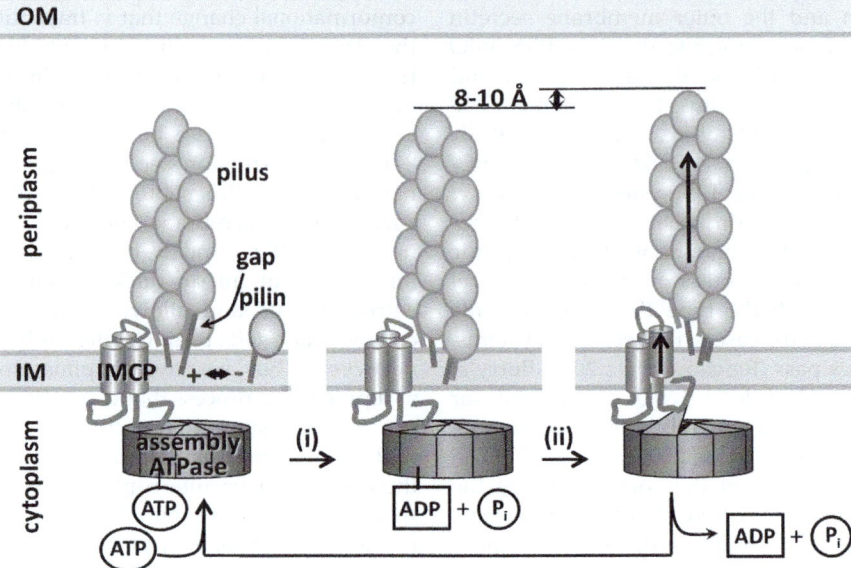

Fig. 1.3. Model for Type IV pilus assembly. In step (i) of the cycle, a single pilin subunit docks into a gap in the growing pilus filament, attracted by complementarity between its negatively-charged Glu5 and the positively-charged main chain amine on the N-terminal residue (N1+) of the terminal pilin subunit in the growing filament. ATP binds to the assembly ATPase and is hydrolysed to ADP and inorganic phosphate (Pi). In step (ii), ATP is released by the assembly ATPase, inducing a conformational change that is transmitted to the inner membrane core protein (IMCP), which in turn extrudes the pilus filament a short distance out of the membrane, opening up a new gap ~120° around the base of the filament for a new subunit to dock. Only one of three predicted IMCPs is shown. IM, inner membrane; OM, outer membrane. (Modified from *Molecular Cell*, Vol. 23, Lisa Craig, Niels Volkmann, Andrew S. Arvai, Michael E. Pique, Mark Yeager, Edward H. Egelman and John A. Tainer, Type IV Pilus Structure by Cryo-Electron Microscopy and Crystallography: Implications for Pilus Assembly and Functions, Pages 651–662, 2006, with permission from Elsevier.)

1987; Kirn *et al.*, 2000; Lim *et al.*, 2010). Charged residues located on the TcpA D-region (Fig. 1.2A), which protrudes from the TCP surface, have been implicated in pilus:pilus interactions, as have residues in the cavity between the subunits. D-region residue Ala156 and cavity residue Asp113 of *V. cholerae* strain O395 of the classical biotype are particularly important in these inter-actions. We have proposed that pili interact by intercalation of the protruding D-regions into the cavities of adjacent pili (Lim *et al.*, 2010).

A second function of TCP is as a secretion organelle. TCP are required for *V. cholerae* to export TcpF, a 37 kDa protein of unknown function. TcpF is essential for *V. cholerae*

colonization of the infant mouse, as a *V. cholerae tcpF* deletion strain has a five-log decrease in colonization index relative to the wild-type parental strain similar to *tcpA* mutant (Kirn *et al.*, 2003). The wild-type strain is unable to rescue the Δ*tcpF* the phenotype of a mutant in coinfection assays, demonstrating that TcpF does not act *in trans*. Yet TcpF also does not appear to associate directly with the pili. TcpF is encoded in the *tcp* operon but is not required for TCP assembly or for auto-agglutination. TcpF has no identifiable amino acid sequence homology with other known proteins. The 2.6 Å crystal structure of TcpF reveals a bilobed protein with a flexible linker and a pronounced cleft but shows little structural homology to other proteins of known structure, which might have provided clues to its functions in *V. cholerae* colonization (Megli *et al.*, 2011). TcpF is synthesized in the cytoplasm and transported into the periplasm via the Sec secretion apparatus. Transport across the outer membrane into the extra-cellular space requires a functional TCP (Kirn *et al.*, 2003). TcpF is presumably extruded through the outer membrane secretin, TcpC, via a piston-like motion of TCP, as is proposed for the type II secretion system. Importantly, antibodies against TcpF are protective when co-administered to mouse pups along with *V. cholerae* or when TcpF is used as an im-munogen in combination with the cholera toxin B subunit, demonstrating its potential value in a multisubunit vaccine (Kirn *et al.*, 2003; Megli *et al.*, 2011; Muse *et al.*, 2012; Price and Holmes, 2012).

A third activity of TCP does not feature directly in colonization, but was critical to the evolution of *V. cholerae* as a deadly human pathogen. TCP is the primary receptor for the filamentous bacteriophage CTXϕ, which forms a lysogen (Waldor and Mekalanos, 1996). The CTXϕ phage carries within its genome the genes encoding the two cholera toxin subunits, *ctxA* and *ctxB*. CTXϕ binds initially to TCP, followed by a second inter-action with TolA in the *V. cholerae* periplasm, whereupon the phage uncoats and injects its circular single-stranded genome into the cytoplasm. A complementary copy of the phage genome is synthesized and this double-stranded DNA integrates into the large or sometimes both the large and small chromosomes of the bacterium to form a lysogen (Waldor and Friedman, 2005; Kimsey and Waldor, 2009). Under optimal conditions, such as those in the human or infant mouse intestine, the cholera toxin A and B subunits are synthesized from the CTXϕ element and the TCP proteins are expressed from the *tcp* operon on VPI-1, with the help of the transcriptional activator, ToxT. Thus, CTXϕ infection of *V. cholerae* imparts toxigenicity on this marine microbe. Of the 200 or more known *V. cholerae* serogroups, only two, O1 and O139, possess the CTXϕ element and express cholera toxin. Acquisition of these genes was made possible by the TCP receptor. The ability to secrete cholera toxin affords an enormous selective advantage on these serogroups as this toxin induces voluminous diarrhoea in cholera patients, which allows the many millions of new *V. cholerae* produced during infection to disseminate into the environment and infect other human hosts.

In organisms other than *V. cholerae*, type IV pili have additional functions such as twitching motility and DNA uptake. These functions, which are associated with type IV pilus retraction, have not been demonstrated for *V. cholerae* TCP, and no retraction ATPase has been identified for the TCP system. However, we observe that all known TCP activities – autoagglutination, TcpF secretion and CTXϕ uptake – appear to utilize a dynamic pilus that cycles between extension and retraction. For instance, *V. cholerae* aggregates are remarkably compact and well-defined, which is surprising given that the pili are many times the length of the *V. cholerae* cells and might therefore be expected to form loosely associated bacterial clusters. Further-more, *V. cholerae tcpA* mutants have been identified that do not autoagglutinate, yet the TCP form large bundles as observed by electron microscopy (Kirn *et al.*, 2000; Lim *et al.*, 2010). This suggests that there is more to microcolony formation than simple pilus:pilus interactions. Such interactions followed by pilus retraction would bring cells into close proximity and protect them from antimicrobial agents produced by the host, such as bile and perhaps serum complement. The second pilus function, TcpF secretion, is thought to occur in a manner similar to that of type II secretion, where the pilus grows across the periplasm and drives the substrate through the outer membrane secretin. Unlike the periplasmic pseudopilus of the type II secretion system, the TCP filament would continue to grow across the outer membrane after pushing TcpF through. But this model seems inefficient, as only one TcpF molecule would be secreted for every surface-displayed pilus, after which the secretin would be blocked by the presence of the pilus. We propose that TCP can dynamically assemble and disassemble, allowing each TCP apparatus to export many molecules of TcpF. The third TCP activity, CTXϕ uptake, is most compelling in its requirement for pilus retraction. CTXφ infection of *V. cholerae* has strong parallels with that of the Ff filamentous phage infection of *E. coli*. Ff phage such as M13 and Fd use their minor tip protein, pIII, to bind to the *E. coli* F pilus, which retracts into the cell, at which point pIII then binds to TolA in the *E. coli* periplasm (Gray *et al.*, 1981; Russel *et al.*, 1988; Deng and Perham, 2002; Clarke *et al.*, 2008). Similarly, CTXφ initially utilizes its pIII tip protein to first bind to TCP, followed by its interaction with TolA (Waldor and Mekalanos, 1996; Heilpern and Waldor, 2000, 2003; Ford *et al.*, 2012). We have proposed that TCP retract spontaneously, drawing the tip of bound CTXφ, which has a diameter similar to the pilus, into the periplasm via the secretin channel, where it is then exposed to TolA (Ford *et al.*, 2012). Type IV pilus retraction is a necessary step for uptake of the PO4 phage by *Pseudomonas aeruginosa*, and disruption of the *pilT* gene encoding the retraction ATPase renders the bacteria resistant to phage infection (Bradley, 1973; Whitchurch and Mattick, 1994). Pilus retraction represents a new avenue of investigation into antibiotic development because it can be exploited as a highly specific uptake mechanism for antimicrobial agents.

How *V. cholerae* TCP retraction might occur in the absence of a retraction ATPase is not known, but it is important to point out that (i) the mechanism of retraction in the presence of a retraction ATPase is also not understood, and (ii) the type II secretion system and the *E. coli* F pilus also appear to be retractile, yet neither of these systems possesses a retraction ATPase. According to our type IV pilus assembly model, pilus extension is processive, and will proceed as long as there is a steady supply of pilin subunits and ATP. Interruption in the supply of either component, or inhibition of the assembly ATPase, possibly by the retraction ATPase, would block pilus extension and lead to spontaneous and processive disassembly. This is because the last subunit added to the growing filament cannot be stabilized by the ratcheting effect of filament extrusion out of the membrane, and thus would diffuse back into the membrane, causing the filament to collapse into the membrane and allow the next subunit to diffuse away, etc. In simple systems like those

of TCP and type II secretion, assembly and disassembly may occur randomly and spontaneously, whereas more complex systems like those that possess a retraction ATPase perhaps require more coordinated retraction of multiple pili to perform functions such as twitching motility.

1.6 TcpA pilin and TcpF as Components of Vaccines

The TCP filaments, the TcpA pilin subunit and the secreted colonization factor, TcpF, show promise as components of multi-subunit vaccines. All are immunogenic in humans, as shown by analysis of sera from convalescing cholera patients and all elicit a protective response against *V. cholerae* in passive immunization studies using animal models (Hang *et al.*, 2003; Attridge *et al.*, 2004). Thus, these proteins may prove efficacious when combined with additional virulence factors or bacterial vaccine formulations. The cholera vaccines that are currently in clinical use are comprised of killed, whole-cell formulations that are administered orally. Dukoral, which has been available since 1991 (Shin *et al.*, 2011), is a mixture of classical and El Tor biotype strains plus recombinant cholera toxin subunit B (CT-B). Dukoral requires two doses and confers about 70% protection. Shanchol, which was licensed in 2009, consists of both classical and El Tor biotypes of the O1 serogroup plus the O139 serogroup and confers protection with an efficacy similar to Dukoral. Shanchol is much less expensive per dose because it does not contain recombinant CT-B. Both vaccines are offered to travellers as well as residents in endemic areas (Mahalanabis *et al.*, 2008). Interestingly, when prototype killed, whole-cell formulations of both Dukoral and Shanchol were tested for the presence of TcpA antigen, none was detected (Sun *et al.*, 1990a; Hauke and Taylor, unpublished). It is highly likely that because TCP expression requires very precise conditions, pili are not expressed (and hence, TcpF is not secreted) under the growth conditions used for vaccine production.

More recently, live attenuated oral vaccines have been developed that are presumed to express TCP because the *V. cholerae* within the formulation are proficient at colonization, a process that requires functional pili. These vaccines are undergoing clinical trials and are not yet available commercially. They include CVD103-HgR, which is highly efficacious in North American volunteers and is licensed for use in travellers but is not approved for use by residents in endemic areas (Cryz *et al.*, 1995). A large field trial for CVD103-HgR was conducted in Indonesia, but was inconclusive due to a low overall incidence of cholera during the field trial that did not allow for enough statistical power to prove efficacy. A second trial in Micronesia showed promising results with 79% efficacy (Richie *et al.*, 2000; Calain *et al.*, 2004). A second live oral candidate is Peru-15, an El Tor biotype derivative that is still undergoing field trials but has thus far proven to be immunogenic and well-tolerated (Chowdhury *et al.*, 2009).

The infant mouse has been used as the experimental model to test the potential of intact TCP, TcpA subunits and TcpF to serve as vaccine components. Its suitability for these studies is based on the fact that *V. cholerae* strains that colonize humans are also able to colonize the infant mouse intestine in a TCP-dependent manner. *V. cholerae tcpA* mutants are unable to colonize either humans or mice, whereas wild-type strains and *mshA* mutants, which are deficient in expression of another type IV pilus, mannose-sensitive haemagglutinin, efficiently colonize in both cases (Taylor *et al.*, 1987; Herrington *et al.*, 1988; Thelin and Taylor, 1996; Tacket *et al.*, 1998). Active vaccination is not feasible in the infant mouse itself, so passive immunization approaches have been taken to study the vaccine efficacy of TCP and its related components. The most common approach, and the first to be used for TCP, is to immunize adult mice or rabbits with intact pili, extract the antisera and mix it with *V. cholerae*, then administer this mixture by oral gavage directly into the stomach of mouse pups. This approach established that effective protection, approaching 100% would be achieved with high levels of infecting organisms (Sun *et al.*, 1990b). Importantly, the protective activity of the antiserum was lost when it was pre-adsorbed with wild-type *V. cholerae*, but was

retained when pre-adsorbed with a *tcpA* null mutant, demonstrating that the anti-TCP antibodies were the protective component of the antisera. A second approach is to administer monoclonal antibodies raised against TCP in place of the polyclonal antisera. This approach established that certain portions of TcpA pilin subunit are more protective than others (Sun *et al.*, 1990a). Subsequent studies using synthetic peptides mapped the most protective epitopes to the C-terminal portion of the pilin, including the D-region that protrudes from the pilus surface, and established the use of the corresponding peptides as effective immunogens (Sun *et al.*, 1991, 1997). The effect of adjuvants on the peptide immunization regimens was also examined. Adult female mice were immunized with anti-TcpA peptide with and without adjuvant and the efficacy of the antibodies was tested by oral challenge of their pups with a *V. cholerae*/antisera mixture (Wu *et al.*, 2001). Addition of adjuvant when immunizing the adult mice correlated with increased protection for the pups.

When considering the utilization of TCP as a cholera vaccine component, it is important to recognize that all TcpA is not created equal. Prior to 1961, all clinical cases of cholera were caused by strains of the classical biotype of the O1 serogroup. Since 1961, the cholera incidence due to El Tor strains from the same serogroup has steadily increased, with a coincident outbreak of the O139 serogroup during the 1990s. Although virtually all clinical isolates are currently of the El Tor biotype, it is prudent to maintain broad vaccine coverage should the classical biotype re-emerge. The classical and El Tor biotypes within the O1 serogroup are distinguished by biochemical differences such as the ability to carry out specific metabolic activities and innate differences in antibiotic sensitivities. The primary amino acid sequences of TcpA from the classical and El Tor biotypes are 82% identical and their structures are very similar, most of the amino acid diversity occurs in the C-terminal region that confers the best protection (Rhine and Taylor, 1994; Lim *et al.*, 2010). *V. cholerae* O139 contains the *tcpA* allele from O1 El Tor.

Throughout the course of TCP vaccine studies it has become evident that protection is much more efficacious when mouse pups are challenged with a *V. cholerae* biotype strain homologous to that used for the immunization than when challenged with a heterologous biotype strain (Sun *et al.*, 1990a; Voss *et al.*, 1996).

Along the same lines as the TCP and TcpA vaccine studies, the secreted colonization factor, TcpF, has also been shown to be protective in the infant mouse cholera model. A 100-fold higher *V. cholerae* inoculum was required to achieve a 50% death rate in infant mice when the bacteria were co-administered with rabbit polyclonal anti-TcpF serum (Kirn and Taylor, 2005). Subsequently, a monoclonal anti-TcpF antibody was identified that confers a level of protection similar to the polyclonal antibody. The crystal structure of TcpF, in combination with mutational analysis, revealed that the protective epitope recognized by the monoclonal antibody lies on the ridge of a cleft in the C-terminal domain of the protein (Megli *et al.*, 2011). The primary sequence of TcpF is the same in both the classical and El Tor biotypes of the O1 serogroup and the O139 serogroup, so its inclusion as a vaccine component should provide universal protection. When the monoclonal anti-TcpF antibody is used in combination with a polyclonal anti-TcpA antibody the level of protection is greater than when either of these is used individually (Megli *et al.*, 2011). Protection conferred by TcpF is also enhanced by co-vaccination with (CT) (Price and Holmes, 2012).

So, how can we take advantage of these findings to potentially improve existing vaccines by incorporating TCP, TcpA and TcpF into cholera vaccine formulations? One way might be to incorporate TcpA and TcpF into a defined component vaccine. Individuals who experience bouts of cholera show long-lasting protection (reviewed in Harris *et al.*, 2012). Although the basis for this protection is not fully understood, it has been suggested to involve responses to CT-B, lipopolysaccharide (LPS) and colonization factors. A combination of factors that have been shown to be protective for *V. cholerae*, such as LPS, TcpA, TcpF and the colonization factor GlcNAc-

binding protein A (GbpA) (Kirn *et al.*, 2005; Muse *et al.*, 2012), could be added to the existing bacterial vaccine formulations. Currently, no formulations of this type are approved for human use. Perhaps the most direct way to include TCP and TcpF in a vaccine is to place the expression of their genes under the control of a regulatory system that can be induced in growth conditions that are not normally conducive to its expression. In this way, TCP can be expressed and TcpF will be secreted under conditions used to grow strains for the preparation of currently approved killed, whole cell or live attenuated vaccines. Such systems are under development.

1.7 Conclusions

The *V. cholerae* TCP are among the simplest of the type IV pilus systems, yet their key functions in colonization of the human intestine plus their central role in evolution of this pathogen make them fascinating and important subjects. The possibility that TCP are retractile even though they lack a retraction ATPase makes them ideal model systems to study this process to derive a unifying mechanism for all type IV pili and the related type II secretion systems. And most importantly, these surface-displayed virulence factors show promise as components to enhance the efficacy of existing cholera vaccines. Their newly recognized retraction feature can be exploited as a highly specific delivery system for anti-Vibrio agents that act upon classical antibiotic targets such as peptidoglycan synthesis enzymes, or upon novel targets such as the type II secretion system or the TCP assembly apparatus itself.

References

Aas, F.E., Wolfgang, M., Frye, S., Dunham, S., Lovold, C. and Koomey, M. (2002) Competence for natural transformation in *Neisseria gonorrhoeae*: components of DNA binding and uptake linked to type IV pilus expression. *Molecular Microbiology* 46, 749–760.

Aas, F.E., Winther-Larsen, H.C., Wolfgang, M., Frye, S., Lovold, C., Roos, N., van Putten, J.P. and Koomey, M. (2007) Substitutions in the N-terminal alpha helical spine of *Neisseria gonorrhoeae* pilin affect type IV pilus assembly, dynamics and associated functions. *Molecular Microbiology* 63, 69–85.

Abendroth, J., Murphy, P., Sandkvist, M., Bagdasarian, M. and Hol, W.G. (2005) The X-ray structure of the type II secretion system complex formed by the N-terminal domain of EpsE and the cytoplasmic domain of EpsL of *Vibrio cholerae*. *Journal of Molecular Biology* 348, 845–855.

Attridge, S.R., Wallerstrom, G., Qadri, F. and Svennerholm, A.M. (2004) Detection of antibodies to toxin-coregulated pili in sera from cholera patients. *Infection and Immunity* 72, 1824–1827.

Ayers, M., Howell, P.L. and Burrows, L.L. (2010) Architecture of the type II secretion and type IV pilus machineries. *Future Microbiology* 5, 1203–1218.

Berry, J.L., Phelan, M.M., Collins, R.F., Adomavicius, T., Tonjum, T., Frye, S.A., Bird, L., Owens, R., Ford, R.C., Lian, L.Y. and Derrick, J.P. (2012) Structure and assembly of a trans-periplasmic channel for type IV pili in *Neisseria meningitidis*. *PLoS Pathogens* 8, e1002923.

Bose, N. and Taylor, R.K. (2005) Identification of a TcpC-TcpQ outer membrane complex involved in the biogenesis of the toxin-coregulated pilus of *Vibrio cholerae*. *Journal of Bacteriology* 187, 2225–2232.

Bradley, D. (1973) A pilus-dependent *Pseudomonas aeruginosa* bacteriophage with a long noncontractile tail. *Virology* 51, 489–492.

Calain, P., Chaine, J.P., Johnson, E., Hawley, M.L., O'Leary, M.J., Oshitani, H. and Chaignat, C.L. (2004) Can oral cholera vaccination play a role in controlling a cholera outbreak? *Vaccine* 22, 2444–2451.

Chowdhury, M.I., Sheikh, A. and Qadri, F. (2009) Development of Peru-15 (CholeraGarde), a live-attenuated oral cholera vaccine: 1991–2009. *Expert Review of Vaccines* 8, 1643–1652.

Clarke, M., Maddera, L., Harris, R.L. and Silverman, P.M. (2008) F-pili dynamics by live-cell imaging. *Proceedings of the National Academy of Sciences USA* 105, 17978–17981.

Craig, L., Taylor, R.K., Pique, M.E., Adair, B.D., Arvai, A.S., Singh, M., Lloyd, S.J., Shin, D.S., Getzoff, E.D., Yeager, M., Forest, K.T. and Tainer, J.A. (2003) Type IV pilin structure and assembly: X-ray and EM analyses of *Vibrio cholerae* toxin-coregulated pilus and *Pseudomonas aeruginosa* PAK pilin. *Molecular Cell* 11, 1139–1150.

Craig, L., Volkmann, N., Arvai, A.S., Pique, M.E., Yeager, M., Egelman, E.H. and Tainer, J.A. (2006) Type IV pilus structure by cryo-electron microscopy and crystallography: implications for pilus assembly and functions. *Molecular Cell* 23, 651–662.

Cryz, S.J. Jr, Kaper, J., Tacket, C., Nataro, J. and Levine, M.M. (1995) *Vibrio cholerae* CVD103-HgR live oral attenuated vaccine: construction, safety, immunogenicity, excretion and non-target effects. *Developments in Biological Standardization* 84, 237–244.

Deng, L.W. and Perham, R.N. (2002) Delineating the site of interaction on the pIII protein of filamentous bacteriophage fd with the F-pilus of *Escherichia coli*. *Journal of Molecular Biology* 319, 603–614.

Ford, C., Kolappan, S., Phan, H.T.H., Winther-Larsen, H.C. and Craig, L. (2012) Crystal structures of CTX-phi pIII unbound and in complex with *Vibrio cholerae* TolA reveal novel interaction interfaces. *Journal of Biological Chemistry* 287, 36258–36262.

Graupner, S., Weger, N., Sohni, M. and Wackernagel, W. (2001) Requirement of novel competence genes pilT and pilU of *Pseudomonas stutzeri* for natural transformation and suppression of *pilT* deficiency by a hexahistidine tag on the type IV pilus protein PilAI. *Journal of Bacteriology* 183, 4694–4701.

Gray, C.W., Brown, R.S. and Marvin, D.A. (1981) Adsorption complex of filamentous fd virus. *Journal of Molecular Biology* 146, 621–627.

Hager, A.J., Bolton, D.L., Pelletier, M.R., Brittnacher, M.J., Gallagher, L.A., Kaul, R., Skerrett, S.J., Miller, S.I. and Guina, T. (2006) Type IV pili-mediated secretion modulates *Francisella* virulence. *Molecular Microbiology* 62, 227–237.

Hang, L., John, M., Asaduzzaman, M., Bridges, E.A., Vanderspurt, C., Kirn, T.J., Taylor, R.K., Hillman, J.D., Progulske-Fox, A., Handfield, M., Ryan, E.T. and Calderwood, S.B. (2003) Use of *in vivo*-induced antigen technology (IVIAT) to identify genes uniquely expressed during human infection with *Vibrio cholerae*. *Proceedings of the National Academy of Sciences USA* 100, 8508–8513.

Harris, J.B., LaRocque, R.C., Qadri, F., Ryan, E.T. and Calderwood, S.B. (2012) Cholera. *Lancet* 379, 2466–2476.

Heilpern, A.J. and Waldor, M.K. (2000) CTXphi infection of *Vibrio cholerae* requires the *tolQRA* gene products. *Journal of Bacteriology* 182, 1739–1747.

Heilpern, A.J. and Waldor, M.K. (2003) pIIICTX, a predicted CTXphi minor coat protein, can expand the host range of coliphage fd to include

Vibrio cholerae. Journal of Bacteriology 185, 1037–1044.

Herrington, D.A., Hall, R.H., Losonsky, G., Mekalanos, J.J., Taylor, R.K. and Levine, M.M. (1988) Toxin, toxin-coregulated pili, and the *toxR* regulon are essential for *Vibrio cholerae* pathogenesis in humans. *Journal of Experimental Medicine* 168, 1487–1492.

Kaper, J.B., Morris, J.G. Jr and Levine, M.M. (1995) Cholera. *Clinical Microbiology Reviews* 8, 48–86.

Kimsey, H.H. and Waldor, M.K. (2009) *Vibrio cholerae* LexA coordinates CTX prophage gene expression. *Journal of Bacteriology* 191, 6788–6795.

Kirn, T.J. and Taylor, R.K. (2005) TcpF is a soluble colonization factor and protective antigen secreted by El Tor and classical O1 and O139 *Vibrio cholerae* serogroups. *Infection and Immunity* 73, 4461–4470.

Kirn, T.J., Lafferty, M.J., Sandoe, C.M. and Taylor, R.K. (2000) Delineation of pilin domains required for bacterial association into microcolonies and intestinal colonization by *Vibrio cholerae. Molecular Microbiology* 35, 896–910.

Kirn, T.J., Bose, N. and Taylor, R.K. (2003) Secretion of a soluble colonization factor by the TCP type 4 pilus biogenesis pathway in *Vibrio cholerae. Molecular Microbiology* 49, 81–92.

Kirn, T.J., Jude, B.A. and Taylor, R.K. (2005) A colonization factor links *Vibrio cholerae* environmental survival and human infection. *Nature* 438, 863–866.

Korotkov, K.V., Krumm, B., Bagdasarian, M. and Hol, W.G. (2006) Structural and functional studies of EpsC, a crucial component of the type 2 secretion system from *Vibrio cholerae. Journal of Molecular Biology* 363, 311–321.

Krebs, S.J. and Taylor, R.K. (2011) Protection and attachment of *Vibrio cholerae* mediated by the toxin-coregulated pilus in the infant mouse model. *Journal of Bacteriology* 193, 5260–5270.

LaPointe, C.F. and Taylor, R.K. (2000) The type 4 prepilin peptidases comprise a novel family of aspartic acid proteases. *Journal of Biological Chemistry* 275, 1502–1510.

Li, J., Egelman, E. and Craig, L. (2012) Electron microscopy reconstruction of the *Vibrio cholerae* toxin coregulated pilus and comparative analysis with the *Neisseria gonorrhoeae* GC pilus. *Journal of Molecular Biology* 418, 47–64.

Lim, M.S., Ng, D., Zong, Z., Arvai, A.S., Taylor, R.K., Tainer, J.A. and Craig, L. (2010) *Vibrio cholerae* El Tor TcpA crystal structure and mechanism for pilus-mediated microcolony formation. *Molecular Microbiology* 77, 755–770.

Mahalanabis, D., Lopez, A.L., Sur, D., Deen, J., Manna, B., Kanungo, S., von Seidlein, L., Carbis, R., Han, S.H., Shin, S.H., Attridge, S., Rao, R., Holmgren, J., Clemens, J. and Bhattacharya, S.K. (2008) A randomized, placebo-controlled trial of the bivalent killed, whole-cell, oral cholera vaccine in adults and children in a cholera endemic area in Kolkata, India. *PloS One* 3, e2323.

Maier, B., Potter, L., So, M., Seifert, H.S. and Sheetz, M.P. (2002) Single pilus motor forces exceed 100 pN. *Proceedings of the National Academy of Sciences USA* 99, 16012–16017.

Megli, C.J., Yuen, A.S., Kolappan, S., Richardson, M.R., Dharmasena, M.N., Krebs, S.J., Taylor, R.K. and Craig, L. (2011) Crystal structure of the *Vibrio cholerae* colonization factor TcpF and identification of a functional immunogenic site. *Journal of Molecular Biology* 409, 146–158.

Melville, S.B. and Craig, L. (2013) Type IV pili in Gram-positive bacteria. *Microbiology and Molecular Biology Reviews* 77.

Merz, A.J., So, M. and Sheetz, M.P. (2000) Pilus retraction powers bacterial twitching motility. *Nature* 407, 98–102.

Misic, A.M., Satyshur, K.A. and Forest, K.T. (2010) *P. aeruginosa* PilT structures with and without nucleotide reveal a dynamic type IV pilus retraction motor. *Journal of Molecular Biology* 400, 1011–1021.

Muse, M., Grandjean, C., Wade, T.K. and Wade, W.F. (2012) A one dose experimental cholera vaccine. *FEMS Immunology and Medical Microbiology* 66, 98–115.

Pelicic, V. (2008) Type IV pili: *e pluribus unum? Molecular Microbiology* 68, 827–837.

Planet, P.J., Kachlany, S.C., DeSalle, R. and Figurski, D.H. (2001) Phylogeny of genes for secretion NTPases: identification of the widespread *tadA* subfamily and development of a diagnostic key for gene classification. *Proceedings of the National Academy of Sciences USA* 98, 2503–2508.

Price, G.A. and Holmes, R.K. (2012) Evaluation of TcpF-A2-CTB chimera and evidence of additive protective efficacy of immunizing with TcpF and CTB in the suckling mouse model of cholera. *PloS One* 7, e42434.

Reichow, S.L., Korotkov, K.V., Hol, W.G. and Gonen, T. (2010) Structure of the cholera toxin secretion channel in its closed state. *Nature Structure and Molecular Biology* 17, 1226–1232.

Rhine, J.A. and Taylor, R.K. (1994) TcpA pilin sequences and colonization requirements for O1 and O139 *Vibrio cholerae. Molecular Microbiology* 13, 1013–1020.

Richie, E.E., Punjabi, N.H., Sidharta, Y.Y., Peetosutan, K.K., Sukandar, M.M., Wasserman, S.S., Lesmana, M.M., Wangsasaputra, F.F., Pandam, S.S., Levine, M.M., O'Hanley, P.P., Cryz, S.J. and Simanjuntak, C.H. (2000) Efficacy trial of single-dose live oral cholera vaccine CVD 103-HgR in North Jakarta, Indonesia, a cholera-endemic area. *Vaccine* 18, 2399–2410.

Robien, M.A., Krumm, B.E., Sandkvist, M. and Hol, W.G. (2003) Crystal structure of the extracellular protein secretion NTPase EpsE of *Vibrio cholerae*. *Journal of Molecular Biology* 333, 657–674.

Rodgers, K., Arvidson, C.G. and Melville, S. (2011) Expression of a *Clostridium perfringens* type IV pilin by *Neisseria gonorrhoeae* mediates adherence to muscle cells. *Infection and Immunity* 79, 3096–3105.

Russel, M., Whirlow, H., Sun, T.P. and Webster, R.E. (1988) Low-frequency infection of F-bacteria by transducing particles of filamentous bacteriophages. *Journal of Bacteriology* 170, 5312–5316.

Satyshur, K.A., Worzalla, G.A., Meyer, L.S., Heiniger, E.K., Aukema, K.G., Misic, A.M. and Forest, K.T. (2007) Crystal structures of the pilus retraction motor PilT suggest large domain movements and subunit cooperation drive motility. *Structure* 15, 363–376.

Sauvonnet, N., Vignon, G., Pugsley, A.P. and Gounon, P. (2000) Pilus formation and protein secretion by the same machinery in *Escherichia coli*. *EMBO Journal* 19, 2221–2228.

Shin, S., Desai, S.N., Sah, B.K. and Clemens, J.D. (2011) Oral vaccines against cholera. *Clinical Infectious Diseases: An Official Publication of the Infectious Diseases Society of America* 52, 1343–1349.

Skerker, J.M. and Berg, H.C. (2001) Direct observation of extension and retraction of type IV pili. *Proceedings of the National Academy of Sciences USA* 98, 6901–6904.

Skorupski, K. and Taylor, R.K. (2013) Toxin and virulence regulation in *Vibrio cholerae*. In: Darwin, A. and Vasil, M. (eds) *Regulation of Bacterial Virulence*. ASM Press, Washington, DC, pp. 241–261.

Strom, M.S., Nunn, D.N. and Lory, S. (1993) A single bifunctional enzyme, PilD, catalyzes cleavage and N-methylation of proteins belonging to the type IV pilin family. *Proceedings of the National Academy of Sciences USA* 90, 2404–2408.

Sun, D., Tillman, D.N., Marion, T.N. and Taylor, R.K. (1990a) Production and characterization of monoclonal antibodies to the toxin coregulated pilus (TCP) of *Vibrio cholerae* that protect against experimental cholera in infant mice. *Serodiagnosis and Immunotherapy in Infectious Disease* 4, 73–81.

Sun, D.X., Mekalanos, J.J. and Taylor, R.K. (1990b) Antibodies directed against the toxin-coregulated pilus isolated from *Vibrio cholerae* provide protection in the infant mouse experimental cholera model. *Journal of Infectious Diseases* 161, 1231–1236.

Sun, D., Seyer, J.M., Kovari, I., Sumrada, R.A. and Taylor, R.K. (1991) Localization of protective epitopes within the pilin subunit of the *Vibrio cholerae* toxin-coregulated pilus. *Infection and Immunity* 59, 114–118.

Sun, D., Lafferty, M.J., Peek, J.A. and Taylor, R.K. (1997) Domains within the *Vibrio cholerae* toxin coregulated pilin subunit that mediate bacterial colonization. *Gene* 192, 79–85.

Tacket, C.O., Taylor, R.K., Losonsky, G., Lim, Y., Nataro, J.P., Kaper, J.B. and Levine, M.M. (1998) Investigation of the roles of toxin-coregulated pili and mannose-sensitive hemagglutinin pili in the pathogenesis of *Vibrio cholerae* O139 infection. *Infection and Immunity* 66, 692–695.

Tammam, S., Sampaleanu, L.M., Koo, J., Sundaram, P., Ayers, M., Chong, P.A., Forman-Kay, J.D., Burrows, L.L. and Howell, P.L. (2011) Characterization of the PilN, PilO and PilP type IVa pilus subcomplex. *Molecular Microbiology* 82, 1496–1514.

Taylor, R.K., Miller, V.L., Furlong, D.B. and Mekalanos, J.J. (1987) Use of *phoA* gene fusions to identify a pilus colonization factor coordinately regulated with cholera toxin. *Proceedings of the National Academy of Sciences USA* 84, 2833–2837.

Thelin, K.H. and Taylor, R.K. (1996) Toxin-coregulated pilus, but not mannose-sensitive hemagglutinin, is required for colonization by *Vibrio cholerae* O1 El Tor biotype and O139 strains. *Infection and Immunity* 64, 2853–2856.

Tripathi, S.A. and Taylor, R.K. (2007) Membrane association and multimerization of TcpT, the cognate ATPase ortholog of the *Vibrio cholerae* toxin-coregulated-pilus biogenesis apparatus. *Journal of Bacteriology* 189, 4401–4409.

Varga, J.J., Nguyen, V., O'Brien, D.K., Rodgers, K., Walker, R.A. and Melville, S.B. (2006) Type IV pili-dependent gliding motility in the Gram-positive pathogen *Clostridium perfringens* and other Clostridia. *Molecular Microbiology* 62, 680–694.

Vignon, G., Kohler, R., Larquet, E., Giroux, S., Prevost, M.C., Roux, P. and Pugsley, A.P. (2003) Type IV-like pili formed by the type II secreton:

specificity, composition, bundling, polar localization, and surface presentation of peptides. *Journal of Bacteriology* 185, 3416–3428.

Voss, E., Manning, P.A. and Attridge, S.R. (1996) The toxin-coregulated pilus is a colonization factor and protective antigen of *Vibrio cholerae* El Tor. *Microbial Pathogenesis* 20, 141–153.

Waldor, M.K. and Friedman D.I. (2005) Phage regulatory circuits and virulence gene expression. *Current Opinion in Microbiology* 8, 459–465.

Waldor, M.K. and Mekalanos, J.J. (1996) Lysogenic conversion by a filamentous phage encoding cholera toxin. *Science* 272, 1910–1914.

Whitchurch, C.B. and Mattick, J.S. (1994) Characterization of a gene, pilU, required for twitching motility but not phage sensitivity in *Pseudomonas aeruginosa*. *Molecular Microbiology* 13, 1079–1091.

Wu, J.Y., Wade, W.F. and Taylor, R.K. (2001) Evaluation of cholera vaccines formulated with toxin-coregulated pilin peptide plus polymer adjuvant in mice. *Infection and Immunity* 69, 7695–7702.

Yamagata, A. and Tainer, J.A. (2007) Hexameric structures of the archaeal secretion ATPase GspE and implications for a universal secretion mechanism. *EMBO Journal* 26, 878–890.

2 Conjugative Pili

Peter J. Christie

Department of Microbiology and Molecular Genetics, University of Texas Medical School at Houston, USA

2.1 Introduction

Bacterial conjugation, the process by which donor cells deliver mobile DNA elements to recipient cells through direct cell-to-cell contact, has been studied for more than 60 years (Alvarez-Martinez and Christie, 2009). Conjugation systems of Gram-negative bacteria are phylogenetically and functionally diverse, but all characterized systems deliver their cargoes intercellularly by elaboration of two surface structures. The first is a translocation channel through which substrates travel across the donor cell envelope. The second is the conjugative pilus, an extracellular filament that functions predominantly or exclusively to initiate interactions with recipient cells. The protein building blocks for the channel and pilus are nearly the same, and models generally depict the two organelles as a single structural entity (Lawley *et al.*, 2004). However, mutations have been isolated that confer the assembly of translocation channels without detectable pili or pili without functional translocation channels. The existence of such 'uncoupling' mutations establishes, first, that the biogenesis pathway for the conjugation apparatus bifurcates at some point for assembly either of the translocation channel or the pilus, and second, that these two organelles contribute in distinct ways to the overall process of conjugation (Christie *et al.*, 2005).

The conjugation systems are ancestral progenitors of a large family of bacterial translocation systems termed the type IV secretion systems (T4SSs) (Cascales and Christie, 2003). These systems, present in nearly all bacterial and some archaeal species, collectively have evolved exceptional versatility in terms of the types of substrates translocated (DNA, effector proteins) and types of target cells to which substrates are delivered (bacteria, fungi, plant, mammal). The three T4SS subfamilies include the conjugation, DNA release and uptake, and effector translocator systems (Christie and Vogel, 2000; Cascales and Christie, 2003). The evolution and functional versatility especially of T4SSs adapted for use during microbial infection are fascinating areas of investigation. We will briefly summarize roles of T4SSs in human disease at the end of this chapter, but interested readers are encouraged to see recent reviews covering this topic in more detail (Nagai and Roy, 2003; Backert and Selbach, 2008).

This chapter will provide an overview of the biogenesis, structure and biological functions of conjugative pili. Although a large number of conjugative pili have been described, the discussion will focus mainly on the two more extensively characterized pili, F-pili elaborated by the *Escherichia coli* F plasmid and P-pili produced by *E. coli* plasmids of the P, N or W incompatibility

(Inc) groups and the *Agrobacterium tumefaciens* VirB/VirD4 T-DNA transfer system (Lawley *et al.*, 2003; Christie *et al.*, 2005; Schroder and Lanka, 2005). At the outset, it is important to note that the nomenclature surrounding this field is varied and often confusing. For example, the conjugative pili are elaborated by T4SSs, yet these pili are termed 'conjugative' and not 'type IV' pili. The conjugative pili and type IV pili are ancestrally unrelated (the latter is related to type II secretion systems or T2SSs), and they display a number of distinct structural and functional features (Ayers *et al.*, 2010; Thanassi *et al.*, 2012). Additionally, most conjugation systems are assembled from a common set of protein 'building blocks' whose names, unfortunately, are specific to that system. This complicates comparisons of the different systems, so here we will use the *A. tumefaciens* VirB/VirD4 nomenclature as a reference point (Alvarez-Martinez and Christie, 2009). Finally, as is also the case for the entire T4SS family, many conjugation systems have acquired novel domains, proteins or protein subassemblies during evolution that endow these systems with specialized functions. The resulting mosaicism of the conjugation machines complicates efforts to present an overarching view of machine architecture (Alvarez-Martinez and Christie, 2009). Here, some novel adaptations will be discussed insofar as they impact the assembly or function of conjugative pili.

2.2 General Overview of Conjugation and Conjugative Pili

In Gram-negative bacteria, conjugation initiates when a donor cell establishes contact with a recipient cell via the conjugative pilus. There are essentially two types of pilus-mediated contact (see Fig. 2.1). The F-pilus displays a dynamic activity whereby it extends and, upon contact with a recipient cell, retracts to bring the two cells into physical juxtaposition (Clarke *et al.*, 2008). P-pili, by contrast, do not seem to retract but instead are produced and released from the cell surface through nonspecific breakage or an active sloughing mechanism. These pili accumulate

in the milieu and form a mesh of pilus polymers that effectively promote nonspecific clumping of donor and recipient cells (Christie, 2004; Schroder and Lanka, 2005).

Pilus-mediated contact with a recipient cell stimulates propagation of a signal to the

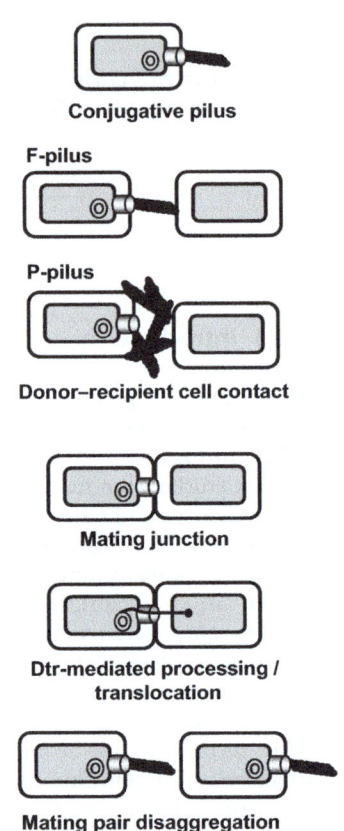

Conjugative pilus

F-pilus

P-pilus

Donor–recipient cell contact

Mating junction

Dtr-mediated processing / translocation

Mating pair disaggregation

Fig. 2.1. Overall mechanism of pilus-mediated conjugative DNA transfer. Gram-negative bacteria elaborate a type IV secretion system (T4SS) composed of a mating channel and conjugative pilus. F-pili extend and retract to bring potential recipient cells into direct cell-to-cell contact, whereas P-pili are released from the cell surface where they form a hydrophobic mesh promoting aggregation of donor and potential recipient cells. Following mating junction formation, the DNA transfer and replication (Dtr) processing factors nick the DNA strand destined for transfer (T-strand) and the nicking enzyme, termed a relaxase, pilots the T-strand into the recipient cell. Upon successful DNA transfer, the mating junction dissociates and the recipient cell carrying the transferred DNA acquires donor status.

donor cell interior to activate the processing of mobile DNA, e.g. a conjugative plasmid or integrative conjugative element (ICE) destined for transfer (Fig. 2.1). The processing proteins are termed the DNA transfer and replication (Dtr) factors, reflecting their shared ancestries and catalytic functions with the rolling circle replicases. The relaxase, together with one or more accessory factors, binds at the mobile element's origin of transfer (*oriT*) sequence to form the relax-osome (de la Cruz *et al.*, 2010). Next, the relaxosome docks with the substrate receptor for the cognate conjugation system. T4SS substrate receptors, also termed type IV coupling proteins (T4CPs), are a conserved family of ATPases associated with all characterized conjugation machines as well as most effector translocator systems (Gomis-Ruth *et al.*, 2001). Engagement of the relax-osome with the T4CP triggers nicking of the DNA strand destined for transfer (T-strand) by a protein termed the relaxase. Upon nicking, the relaxase remains covalently bound to the 5' end of the T-strand, which is then unwound from the nontransferred strand (Zechner *et al.*, 2012). The T4CP delivers the relaxase-T-strand transfer inter-mediate to the transfer channel, also termed the mating channel or pore, for translocation to the donor cell surface and into the recipient cell (Cascales and Christie, 2004).

Various factors regulate the efficiency of DNA transfer. Dynamic extension and retraction of F-pili, for example, enables efficient transfer between donor and recipient cells in liquid (Silverman and Clarke, 2010). The P-pilus-dependent transfer systems function very poorly in liquid, most probably because released P-pili dissipate in the medium and fail to induce donor–recipient cell aggregation (Schroder and Lanka, 2005). Both F- and P-type systems transfer efficiently on solid surfaces, but mechanisms have evolved to prevent redundant transfer of mobile elements into cells bearing a similar element. Surface or entry exclusion (Eex), for example, is manifested by surface proteins that prevent formation of productive mating pairs between cells in a donor population. Additionally, mobile elements are grouped into different Inc systems on the basis of their

replication mechanisms, and during mating the replication of an incoming mobile element is suppressed by a resident element bearing the same Inc system (Lawley *et al.*, 2003).

2.3 Types of Conjugative Pili

The F-type pili are representative of a group of morphologically similar pili encoded by plasmids of the IncF, IncH1 and IncH2 and other incompatibility groups. These pili are typically ~9 nm and flexible, range in length up to 1 micron, and dynamically extend and retract (Fig. 2.2A). The P-type pili by contrast are thicker (9–11 nm), more rigid, and much shorter than F-pili, although length measure-ments are complicated by the fact that isolated pili are typically broken (Fig. 2.2B) (Bradley *et al.*, 1980, Bradley, 1980; Paranchych and Frost, 1988). A third type of pilus-based transfer system that also has been extensively characterized is encoded by the IncI plasmids, as represented by *Pseudomonas aeruginosa* plasmid R64. R64 actually codes for two distinct pili. One pilus, here termed the I-pilus, is elaborated by the same subunits used for assembly of the mating channel. These pili are thick (~9–11 nm) and rigid and confer efficient mating on solid surfaces only. The second pilus is a type IV pilus, which as mentioned above is ancestrally unrelated to the conjugative pili. These pili are thin (~6–7 nm) and flexible, and among their various functions they promote attachment and biofilm formation and a form of motility termed 'twitching motility' (Douzi *et al.*, 2012). Accordingly, the IncI plasmid-encoded type IV pili mediate efficient mating in liquid, most probably due to their capacity to extend and retract dynamically (Horiuchi and Komano, 1998).

2.3.1 F-pili

F-pili assemble by sequential addition of pilin monomers to the base of the growing pilus (Clarke *et al.*, 2008). These pili are composed of a single type of pilin, although the composition and nature of the pilus tip is currently unknown. Early fibre diffraction

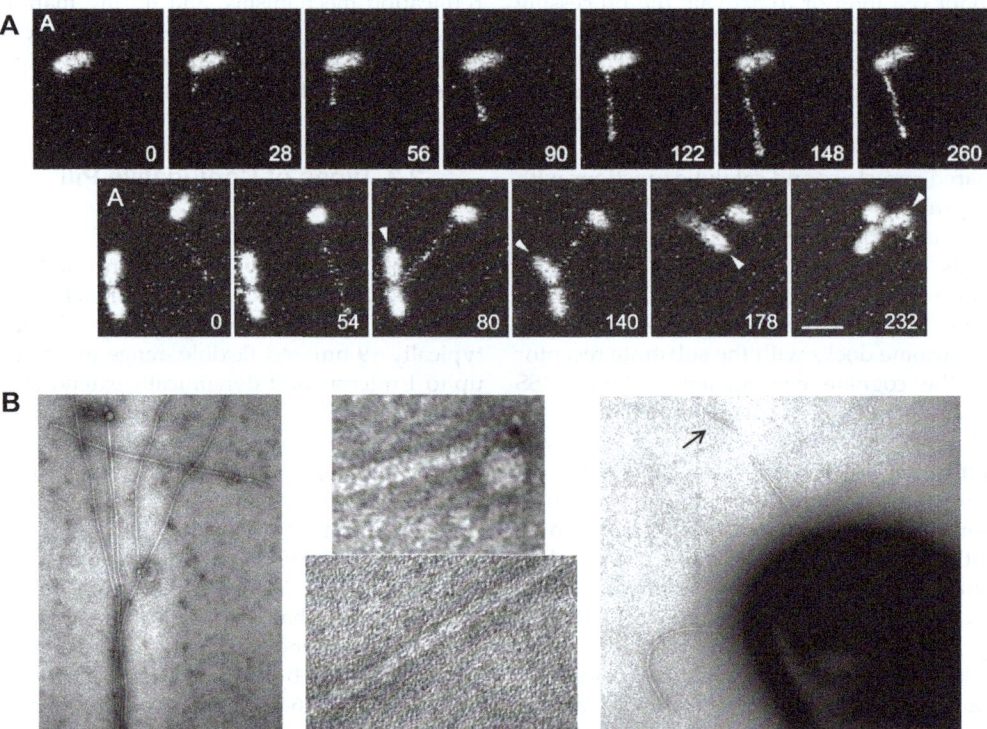

Fig. 2.2. Visualization of F- and P-pili. (A) F-pili were decorated with fluorescent R17 bacteriophage for visualization of F-pilus extension and retraction in real time. Top frames show extension of an F-pilus (thin white filament) from an *E. coli* cell (grey oval) over the indicated times (in seconds) as monitored by fluorescence microscopy. Lower frames show extension (first two frames) and then retraction (subsequent frames) of an F-pilus attached at its tip to a dividing recipient cell, ultimately bringing the donor and recipient cells into direct physical contact. (Images reproduced with permission by P. Silverman.) (B) Electron microscopy images of P-pili elaborated by *A. tumefaciens* VirB/VirD4 T4SS. Left image, P-pilus aggregates in the milieu (magnification 85,000×); Middle images, P-pilus shaft and terminal knob structure (magnification 145,000×), Right image, *A. tumefaciens* cell with polar-localized P-pilus, arrow shows a detached pilus fragment (magnification 30,000×). (Images were taken by E. Cascales and J. Kerr in the Christie laboratory.)

studies established that F-pili are assembled as a one-start helix with a subunit rise of 1.28 nm that was subsequently corrected to a five-start helix when the mass of F-pilin was determined (7.2 kilodaltons (kDa); Silverman, 1997). F-pili are hollow cylinders of ~8.5–9 nm in diameter with an axial lumen of ~3 nm. Intriguingly, recent cryoelectron microscopy (CryoEM) and computational studies showed that F-pili in fact are polymorphic and display two principal helical symmetries. The first is a C4 point symmetry consisting of stacked rings of four radially symmetric subunits. The

adjacent rings were axially spaced at 12.8 Å. A variable inter-ring twist of +/– 20° was also detected, which might account for the capacity of F-pili to withstand torsional stress. The second, a one-start helical symmetry similar to that of the early fibre diffraction studies, displayed a rise of 3.6 units per turn and a pitch of 12.2 Å. Pilus segments with this helical symmetry were thought to be capable of withstanding considerable extension and contraction forces as might be encountered upon engagement of the pilus with recipient cells (Fig. 2.2B) (Wang *et al.*, 2009).

The CryoEM studies provided an architectural view of the F-pilus at a resolution of 13–14 Å. Unfortunately, no X-ray structure exists for the processed mature form of the F-pilin, preventing modelling of pilin packing in the assembled pilus. The F-pilin is termed TraA, and hereafter the source (plasmid or bacterial species) of a given protein will be denoted in subscript, e.g. TraA$_F$. Serological and bacteriophage binding studies have deciphered the general arrangement of TraA$_F$ in the F-pilus. Monoclonal antibodies to N-terminal epitopes failed to bind to TraA$_F$, whereas these antibodies bound to the ends of sonicated pilus fragments suggesting that the N-proximal region of TraA$_F$ lines the base of each layer of pilin in the helix (Frost *et al.*, 1986). By contrast, the C-termini of TraA$_F$ pilins are exposed on the sides of the pilus, as shown both by carboxypeptidase digestion experiments and evidence that F-pili assembled with pilins bearing epitope tags fused to their C termini display these tags along their sides (Frost and Paranchych, 1988; Rondot *et al.*, 1998). Finally, a central hydrophilic domain of the F-pilin (designated domain III; see below) is buried in the pilus and probably lines the inner lumen (Manchak *et al.*, 2002).

2.3.2 P-pili

No high-resolution structures currently exist for P-pili, although various biochemical properties have been defined. Plasmid RP4-encoded P-pili are thick and straight, whereas those elaborated by the *A. tumefaciens* VirB/VirD4 system are more flexuous (Sagulenko *et al.*, 2001; Schroder and Lanka, 2005). P-pili are abundantly present in the extracellular milieu, often as bundles, and are rarely found associated with cells (Fig. 2.2B). Whether this is an artefact accompanying preparation of cells for electron microscopy due to their fragility or is a normally occurring process is not known (Worobec *et al.*, 1986; Sagulenko *et al.*, 2001).

P-type pilins have biochemical properties similar to those described for TraA$_F$ in being small (~7 kDa) proteins with hydrophilic N- and C-termini and two hydrophobic stretches

of ~20–22 residues separated by a small central hydrophilic loop (Kalkum *et al.*, 2002). Recently, accessibility studies of Cys residues introduced along the length of the *A. tumefaciens* VirB2 pilin provided evidence that VirB2$_{At}$ adopts a similar overall architecture as TraA$_F$ in the assembled pilus (Silverman, 1997; Kerr and Christie, 2010). For example, the hydrophobic domains and intervening hydrophilic loops of VirB2$_{At}$ are buried in the pilus, the former probably forming packing interfaces between adjacent pilins or between pilins in adjacent helix stacks and the latter lining the inner lumen. Also, as shown for TraA$_F$, the N-proximal domain of VirB2$_{At}$ is surface-exposed. However, the C terminus of VirB2$_{At}$ seems to be buried and also does not retain function with addition of C-terminal epitopes. Another interesting difference with TraA$_F$ is that a portion of the first hydrophobic domain (designated domain II) of VirB2$_{At}$ is surface-exposed in the assembled pilus. A VirB2$_{At}$ packing geometry that results in patches of surface hydrophobicity might account for the tendency of the P-pili to bundle and induce cellular aggregation (Kerr and Christie, 2010).

Another important distinction between the F- and P-pili derived from an early study in which it was shown that a deletion of the *traC* gene from the IncN plasmid pKM101 can be complemented extracellularly with a helper strain producing *traC* (Winans and Walker, 1985). This finding led to a proposal that TraC$_{pKM101}$ is surface-exposed and functions to mediate contact between pKM101-carrying donor and recipient cells. More recently, the TraC$_{pKM101}$ homologue VirB5$_{At}$ was shown by immunoelectron microscopy to localize at the tip of the VirB/VirD4-encoded pilus (Aly and Baron, 2007). The X-ray structure of TraC$_{pKM101}$ was solved and showed that TraC$_{pKM101}$ comprises a 3-helix bundle flanked by a smaller globular part (Yeo *et al.*, 2003). Structure-function analyses identified residues important for DNA transfer and for binding of pilus-specific bacteriophages. Precisely how VirB5$_{At}$ interacts with VirB2$_{At}$ in the assembled pilus is unknown, but the present findings point to a role for the TraC/VirB5-like subunits in pilus-mediated establishment of donor–recipient cell contacts.

2.3.3 Other conjugative pilus substructures

Other interesting features of conjugative pili have been identified. For example, some pili were found to separate into 'protofilaments', suggesting the possibility that conjugative pili might consist of intertwined ropes of filaments as opposed to the packing of pilins into a single helical fibre (Brinton, 1971). Along these lines, studies showed that cells containing two F-like plasmids, F and R1, assemble 'mixed pili' composed of pilin subunits from both plasmids. Intriguingly, antibodies specific to each pilin type lined up along the length of the mixed pilus consistent with a 'protofilament' architecture (Lawn et al., 1971). In light of the recent structural data (Wang et al., 2009), it is intriguing to ask whether F-pili are restricted to assembly as one fibre with structural features described by Wang et al. (2009) or they might also assemble as a mixture of thinner protofilaments.

Isolated F- and P-pili also have been shown to possess knob structures at their ends (Fig. 2.2B) (Frost et al., 1986; Worobec et al., 1986). Evidence was presented that the knob structures consist of membranous material derived either from donor or recipient cells. The knob structures also consist of pilin subunits, which is interesting in view of evidence that the $TraA_F$ and $VirB2_{At}$ pilins were found to co-fractionate not only with inner membrane but also with outer membrane material (Paiva et al., 1992; Christie, P.J., unpublished data). As discussed further below, the outer membrane-associated pilin potentially corresponds to an intermediate in the pilus assembly pathway or might be involved in mating junction formation.

2.4 Conjugative Pilins and Early Pilin Processing Reactions

F- and P-pili are synthesized as pro-proteins with unusually long (~30–50 residues) leader peptides that are cleaved upon insertion into the inner membrane (Kalkum et al., 2002; Silverman and Clarke, 2010). $TraA_F$ pro-pilin inserts into the membrane by a mechanism dependent on the proton motive force, but independent of the general secretory (Sec) pathway (Majdalani and Ippen-Ihler, 1996). The F-plasmid-encoded membrane protein TraQ also is required for correct orientation and stabilization of $TraA_F$ pro-pilin in the membrane (Maneewannakul et al., 1993).

Upon membrane insertion, the pro-pilins are processed further to yield a membrane pool of mature pilin subunits (Fig. 2.3A). The $TraA_F$ pro-pilin is cleaved of its leader sequence by LepB signal peptidase, and its N terminus is then acetylated by the action of another F plasmid-encoded protein $TraX_F$ (Maneewannakul et al., 1995). Acetylation is not required for assembly of F-pili, but does endow F-pili with certain biochemical properties such as sensitivity to binding of certain RNA bacteriophages along their lengths (Silverman and Clarke, 2010). Upon leader sequence cleavage, the P-pilins can undergo additional proteolytic processing of the C-terminus as shown for $TrbC_{RP4}$ (Haase and Lanka, 1997; Eisenbrandt et al., 1999). Most intriguingly, however, P-pilus maturation involves a head-to-tail cyclization reaction whereby the N- and C-termini of each processed pilin is covalently joined (Eisenbrandt et al., 1999). RP4-encoded TraF catalyses cyclization of $TrbC_{RP4}$, whereas the cyclizing enzyme has not been identified for $VirB2_{At}$ (Eisenbrandt et al., 2000). Cyclization stabilizes the P-type pilins in the membrane and appears to be essential for pilus assembly.

Processed forms of the conjugative pilins typically are composed of alternating stretches of hydrophilic and hydrophobic residues along the length of the protein. For Tra_{AF}, the hydrophilic segments were designated as domains I (residues 1–28) and III (residues 46–49) and hydrophobic segments as domains II (residues 29–45) and IV (residues 50–70) (Silverman, 1997). The P-pilins $TrbC_{RP4}$ and $VirB2_{At}$ have a similar arrangement with the exception that they each contain a short stretch of hydrophilic C-terminal residues (Eisenbrandt et al., 1999; Kerr and Christie, 2010). In the cyclized pilin, however, the N- and C-terminal segments of these pilins are joined, resulting in an overall 4-domain architecture closely resembling that of $TraA_F$ (Eisenbrandt et al., 1999). When inserted into the inner membrane, TraA$_F$, TrbC$_{RP4}$ and VirB2$_{At}$ adopt similar topologies whereby the N/C-terminal domain I is located in the periplasm, the small central domain III is in the cytoplasm and the hydrophobic domains II and IV span the inner membrane (Fig. 2.3A) (Kalkum *et al.*, 2002; Lawley *et al.*, 2003; Kerr and Christie, 2010). As might be expected, mutations in the central loop or the hydrophobic domains that disrupt membrane insertion block further pilus morphogenesis.

The inner membrane pool of TraA$_F$ is estimated at ~100,000 monomers (Silverman, 1997), and a central question is how these pilins are extracted from the membrane to build the pilus. In the Cys-accessibility study mentioned above (Kerr and Christie, 2010), the inner membrane topology of VirB2$_{At}$ was defined and evidence was also presented that the VirB4$_{At}$ ATPase mediates release of VirB2$_{At}$ from the membrane. VirB4$_{At}$ also was shown to release VirB2$_{At}$ into the periplasm by osmotic shock experiments. VirB4$_{At}$

formed an immunoprecipitable complex with VirB2$_{At}$, suggestive of a direct effect of this ATPase on the VirB2$_{At}$ topological state. Finally, a Walker A nucleotide triphosphate binding site mutation abolished all of the observed VirB4-mediated activities. Taken together, these findings prompted a model in which VirB4 catalyses the dislocation of VirB2$_{At}$ monomers from the inner membrane by an ATP-dependent mechanism, presumptively for use in building the P-pilus (Fig. 2.3A). VirB4-like subunits are signatures of all bacterial conjugation machines (Alvarez-Martinez and Christie, 2009), and thus might generally function as pilin dislocases early during morphogenesis of conjugative pili.

2.5 Latter-stage Pilus Assembly

Once pilins are extracted from the membrane, how and where on the cell envelope do they polymerize? To address this question, it is

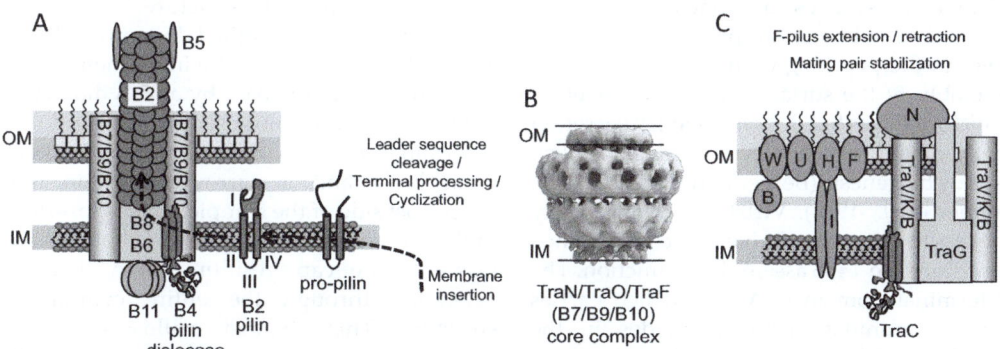

Fig. 2.3. Biogenesis of conjugative pili. (A) A model depicting the assembly pathway for P-pili. The *A. tumefaciens* VirB7, VirB9 and VirB10 subunits form a structural scaffold termed the core complex for pilus assembly across the cell envelope. Pilus polymerization is postulated to initiate on an inner membrane platform composed at least in part of VirB6 and VirB8. VirB2 pilin is inserted into the membrane and processed in several steps to yield a pool of cyclized pilins. Orientation of the two hydrophobic domains (II and IV) and hydrophilic domains (I and IV) of VIrB2 are shown (see text for details). VirB4, aided by VirB11, catalyses dislocation of mature pilins and feeds the pilin monomers into the site of pilus assembly within the core complex. IM, inner membrane; OM, outer membrane. (B) A CryoEM structure of the pKM101 core complex. Both P- and F-pili are postulated to assemble within or at the exterior face of the core complex. (Reproduced with permission by G. Waksman.) (C) F-pilus specific proteins. A TraW, -U, -H, F and TrbI interaction group is (left) required for F-pilus extension and retraction and mating pair formation. TraN interacts with the putative TraV, -K, -B core complex to mediate mating pair stabilization. The C-terminus of VirB6-like TraG is also thought to participate in mating pair stabilization as well as entry exclusion. The only ATPase required for F-pilus biogenesis is VirB4-like TraC. Homologues of VirB1 hydrolase, VirB3 and VirB5 are also required for F-pilus biogenesis (not shown).

first necessary to summarize features of other subunits and subcomplexes required for pilus biogenesis. The *A. tumefaciens* VirB/VirD4 system will serve as a reference for this discussion because VirB subunits are widely conserved among conjugation systems of Gram-negative bacteria (Alvarez-Martinez and Christie, 2009). The 11 VirB subunits and the VirD4 T4CP coordinate assembly of the substrate translocation channel, whereas only the VirB subunits are needed for pilus assembly (Figs 2.3A, B). Contributions of the VirB subunits to pilus biogenesis can be summarized as follows.

2.5.1 VirB1 cell wall hydrolase

VirB1 is a periplasmic protein that harbours a lysozyme-like structural fold. It belongs to a large superfamily of proteins that are associated with bacterial surface structures, including T4SSs, the types II and III secretion systems, type IV pili, DNA uptake systems, flagella and bacteriophage entry systems. In all of these systems, associated lytic trans-glycosylase subunits are proposed to punch holes in the peptidoglycan cell wall to allow assembly of the surface structures. Deletion of *virB1* reduces but does not abolish secretion in *A. tumefaciens*; however, it does abolish T-pilus biogenesis (Berger and Christie, 1994; Fullner *et al.*, 1996). VirB1 is proteolytically processed into two halves, each of which contributes to T4SS assembly or function. The N-terminal domain of VirB1, which carries the muraminidase activity, resides in the periplasm, where it presumably degrades peptidoglycan. The C-terminal domain, known as VirB1*, is secreted into the extra-cellular milieu and is thought to be involved in T-pilus biogenesis (Zupan *et al.*, 2007; Fronzes *et al.*, 2009a).

2.5.2 A pilus structural scaffold – the VirB7, VirB9 and VirB10 core complex

Assembly of P-pili and probably also F-pili requires the elaboration of a cell-envelope-spanning complex termed the core complex. Early studies of the *A. tumefaciens* VirB/VirD4

system established that two outer membrane-associated subunits, lipoprotein $VirB7_{At}$ and $VirB9_{At}$ form a disulfide crosslinked dimer that also interacts with and stabilizes the bitopic subunit VirB10, resulting in formation of an envelope-spanning complex (see Christie *et al.*, 2005). This VirB7/VirB/VirB10 complex, termed the core complex, is in-trinsically stable and it also exerts stabilizing effects on several other VirB channel subunits, suggesting that this complex forms an envelope-spanning structural scaffold early during machine biogenesis.

Recently, a core complex composed of homologs of these subunits was visualized by CryoEM, and a portion was further solved by X-ray crystallography. The pKM101 core complex, composed of 14 copies of each of the VirB7-like TraN, VirB9-like TraO and VirB10-like TraF subunits, is a 1.05 megadalton (MDa) structure, 185 Å wide and 185 Å long, with the potential of spanning the entire Gram-negative bacterial cell envelope (Fig. 2.3B) (Fronzes *et al.*, 2009b). It is composed of two layers (I- and O-layers) that form a double-walled ring-like structure. The I-layer, composed of the N-terminal domains of TraO and TraF, is anchored in the inner membrane and is opened at the base by a 55 Å diameter hole. The O-layer, composed of TraN and the C-terminal domains of TraO and TraF, has a main body and a narrower cap on the outermost side of the complex that is inserted in the outer membrane. A narrow hole (10 Å) exists in the cap, resulting in a channel extending through the entire cylindrical structure. This channel could allow for substrate passage or exchange of small molecule signals between the cytoplasm and the extracellular milieu.

A crystal structure of the entire O-layer reveals that VirB10/TraF forms the outer membrane channel (Chandran *et al.*, 2009). Specifically, 14 copies of a VirB10 domain comprised of two α-helices separated by a loop, termed the antennae projection or AP, form the outer membrane channel. VirB10/TraF subunits are thus unique among described bacterial proteins in spanning the entire cell envelope; the N-terminal TM domain crosses the inner membrane, a Pro-rich region and β-barrel domain extend across the periplasm, and the C-terminal α-helical AP forms an outer membrane pore of 32 Å in diameter (Chandran *et al.*, 2009; Jakubowski *et al.*, 2009).

2.5.3 Other P-pilus assembly factors

Other VirB subunits that are also required for biogenesis of the pilus include the two ATPases, VirB4 and VirB11, and several inner membrane subunits including VirB3, VirB6 and VirB8 (Fig. 2.3A) (Christie *et al.*, 2005). In the CryoEM structure of the pKM101 core complex, the 14 N-terminal N-proximal transmembrane domains of VirB10-like TraF$_{pKM101}$ form a ring of ~55 Å in diameter in the inner membrane (Fronzes *et al.*, 2009b). Therefore, a current model postulates that the ATPases and inner membrane subunits assemble within this ring to form a pilus assembly platform (Fig. 2.3A) (Christie, 2009; Wallden *et al.*, 2010). In agreement with this prediction, recently it was shown by CryoEM that VirB4-like TraB$_{pKM101}$ associates with the side of the pKM101 core complex (Wallden *et al.*, 2012). Although a monomeric form of TraB$_{pKM101}$ associated with the core, a recent crystal structure of VirB4-like TrwER$_{388}$ was solved and shown to assemble as a homo-hexamer (Pena *et al.*, 2012). In modelling studies, TrwE was placed in the centre of the core's inner membrane ring. Regardless of the precise physical relationship, in light of evidence that VirB4$_{At}$ functions as a pilin dislocase (Kerr and Christie, 2010), it is appealing to suggest that the VirB4–core complex interaction serves to promote delivery of membrane-extracted pilin mono-mers into the core's central chamber. Once in the chamber, pilins would then polymerize to build the conjugative pilus (Fig. 2.3A) (Christie, 2009).

Another protein that is critical for bio-genesis of P-pili is the VirB11 ATPase. VirB11 ATPases are members of a large family of structurally conserved secretion ATPases that also are associated with the types II, III and VI secretion systems (Savvides, 2007). Like the other members of this superfamily, VirB11 subunits assemble as double-stacked homo-hexameric rings formed respectively by

N- and C-terminal domains. The VirB11 hexamer is ~110 Å in diameter and ~70 Å in height with a central chamber of ~50 Å in diameter (Yeo *et al.*, 2000; Hare *et al.*, 2006). VirB11 subunits undergo profound structural changes upon binding of ATP, as shown by a combination of X-ray crystallography and analytical ultracentrifugation (Savvides, 2007). In the absence of nucleotide, the N-terminal domains exhibit rigid-body con-formations, and nucleotide binding 'locks' the hexamer into a symmetric and compact structure. Based on these structural findings, it was proposed that VirB11 subunits use the mechanical leverage generated by nucleotide-dependent conformational changes to drive substrate export or assembly of the trans-location channel or conjugative pilus. VirB11's role in assembly of the *A. tumefaciens* P-pilus was confirmed by mutational and protein–protein interaction studies (Sagulenko *et al.*, 2001; Atmakuri *et al.*, 2004). Moreover, in the VirB2$_{At}$ Cys-accessibility study, although VirB4 alone could mediate structural changes in the topological state of membrane-integrated pilin, VirB11 also contributed to structural changes in the pilin (Kerr and Christie, 2010). These findings support a general proposal that VirB4 and VirB11 might coordinate their activities for membrane extraction and delivery of VirB2$_{At}$ monomers into the core's chamber (Fig. 2.3A).

The core's chamber is approximately 100 Å in diameter, sufficiently large to house the conjugative pilus. However, the outer membrane ring formed by TraF's AP domain is at most only 32 Å and clearly not large enough to accommodate pili (Fig. 2.3B) (Fronzes *et al.*, 2009a). If pili assemble from a platform of subunits located at the inner membrane, pilus extension through the chamber and across the outer membrane must induce gross structural changes in the distal portion of the core complex. A model for pilus assembly from an inner membrane platform is reminiscent of that described for type IV pili (Douzi *et al.*, 2012). For F-pilus biogenesis, this is an appealing model because F-pili, like type IV pili, extend and retract dynamically through recruitment from and regeneration of an inner membrane pilin pool. Another model proposed that pili assemble at the outer membrane, on a platform formed by the AP domain of the core complex (Yuan *et al.*, 2005; Christie, 2009). According to this model, pilin subunits would be chaperoned within the core's chamber or directly across the periplasm where they would nucleate upon contact with the AP domain. This assembly pathway would be more reminiscent of another family of extracellular filaments termed P or type I fimbriae that undergo polymerization upon delivery through an outer membrane pore-forming protein termed the usher (Busch and Waksman, 2012). It is somewhat surprising that, despite the recent advances in our structural definition of core complex and the F-pilus, the molecular details of pilus assembly across the cell envelope are still unknown.

2.5.4 The F-pilus assembly machine

F-type transfer systems code for homologues of most of the VirB subunits as well as several other F-specific factors (Fig. 2.3C) (Lawley *et al.*, 2003). Importantly, F-pilus assembly requires TraV, TraK and TraB, which are homologues of the VirB7, VirB9 and VirB10 core complex subunits (Harris *et al.*, 2001). F-pilus assembly therefore likely relies on a core complex structurally similar to that solved for the pKM101 conjugation system. Other VirB homologues required for pilus assembly include the VirB4-like ATPase TraC and VirB3-like TraL, whereas another subunit, TraG, seems to be a chimera of VirB6- and VirB8-like proteins (Lawley *et al.*, 2003; Silverman and Clarke, 2010). Also of interest, the F transfer system codes for VirB5-like TraE, which unlike $VirB5_{At}$ localizes at the inner membrane as opposed to the F-pilus tip (Lawley *et al.*, 2003). F-type systems also code for a VirB1-like transglycosylase (termed Orf169 or P19), which like $VirB1_{At}$ is dispensable for elaboration of the translocation channel.

The P- and F-type systems thus share a common ancestry and VirB-mediated architectural and functional features. The F-type systems, however, also are more complex than P-type systems by virtue of having

acquired additional novel domains and protein subassemblies during evolution (Fig. 2.3C). These systems, for example, code for an outer membrane protein designated TraN. TraN family proteins are large (~600–1200 residues) cysteine-rich proteins thought to function as adhesins for formation of stable donor–recipient cell mating pairs. TraN interacts with the TraV/K/B core complex through the TraV protein (Klimke *et al.*, 2005). The F-type systems also code for a unique set of proteins, TraF, -H, -U, -W and TrbB, -I (Harris and Silverman, 2004; Arutyunov *et al.*, 2010; Silverman and Clarke, 2010). These proteins form an interaction network distinct from the core complex assembled by the VirB-like subunits. The TraF/H/U/W/B/I complex is implicated in the unique capacity of F-type conjugative pili to dynamically extend and retract and confer efficient mating in liquid media. TraF, -H, -U and -W localize at the outer membrane and, like TraN, the TraH and TraU subunits possess a large number of cysteine residues. TraF and the periplasmic protein TrbB possess thioredoxin folds and appear to act in concert with the Dsb proteins to promote the formation or maintenance of disulfide bonds in the Cys-rich TraN, TraH and TraU subunits (Elton *et al.*, 2005; Klimke *et al.*, 2005). TrbI spans the inner membrane and periplasm, making contact with the outer membrane Tra proteins. TrbI mutations do not abolish pilus production but do disrupt pilus retraction (Maneewannakul *et al.*, 1992).

Finally, F-type systems lack a homologue for the VirB11 ATPase (Fig. 2.3C) (Lawley *et al.*, 2003). The F-type systems therefore have only one ATPase, VirB4-like TraC, which is required for production of the F-pilus. Accordingly, TraC must supply all ATP energy-dependent reactions associated with F-pilus extension/retraction dynamics.

Despite holes in our knowledge of pilus assembly across the cell envelope, a recent study provided valuable new insights into F-pilus extension and retraction dynamics in real time in living cells (Clarke *et al.*, 2008). By use of fluorescently labelled phage R17, which binds along the sides of F-pili, it was shown that F-pili extend and retract in an ongoing process regardless of the presence of recipient cells in the vicinity (Fig. 2.2A). Evidence was presented for pilus polymerization by addition of pilins to the base of the pilus, and retraction requiring a time frame of ~5 minutes for completion. Additionally, cells were seen to elaborate new pili prior to completion of retraction of older pili and the number of new pili was not correlated with the number of pre-existing pili, suggesting that initiation and retraction events are randomly timed and not coordinated.

These studies also presented evidence for a twisting or spiral motion of F-pili during extension and retraction (Clarke *et al.*, 2008). These observations supported an early finding using tethered donor cells that F-pili rotate along their longitudinal axis during retraction, causing a bound recipient cell to spin as it is drawn closer to the donor cell. It was proposed that a rotary motion of flexible pili during extension and retraction allows the pili to 'sweep' a large volume around the donor cell. Accordingly, F-pili can be viewed as sensory organelles whose dynamic activities enhance the probability of a productive encounter with a recipient cell. This sensory function would be most useful for liquid matings, whereby the sweeping motion in three dimensions accompanying pilus extension and retraction would enhance F-pilus search efficiency.

The notion that conjugative pili function as sensory nanowires is further supported by two lines of evidence suggesting that F-pili are capable of transducing a signal into the donor cell to activate translocation. First, in the early 1970s it was shown that F-pilus contact with recipient cells was essential for detection of synthesis of the second-strand of F-plasmid to replace the displaced T-strand. This finding indicated that a contact-dependent signal was necessary for initiation of the *oriT*-nicking and T-strand displacement reactions in the donor cell (Ippen-Ihler and Maneewannakul, 1991). Second, recently it was shown that bacteriophage R17, whose receptor is F-pilus, enters plasmid-carrying donor cells only if the relaxosome is docked with the TraD T4CP (Lang *et al.*, 2011). Phage uptake also was found to require *oriT*-nicking, establishing that the relaxosome must be

catalytically active. In this system, the TraD T4CP does not display detectable ATPase activity even when bound by relaxosome. Binding of phage R17 to the conjugative pilus was found to stimulate TraD ATPase activity, in turn resulting in phage uptake (Lang *et al.*, 2011). Thus, on the one hand, phage binding to the F-type pilus requires relaxosome docking at the T4CP to signal that the conjugative channel is 'primed' for translocation of a nucleic acid substrate. On the other hand, phage binding stimulates T4CP ATP hydrolysis, which is critical for delivery of the nucleic acid substrate through the channel. The authors suggest that a combination of contact-mediated extracellular signals (pilus-mediated recipient cell contact, phage binding to the pilus) and intracellular signals (relaxosome–T4CP engagement, T4CP ATP hydrolysis) together activate the conjugative channel either for export of the natural DNA substrate or import of the parasitic phage.

2.6 Contributions of Conjugative Pili and Pilins to Formation of the Mating Junction

For F-type systems, DNA transfer is detectable in less than 10 minutes upon mixing of donor and recipient cells, with complete conversion of a recipient cell population to donor status within 30 minutes. Transfer rates for the F plasmid were estimated at about 45 kilobases/minute in both liquid and solid surface matings (Wilkins and Frost, 2001). The P-type systems are equally efficient provided that mating is done on solid surfaces. A substantial body of evidence suggests that the pili are not specificity determinants, that is, they do not bind specific receptors on the surfaces of recipient cells. Instead, they mediate a nonspecific interaction that is characterized as a loose association, e.g. an interaction easily disrupted by vortexing or other mild shear forces. This loose association is converted to a tight mating junction through the actions of the TraN and TraG mating pair stabilization proteins (Lawley *et al.*, 2003). A tight mating junction is characterized by its resistance to disruption by high shear forces. Studies have attempted without much success to identify the surface components on recipient cells that are required for establishment of a tight mating junction. In general, two classes of mutations have been isolated, those in the outer membrane protein OmpA and those in the lipopolysaccharide (LPS) biosynthetic pathway (Anthony et al., 1994). For the F-systems, TraN was shown to bind both OmpA and LPS and thus is an important surface component of the donor cell for formation of tight mating junctions (Klimke et al., 2005).

The nature and composition of the mating junction is entirely unknown at this time. For RP4-carrying (P-type) donor cells, mating junctions were detected at poles and along the lengths of cells by electron microscopy. Junctions at cell poles were estimated to be ~150–200 nm in length, whereas those along the cell body extended up to 1500 nm in length or up to half the lengths of the cells (Samuels et al., 2000). Intriguingly, neither the core complex nor any other structures have been detected at the mating junction (Samuels et al., 2000). These findings have supported proposals that surface adhesins, e.g. TraN, or outer-membrane-bound forms of pilins mediate mating junction formation. The inability to detect mating channels also raises the intriguing possibility that these channels form only transiently upon surface contact by donor and recipient cells.

Interestingly, there is some evidence for DNA transfer through the F-pilus. For example, DNA transfer was shown to occur between cells that were separated by a filter (Harrington and Rogerson, 1990). More recently, this concept was advanced with a novel mating assay enabling visualization of DNA transfer by fluorescence microscopy (Babic et al., 2008). In a mixed population of donor and recipient cells fixed on a glass slide, DNA transfer could be detected in the absence of apparent donor–recipient cell contact. Substrate transfer through an extended F-pili is formally possible in view of the recent CryoEM study confirming that the central lumen of the F-pilus is ~3 nm, a diameter sufficiently large to accommodate ssDNA and unfolded protein(s) (Wang et al., 2009). It is important to note, however, that the reported frequencies for DNA transfer at a distance are far below known F-plasmid transfer rates. Thus, while this mechanism of transfer might occur infrequently, the evidence still weighs heavily in favour of the notion that the conjugative pilus functions as an attachment organelle to initiate the donor–recipient contact and not as a conduit for transmission of DNA substrates.

2.7 Beyond Gene Transfer: Conjugative Pili and Biofilm Formation

Conjugative DNA transfer is a principal mechanism underlying genome plasticity and evolution. On a more immediate time scale, conjugation enables adaptation and survival of bacterial hosts to changing environmental conditions (for reviews on this topic, see Phillips and Funnell, 2004). Although mutations enabling DNA transfer in the absence of detectable pili can be isolated in the laboratory, natural conjugation systems with this Tra[+], Pil[-] phenotype are not readily identified, suggesting that there is a strong selective pressure to retain this organelle among the conjugative machinery of Gram-negative cells. F-pili clearly confer enhanced DNA transfer frequencies in liquid environments. Moreover, the capacity of both F- and P-pili to promote nonspecific aggregation can confer several selective advantages. For example, conjugative pili have been shown to promote formation of biofilms by facilitating attachment of donor cells to both biotic and abiotic surfaces (Ghigo, 2001; Molin and Tolker-Nielsen, 2003; Reisner et al., 2003, 2006). As biofilms provide some protection against predation and fluctuations in natural environmental conditions through transmission of fitness traits, e.g. antibiotic resistance genes, conjugative pili clearly can provide a direct selective advantage to bacteria in such settings. The contribution of conjugative pili to colonization of human tissues has not been extensively studied, although the demonstrated contributions of conjugative pili to attachment, biofilm formation and dissemination of adaptive traits establishes a role for these organelles as pathogenicity determinants (Ong *et al.*, 2009).

Conjugative pili also promote biofilm formation in ways other than by mediating attachment and aggregation. For example, several studies have presented evidence that biogenesis of the conjugative pilus indirectly induces biofilm formation via activation of envelope stress responses. Elaboration of the F transfer system was found to sensitize cells to bile salts and SDS, thus, it was postulated that the translocation channel itself might mediate uptake of envelope-damaging compounds resulting in the envelope stress response (Bidlack and Silverman, 2004). It has also been proposed that a large number of assembly intermediates or dead-end complexes form during machine biogenesis that induce the stress response. A recent microarray analysis further established a causal relationship between biogenesis of the conjugative pilus machinery, induction of envelope-stress response systems, repression of motility genes, and upregulation of biofilm formation genes (Yang *et al.*, 2008). While most studies exploring the contribution of conjugative pili to biofilm formation have focused on derepressed systems that constitutively express F-type pili at all times, studies of non-derepressed F systems corroborate the general finding that production of conjugative pili stimulate biofilm formation. Cells carrying these natural F-type systems typically elaborate very few F-pili and conjugate at much lower frequencies than the derepressed systems. Even so, it was shown that elaboration of F-pili by natural systems correlates with increased expression of genes involved in colonic acid and curli production, both of which were linked to maturation of dense biofilms. The natural F-type systems were shown to activate expression of the genes encoding the EnvZ/OmpR two-component regulatory system, which in turn regulates production of curli (May and Okabe, 2008). Thus, even conjugation systems mediating low-frequency DNA transfer in the environment can strongly impact biofilm formation indirectly through induction of envelope stress response systems.

2.8 Adapted Conjugation Systems: The T4SS Effector Translocators

As mentioned in the Introduction, the conjugation systems are the ancestral progenitors of the larger T4SS superfamily. Accordingly, most T4SSs of the effector translocator subfamily employ a conjugation-like mechanism to deliver their protein cargoes to eukaryotic target cells. The relatedness of conjugation and effector translocator subfamilies is highlighted by the 'dual function' systems. These systems have evolved specialized functions as effector translocators during the course of infection, but also have retained their ancestral conjugation function vis-à-vis the capacity to translocate DNA substrates intercellularly. The *A. tumefaciens* VirB/VirD4 system is the best known of these systems. This T4SS delivers protein and DNA effector molecules to plant cells to incite Crown Gall disease, but it also can mobilize the transfer of a small, nonself-transmissible IncQ plasmid to other *A. tumefaciens* cells (Cascales and Christie, 2003). In this system, the P-pilus initiates contacts with plant tissue, presumptively followed by formation of a mating junction between bacterial and plant membranes.

Two other 'dual functions' have been described. In *Legionella pneumophila*, the Dot/Icm T4SS is now known to deliver an impressive 250 or more effector proteins to eukaryotic target cells during the course of infection (Zhu *et al.*, 2011). However, the Dot/Icm system also can mobilize the transfer of IncQ plasmid to bacterial recipients by a conjugative mechanism (Vogel *et al.*, 1998). Interestingly, the Dot/Icm system is related to the I-type translocation systems mentioned earlier, but unlike these systems, the Dot/Icm system does not appear to elaborate conjugative pili. Rather, it elaborates a fibrous mesh that covers the cell surface. This fibrous mesh is composed of DotO and DotH subunits and is thought to facilitate specific stages of the *L. pneumophila* infection cycle (Watarai *et al.*, 2000). This mesh likely also mediates interbacterial DNA transfer, making this one of the few known conjugation systems in Gram-negative bacteria capable of mediating DNA transfer in the absence of a conjugative pilus.

Bartonella henselae uses a VirB/VirD4-like T4SS to translocate seven *Bartonella* effector proteins (Beps) into human endothelial cells. Noted similarities between the translocation signals (BID domains) and signals associated with conjugative relaxases prompted a study of whether *Bartonella* spp. could mobilize the transfer of DNA into human cells. Very intriguingly, *B. henselae* was shown to deliver DNA through its VirB/VirD4 system into human cells, where the DNA recombined into the human genome (Schroder *et al.*, 2011; Llosa *et al.*, 2012). The VirB/VirD4 system has not been shown to elaborate a pilus structure, although this is formally possible. However, *Bartonella* spp. also code for a second T4SS, designated Trw for its strong relatedness (up to 80% amino acid sequence identity in individual components) to the Trw conjugation system encoded by *E. coli* plasmid R388. Indeed, several *trw* genes from *Bartonella* were shown to complement the respective *trw* gene deletions in R388 (Seubert *et al.*, 2003; de Paz *et al.*, 2005), underscoring the high degree of structural and functional conservation of individual subunits of these functionally diversified T4SSs. In contrast to the R388 system, however, the *Bartonella* Trw system lacks a T4CP and it codes for multiple copies of homologs of VirB2 and VirB5, VirB6, and VirB7. The *Bartonella* Trw system thus has probably lost its capacity to translocate substrates and instead displays variant forms of surface-exposed pili or pilins. The Trw system is essential for erythrocyte invasion, and it is postulated that the variant pili might facilitate interactions with different erythrocyte receptors, either within the reservoir host population (e.g. different blood group antigens) or among different reservoir hosts (Dehio, 2008).

Other T4SSs of the effector translocator family have not been shown to mediate DNA transfer, but nevertheless possess the machine infrastructure for elaboration of a pilus organelle or, alternatively, pilins or other adhesins on the cell surface. As with the *Bartonella* Trw system, *Rickettsial* spp. elaborate T4SSs with multiple copies of the VirB2 and VirB5 homologues that, if displayed on the cell surface, might facilitate attachment to novel cell types or mediate evasion of the host immune system (Lockwood *et al.*, 2011). In *Helicobacter pylori*, the Cag pathogenicity island mediates the formation of sheathed appendages of 100 to 200 nm in length (Rohde *et al.*, 2003). The needle portion of this structure has a diameter of 40 nm, and the sheathed structure has a diameter of 70 nm. Subunits associated with this sheathed structure include HP0527 (VirB10-like domain), HP0532 (VirB7-like lipoprotein), HP0528 (VirB9-like) and, at its tip, the CagA effector protein (Rohde *et al.*, 2003; Tanaka *et al.*, 2003; Tegtmeyer *et al.*, 2011). *H. pylori* also appears to elaborate another type of Cag-dependent organelle structurally more similar to conjugative pili; these organelles are composed of HP0546 (VirB2$_{At}$-like) and CagL (VirB5$_{At}$-like) (Terradot and Waksman, 2011). Interestingly, CagL carries an Arg-Gly-Asp (RGD) motif, which was shown to mediate binding to and activation of $\beta 1$ integrin receptors on mammalian epithelial cells (Backert *et al.*, 2008). A recent study further showed that CagL and two CagL homologues, CagH and CagI, interact with each other and are important for pilus assembly (Shaffer *et al.*, 2011). The relationship between the sheathed structure and the VirB2/VirB5-like pilus is not clear at this time.

2.9 Conclusions

This chapter has attempted to integrate research findings for conjugative pili over the past six decades with the more recent structural data generated with state-of-the-art fluorescence and electron microscopy techniques. The continued application of both classical genetic/biochemical approaches and modern high-resolution imaging technologies is essential for developing a complete mechanistic and structural picture of the process of conjugation. The high-resolution studies promise to answer many of the questions still surrounding these fascinating organelles. For example, what is the physical relationship of the pilus with the translocation channel and the role of the core complex in assembly of both structures? What are the structures of pilins and how do they assemble in the pilus? And what are the architectures of mating junctions and the roles of conjugative pili in establishment of these junctions? While answering these questions will generate a gratifying visual blueprint, the classical approaches together with modern cell biological techniques are needed decipher the underlying mechanistic details. These approaches ultimately will answer questions like: how do the energy-powering ATPase subunits and the proton motive force contribute to pilus morphogenesis? How is F-pilus extension and retraction mechanistically achieved? And what are the intra- and extracellular signals and the mechanisms of signal transduction that lead to machine assembly and function? Finally, it is intriguing to recall that conjugation machines are ancestrally and mechanistically related to the effector translocator systems, and that both of these T4SS subfamilies have acquired novel domains, proteins and protein subassemblies during evolution. Understanding this mosaicism in structural and functional detail remains an exciting challenge that assuredly will occupy the attention of many researchers in this field for the foreseeable future.

References

Alvarez-Martinez, C.E. and Christie, P.J. (2009) Biological diversity of prokaryotic type IV secretion systems. *Microbiology and Molecular Biology Reviews* 73, 775–808.

Aly, K.A. and Baron, C. (2007) The VirB5 protein localizes to the T-pilus tips in *Agrobacterium tumefaciens*. *Microbiology* 153, 3766–3775.

Anthony, K.G., Sherburne, C., Sherburne, R. and Frost, L.S. (1994) The role of the pilus in recipient cell recognition during bacterial conjugation mediated by F-like plasmids. *Molecular Microbiology* 13, 939–953.

Arutyunov, D., Arenson, B., Manchak, J. and Frost, L.S. (2010) F plasmid TraF and TraH are components of an outer membrane complex involved in conjugation. *Journal of Bacteriology* 192, 1730–1744.

Atmakuri, K., Cascales, E. and Christie, P.J. (2004) Energetic components VirD4, VirB11 and VirB4 mediate early DNA transfer reactions required for bacterial type IV secretion. *Molecular Microbiology* 54, 1199–1211.

Ayers, M., Howell, P.L. and Burrows, L.L. (2010)

Architecture of the type II secretion and type IV pilus machineries. *Future Microbiology* 5, 1203–1218.

Babic, A., Lindner, A.B., Vulic, M., Stewart, E.J. and Radman, M. (2008) Direct visualization of horizontal gene transfer. *Science* 319, 1533–1536.

Backert, S. and Selbach, M. (2008) Role of type IV secretion in *Helicobacter pylori* pathogenesis. *Cellular Microbiology* 10, 1573–1581.

Backert, S., Fronzes, R. and Waksman, G. (2008) VirB2 and VirB5 proteins: specialized adhesins in bacterial type-IV secretion systems? *Trends in Microbiology* 16, 409–413.

Berger, B.R. and Christie, P.J. (1994) Genetic complementation analysis of the *Agrobacterium tumefaciens virB* operon: *virB2* through *virB11* are essential virulence genes. *Journal of Bacteriology* 176, 3646–3660.

Bidlack, J.E. and Silverman, P.M. (2004) An active type IV secretion system encoded by the F plasmid sensitizes *Escherichia coli* to bile salts. *Journal of Bacteriology* 186, 5202–5209.

Bradley, D.E. (1980) Morphological and serological relationships of conjugative pili. *Plasmid* 4, 155–169.

Bradley, D.E., Taylor, D.E. and Cohen, D.R. (1980) Specification of surface mating systems among conjugative drug resistance plasmids in *Escherichia coli* K-12. *Journal of Bacteriology* 143, 1466–1470.

Brinton, C.C. Jr (1971) The properties of sex pili, the viral nature of 'conjugal' genetic transfer systems, and some possible approaches to the control of bacterial drug resistance. *CRC Critical Reviews in Microbiology* 1, 105–160.

Busch, A. and Waksman, G. (2012) Chaperone-usher pathways: diversity and pilus assembly mechanism. *Philosophical Transactions of the Royal Society B: Biological Sciences* 367, 1112–1122.

Cascales, E. and Christie, P.J. (2003) The versatile bacterial type IV secretion systems. *Nature Reviews Microbiology* 1, 137–150.

Cascales, E. and Christie, P.J. (2004) Definition of a bacterial type IV secretion pathway for a DNA substrate. *Science* 304, 1170–1173.

Chandran, V., Fronzes, R., Duquerroy, S., Cronin, N., Navaza, J. and Waksman, G. (2009) Structure of the outer membrane complex of a type IV secretion system. *Nature* 462, 1011–1015.

Christie, P.J. (2004) Type IV secretion: the *Agrobacterium* VirB/D4 and related conjugation systems. *Biochimica et Biophysica Acta* 1694, 219–234.

Christie, P.J. (2009) Structural biology: translocation chamber's secrets. *Nature* 462, 992–994.

Christie, P.J. and Vogel, J.P. (2000) Bacterial type IV secretion: conjugation systems adapted to deliver effector molecules to host cells. *Trends in Microbiology* 8, 354–360.

Christie, P.J., Atmakuri, K., Krishnamoorthy, V., Jakubowski, S. and Cascales, E. (2005) Biogenesis, architecture, and function of bacterial type IV secretion systems. *Annual Review of Microbiology* 59, 451–485.

Clarke, M., Maddera, L., Harris, R.L. and Silverman, P.M. (2008) F-pili dynamics by live-cell imaging. *Proceedings of the National Academy of Sciences of the USA* 105, 17978–17981.

De La Cruz, F., Frost, L.S., Meyer, R.J. and Zechner, E.L. (2010) Conjugative DNA metabolism in Gram-negative bacteria. *FEMS Microbiology Reviews* 34, 18–40.

De Paz, H.D., Sangari, F.J., Bolland, S., Garcia-Lobo, J.M., Dehio, C., De La Cruz, F. and Llosa, M. (2005) Functional interactions between type IV secretion systems involved in DNA transfer and virulence. *Microbiology* 151, 3505–3516.

Dehio, C. (2008) Infection-associated type IV secretion systems of *Bartonella* and their diverse roles in host cell interaction. *Cellular Microbiology* 10, 1591–1598.

Douzi, B., Filloux, A. and Voulhoux, R. (2012) On the path to uncover the bacterial type II secretion system. *Philosophical Transactions of the Royal Society B: Biological Sciences* 367, 1059–1072.

Eisenbrandt, R., Kalkum, M., Lai, E.M., Lurz, R., Kado, C.I. and Lanka, E. (1999) Conjugative pili of IncP plasmids, and the Ti plasmid T pilus are composed of cyclic subunits. *Journal of Biological Chemistry* 274, 22548–22555.

Eisenbrandt, R., Kalkum, M., Lurz, R. and Lanka, E. (2000) Maturation of IncP pilin precursors resembles the catalytic dyad-like mechanism of leader peptidases. *Journal of Bacteriology* 182, 6751–6761.

Elton, T.C., Holland, S.J., Frost, L.S. and Hazes, B. (2005) F-like type IV secretion systems encode proteins with thioredoxin folds that are putative DsbC homologues. *Journal of Bacteriology* 187, 8267–8277.

Fronzes, R., Christie, P.J. and Waksman, G. (2009a) The structural biology of type IV secretion systems. *Nature Reviews Microbiology* 7, 703–714.

Fronzes, R., Schafer, E., Wang, L., Saibil, H.R., Orlova, E.V. and Waksman, G. (2009b) Structure of a type IV secretion system core complex. *Science* 323, 266–268.

Frost, L.S. and Paranchych, W. (1988) DNA sequence analysis of point mutations in traA, the F pilin gene, reveal two domains involved in F-specific bacteriophage attachment. *Molecular and General Genetics* 213, 134–149.

Frost, L.S., Lee, J.S., Scraba, D.G. and Paranchych, W. (1986) Two monoclonal antibodies specific for different epitopes within the amino-terminal region of F pilin. *Journal of Bacteriology* 168, 192–198.

Fullner, K.J., Lara, J.C. and Nester, E.W. (1996) Pilus assembly by Agrobacterium T-DNA transfer genes. *Science* 273, 1107–1109.

Ghigo, J.M. (2001) Natural conjugative plasmids induce bacterial biofilm development. *Nature* 412, 442–445.

Gomis-Ruth, F.X., Moncalian, G., Perez-Luque, R., Gonzalez, A., Cabezon, E., De La Cruz, F. and Coll, M. (2001) The bacterial conjugation protein TrwB resembles ring helicases and F1-ATPase. *Nature*, 409, 637–641.

Haase, J. and Lanka, E. (1997) A specific protease encoded by the conjugative DNA transfer systems of IncP and Ti plasmids is essential for pilus synthesis. *Journal of Bacteriology* 179, 5728–5735.

Hare, S., Bayliss, R., Baron, C. and Waksman, G. (2006) A large domain swap in the VirB11 ATPase of *Brucella suis* leaves the hexameric assembly intact. *Journal of Molecular Biology* 360, 56–66.

Harrington, L.C. and Rogerson, A.C. (1990) The F pilus of *Escherichia coli* appears to support stable DNA transfer in the absence of wall-to-wall contact between cells. *Journal of Bacteriology* 172, 7263–7274.

Harris, R.L. and Silverman, P.M. (2004) Tra proteins characteristic of F-like type IV secretion systems constitute an interaction group by yeast two-hybrid analysis. *Journal of Bacteriology* 186, 5480–5485.

Harris, R.L., Hombs, V. and Silverman, P.M. (2001) Evidence that F-plasmid proteins TraV, TraK and TraB assemble into an envelope-spanning structure in *Escherichia coli*. *Molecular Microbiology* 42, 757–766.

Horiuchi, T. and Komano, T. (1998) Mutational analysis of plasmid R64 thin pilus prepilin: the entire prepilin sequence is required for processing by type IV prepilin peptidase. *Journal of Bacteriology* 180, 4613–4620.

Ippen-Ihler, K. and Maneewannakul, S. (1991) Conjugation among enteric bacteria: mating systems dependent on expression of pili. In: Dworkin, M. (ed.) *Microbial Cell–Cell Interactions*. ASM, Washington, DC, pp. 35–69.

Jakubowski, S.J., Kerr, J.E., Garza, I., Krishnamoorthy, V., Bayliss, R., Waksman, G. and Christie, P.J. (2009) *Agrobacterium* VirB10 domain requirements for type IV secretion and

T pilus biogenesis. *Molecular Microbiology* 71, 779–794.

Kalkum, M., Eisenbrandt, R., Lurz, R. and Lanka, E. (2002) Tying rings for sex. *Trends in Microbiology* 10, 382–387.

Kerr, J.E. and Christie, P.J. (2010) Evidence for VirB4-mediated dislocation of membrane-integrated VirB2 pilin during biogenesis of the *Agrobacterium* VirB/VirD4 type IV secretion system. *Journal of Bacteriology* 192, 4923–4934.

Klimke, W.A., Rypien, C.D., Klinger, B., Kennedy, R.A., Rodriguez-Maillard, J.M. and Frost, L.S. (2005) The mating pair stabilization protein, TraN, of the F plasmid is an outer-membrane protein with two regions that are important for its function in conjugation. *Microbiology* 151, 3527–3540.

Lang, S., Kirchberger, P.C., Gruber, C.J., Redzej, A., Raffl, S., Zellnig, G., Zangger, K. and Zechner, E.L. (2011) An activation domain of plasmid R1 TraI protein delineates stages of gene transfer initiation. *Molecular Microbiology* 82, 1071–1085.

Lawley, T.D., Klimke, W.A., Gubbins, M.J. and Frost, L.S. (2003) F factor conjugation is a true type IV secretion system. *FEMS Microbiology Letters* 224, 1–15.

Lawley, T., Wilkins, B.M. and Frost, L.S. (2004) Conjugation in Gram-negative bacteria. In: Funnell, B.E. and Phillips, G.J. (eds) *Plasmid Biology*. ASM Press, Washington, DC, pp. 203–226.

Lawn, A.M., Meynell, E. and Cooke, M. (1971) Mixed infections with bacterial sex factors: sex pili of pure and mixed phenotype. *Annales de l'Institut Pasteur* 120, 3–8.

Llosa, M., Schroder, G. and Dehio, C. (2012) New perspectives into bacterial DNA transfer to human cells. *Trends in Microbiology* 20, 355–359.

Lockwood, S., Voth, D.E., Brayton, K.A., Beare, P.A., Brown, W.C., Heinzen, R.A. and Broschat, S.L. (2011) Identification of *Anaplasma marginale* type IV secretion system effector proteins. *PLoS ONE* 6, e27724.

Majdalani, N. and Ippen-Ihler, K. (1996) Membrane insertion of the F-pilin subunit is Sec independent but requires leader peptidase B and the proton motive force. *Journal of Bacteriology* 178, 3742–3747.

Manchak, J., Anthony, K.G. and Frost, L.S. (2002) Mutational analysis of F-pilin reveals domains for pilus assembly, phage infection and DNA transfer. *Molecular Microbiology* 43, 195–205.

Maneewannakul, S., Maneewannakul, K. and Ippen-Ihler, K. (1992) Characterization, locali-zation, and sequence of F transfer region products: the pilus assembly gene product TraW and a new product, TrbI. *Journal of Bacteriology* 174, 5567–5574.

Maneewannakul, K., Maneewannakul, S. and Ippen-Ihler, K. (1993) Synthesis of F pilin. *Journal of Bacteriology* 175, 1384–1391.

Maneewannakul, K., Maneewannakul, S. and Ippen-Ihler, K. (1995) Characterization of *traX*, the F plasmid locus required for acetylation of F-pilin subunits. *Journal of Bacteriology* 177, 2957–2964.

May, T. and Okabe, S. (2008) *Escherichia coli* harboring a natural IncF conjugative F plasmid develops complex mature biofilms by stimulating synthesis of colanic acid and curli. *Journal of Bacteriology* 190, 7479–7490.

Molin, S. and Tolker-Nielsen, T. (2003) Gene transfer occurs with enhanced efficiency in biofilms and induces enhanced stabilisation of the biofilm structure. *Current Opinion in Biotechnology* 14, 255–261.

Nagai, H. and Roy, C.R. (2003) Show me the substrates: modulation of host cell function by type IV secretion systems. *Cellular Microbiology* 5, 373–383.

Ong, C.L., Beatson, S.A., McEwan, A.G. and Schembri, M.A. (2009) Conjugative plasmid transfer and adhesion dynamics in an *Escherichia coli* biofilm. *Applied and Environmental Microbiology* 75, 6783–6791.

Paiva, W.D., Grossman, T. and Silverman, P.M. (1992) Characterization of F-pilin as an inner membrane component of *Escherichia coli* K12. *Journal of Biological Chemistry* 267, 26191–26197.

Paranchych, W. and Frost, L.S. (1988) The physiology and biochemistry of pili. *Advances in Microbial Physiology* 29, 53–114.

Pena, A., Matilla, I., Martin-Benito, J., Valpuesta, J.M., Carrascosa, J.L., De La Cruz, F., Cabezon, E. and Arechaga, I. (2012) The hexameric structure of a conjugative VirB4 protein ATPase provides new insights for a functional and phylogenetic relationship with DNA translocases. *Journal of Biological Chemistry* 287, 39925–39932.

Phillips, G.J. and Funnell, B.E. (eds) (2004) *Plasmid Biology*. ASM Press, Washington DC.

Reisner, A., Haagensen, J.A., Schembri, M.A., Zechner, E.L. and Molin, S. (2003) Development and maturation of *Escherichia coli* K-12 biofilms. *Molecular Microbiology* 48, 933–946.

Reisner, A., Holler, B.M., Molin, S. and Zechner, E.L. (2006) Synergistic effects in mixed *Escherichia coli* biofilms: conjugative plasmid transfer drives biofilm expansion. *Journal of Bacteriology* 188, 3582–3588.

Rohde, M., Puls, J., Buhrdorf, R., Fischer, W. and Haas, R. (2003) A novel sheathed surface organelle of the *Helicobacter pylori cag* type IV secretion system. *Molecular Microbiology* 49, 219–234.

Rondot, S., Anthony, K.G., Dubel, S., Ida, N., Wiemann, S., Beyreuther, K., Frost, L.S., Little, M. and Breitling, F. (1998) Epitopes fused to F-pilin are incorporated into functional recombinant pili. *Journal of Molecular Biology* 279, 589–603.

Sagulenko, E., Sagulenko, V., Chen, J. and Christie, P.J. (2001) Role of *Agrobacterium* VirB11 ATPase in T-pilus assembly and substrate selection. *Journal of Bacteriology* 183, 5813–5825.

Samuels, A.L., Lanka, E. and Davies, J.E. (2000) Conjugative junctions in RP4-mediated mating of *Escherichia coli. Journal of Bacteriology* 182, 2709–2715.

Savvides, S.N. (2007) Secretion superfamily ATPases swing big. *Structure* 15, 255–257.

Schroder, G. and Lanka, E. (2005) The mating pair formation system of conjugative plasmids: a versatile secretion machinery for transfer of proteins and DNA. *Plasmid* 54, 1–25.

Schroder, G., Schuelein, R., Quebatte, M. and Dehio, C. (2011) Conjugative DNA transfer into human cells by the VirB/VirD4 type IV secretion system of the bacterial pathogen *Bartonella henselae. Proceedings of the National Academy of Sciences of the USA* 108, 14643–14648.

Seubert, A., Hiestand, R., De La Cruz, F. and Dehio, C. (2003) A bacterial conjugation machinery recruited for pathogenesis. *Molecular Microbiology* 49, 1253–1266.

Shaffer, C.L., Gaddy, J.A., Loh, J.T., Johnson, E.M., Hill, S., Hennig, E.E., McClain, M.S., McDonald, W.H. and Cover, T.L. (2011) *Helicobacter pylori* exploits a unique repertoire of type IV secretion system components for pilus assembly at the bacteria–host cell interface. *PLoS Pathogens* 7, e1002237.

Silverman, P.M. (1997) Towards a structural biology of bacterial conjugation. *Molecular Microbiology* 23, 423–429.

Silverman, P.M. and Clarke, M.B. (2010) New insights into F-pilus structure, dynamics, and function. *Integrative Biology* 2, 25–31.

Tanaka, J., Suzuki, T., Mimuro, H. and Sasakawa, C. (2003) Structural definition on the surface of *Helicobacter pylori* type IV secretion apparatus. *Cellular Microbiology* 5, 395–404.

Tegtmeyer, N., Wessler, S. and Backert, S. (2011) Role of the cag-pathogenicity island encoded type IV secretion system in *Helicobacter pylori* pathogenesis. *FEBS Journal* 278, 1190–1202.

Terradot, L. and Waksman, G. (2011) Architecture of the *Helicobacter pylori* Cag-type IV secretion system. *FEBS Journal* 278, 1213–1222.

Thanassi, D.G., Bliska, J.B. and Christie, P.J. (2012) Surface organelles assembled by secretion systems of Gram-negative bacteria: diversity in structure and function. *FEMS Microbiology Review* 36, 1046–1082.

Vogel, J.P., Andrews, H.L., Wong, S.K. and Isberg, R.R. (1998) Conjugative transfer by the virulence system of *Legionella pneumophila. Science* 279, 873–876.

Wallden, K., Rivera-Calzada, A. and Waksman, G. (2010) Type IV secretion systems: versatility and diversity in function. *Cellular Microbiology* 12, 1203–1212.

Wallden, K., Williams, R., Yan, J., Lian, P.W., Wang, L., Thalassinos, K., Orlova, E.V. and Waksman, G. (2012) Structure of the VirB4 ATPase, alone and bound to the core complex of a type IV secretion system. *Proceedings of the National Academy of Sciences of the USA* 109, 11348–11353.

Wang, Y.A., Yu, X., Silverman, P.M., Harris, R.L. and Egelman, E.H. (2009) The structure of F-pili. *Journal of Molecular Biology* 385, 22–29.

Watarai, M., Andrews, H.L. and Isberg, R. (2000) Formation of a fibrous structure on the surface of *Legionella pneumophila* associated with exposure of DotH and DotO proteins after intracellular growth. *Molecular Microbiology* 39, 313–329.

Wilkins, B.M. and Frost, L.S. (2001) Mechanisms of genetic exchange between bacteria. In: Sussman, M. (ed.) *Molecular Medical Microbiology.* Academic Press, London, pp. 355–400.

Winans, S.C. and Walker, G.C. (1985) Conjugal transfer system of the IncN plasmid pKM101. *Journal of Bacteriology* 161, 402–410.

Worobec, E.A., Frost, L.S., Pieroni, P., Armstrong, G.D., Hodges, R.S., Parker, J.M., Finlay, B.B. and Paranchych, W. (1986) Location of the antigenic determinants of conjugative F-like pili. *Journal of Bacteriology* 167, 660–665.

Yang, X., Ma, Q. and Wood, T.K. (2008) The R1 conjugative plasmid increases *Escherichia coli* biofilm formation through an envelope stress response. *Applied Environmental Microbiology* 74, 2690–2699.

Yeo, H.J., Savvides, S.N., Herr, A.B., Lanka, E. and Waksman, G. (2000) Crystal structure of the hexameric traffic ATPase of the *Helicobacter pylori* type IV secretion system. *Molecular Cell* 6, 1461–1472.

Yeo, H.-J., Yuan, Q., Beck, M.R., Baron, C. and Waksman, G. (2003) Structural and functional characterization of the VirB5 protein from the type IV secretion system encoded by the conjugative plasmid pKM101. *Proceedings of the National Academy of Sciences of the USA* 100, 15947–15952.

Yuan, Q., Carle, A., Gao, C., Sivanesan, D., Aly, K.A., Hoppner, C., Krall, L., Domke, N. and Baron, C. (2005) Identification of the VirB4-VirB8-VirB5-VirB2 pilus assembly sequence of type IV secretion systems. *Journal of Biological Chemistry* 280, 26349–26359.

Zechner, E.L., Lang, S. and Schildbach, J.F. (2012) Assembly and mechanisms of bacterial type IV secretion machines. *Philosophical Transactions of the Royal Society B: Biological Sciences* 367, 1073–1087.

Zhu, W., Banga, S., Tan, Y., Zheng, C., Stephenson, R., Gately, J. and Luo, Z.Q. (2011) Comprehensive identification of protein substrates of the Dot/Icm type IV transporter of *Legionella pneumophila*. *PLoS ONE* 6, e17638.

Zupan, J., Hackworth, C.A., Aguilar, J., Ward, D. and Zambryski, P. (2007) VirB1* promotes T-pilus formation in the vir-Type IV secretion system of *Agrobacterium tumefaciens*. *Journal of Bacteriology* 189, 6551–6563.

3 Pilus Biogenesis by the Chaperone–Usher Pathway

Gilles Phan and Gabriel Waksman

Institute of Structural and Molecular Biology, University College London and Birkbeck College, London, UK

3.1 Introduction

Gram-negative bacteria display a variety of multi-subunit fibrillar structures on their cell surface called 'pili' or 'fimbriae'. Originally based on the filament morphology and the biogenesis machinery, pili are currently classified into five different families (Fronzes *et al.*, 2008; Thanassi *et al.*, 2012): (i) the curli, (ii) the type IV pili, (iii) the type III secretion system needle, (iv) the type IV secretion system pili and (v) the chaperone–usher (CU) pili. Each family is involved in broad and diverse functions such as export of virulence factors (type III secretion needle and type IV secretion system pili), DNA uptake and transformation (type IV secretion system pili and the non-related type IV pili), twitching motility (type IV pili) or adhesion and invasion of host cells (curli, type IV pili and CU pili). Among pathogenic bacteria, pili are often essential virulence factors mediating attachment to target cells, evasion from the host immune systems or bacterial aggregation into a protective biofilm. Therefore, pili are attractive targets for drug development (Cusumano and Hultgren, 2009; Durand *et al.*, 2009).

This review focuses on the biogenesis machinery of one of the most widespread and important microbial adhesion factor, the CU

pili (Nuccio and Baumler, 2007; Waksman and Hultgren, 2009; Zav'yalov *et al.*, 2010). Described for the first time by Duguid *et al.* (1955) as surface appendages that allow *Escherichia coli* to bind and agglutinate erythrocytes, CU pili are made of different subunits that polymerize to either form thin flexible fibres or rigid long filaments. The main function of the CU pili is the adhesion to the host cells, mediated by a specialized pilus subunit, the adhesin, often containing a lectin domain that specifically recognizes the lining sugars of the target tissue, thereby defining the host tropism. Generally clustered under the same promoter, all the necessary components for the CU pili biogenesis are encoded within the same operon and it has been shown that multiple CU pathway clusters may be present within the same genome, extending the range of receptors the same bacterium might be able to adhere to (Korea *et al.*, 2010; Felek *et al.*, 2011).

As indicated by its name, CU pili biogenesis involved a duet of specialized proteins: (i) a periplasmic chaperone that stabilizes newly synthesized pilus subunits and (ii) an outer-membrane transporter, the usher, that catalyses subunit polymerization and translocation through the cell envelope. CU pathways have been under extensive investigation, leading to the emergence of

© CAB International 2014. *Bacterial Pili: Structure, Synthesis and Role in Disease* (eds M.A. Barocchi and J.L. Telford)

three model systems: the type 1 pili (Fim system, mainly responsible for bladder infection), the P pili (Pap system, mainly involved in kidney infection), both from the uropathogenic *E. coli* (UPEC), and the F1 pili, commonly named F1 capsule (Caf system) from the plague agent *Yersina pestis* (Fig. 3.1).

One major challenge to secretion by Gram-negative bacteria is the presence of two lipoid membranes separated by a periplasmic space devoid of classical cellular energy fuel such as ATP or the proton motive force. The strategy adopted by the CU pathway is a two-step secretion mechanism (Rego *et al.*, 2010). First, pilus subunits are secreted through the inner membrane by the general secretory pathway (GSP) which is an ATP-dependent process (Stathopoulos *et al.*, 2000; Driessen and Nouwen, 2008; Chatzi *et al.*, 2013). Then, once released and appropriately capped by the periplasmic chaperone, the newly formed

chaperone:subunit complexes are ferried to the outer membrane usher, which catalyses their subsequent assembly. This process is energetically driven by the intrinsic folding energy of the subunit, which transitions from a high-energy state stabilized by the chaperone to a low-energy state upon chaperone release and incorporation into the growing pilus (Busch and Waksman, 2012).

3.2 Structure, Function and Polymerization of the CU Pili

CU pili are non-covalent polymers of different subunits, the composition and arrangement of which vary from one system to another. As a result, clear morphological differences appear between UPEC and F1 pili. P and type 1 pili are composed of two subassemblies: (i) a long right-handed helical

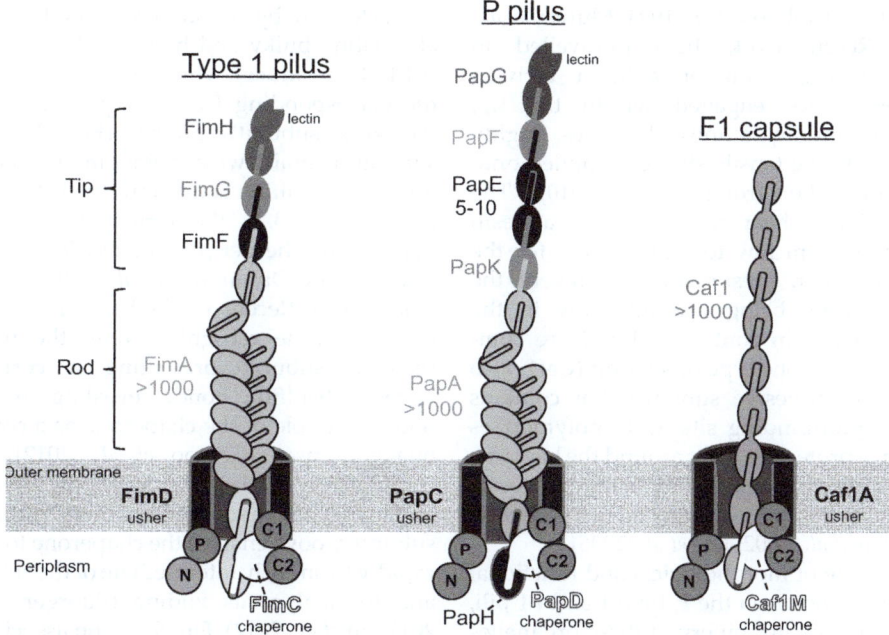

Fig. 3.1. Schematic representation of three different chaperone–usher pili systems. The type 1 pili and the P pili from uropathogenic *Escherichia coli*, and the F1 capsule from *Yersina pestis*. Each pilus is built up with different combination of subunits (Fim, Pap or Caf), represented in different shades of grey. The type 1 and P pili are made of two distinct regions: a flexible end tip and a rigid long rod. The pili expose their adhesion subunit (Adhesin) at the cell surface and are anchored at the bacterial outer membrane by the usher. The usher is constituted by a transmembrane pore domain and four soluble domains: the N-terminal domain (N), the plug domain (P), the C-terminal domains (C1 and C2).

rod (approximately 7 nm in diameter and 2 μm long) extending from the bacterial outer membrane, and (ii) a flexible end tip (approximately 2 nm in diameter and 15 nm in length) that carries the adhesion activity, i.e. the adhesin subunit (Hahn *et al.*, 2002; Mu and Bullitt, 2006). In contrast, the F1 capsule is made of shorter, linear and flexible polymers (2 nm in diameter) that tend to aggregate with each other to form a dense and amorphous coating (Miller *et al.*, 1998; Zavialov *et al.*, 2003). The helical rod of the P and type 1 pili are built up of thousands copies of a major subunit (respectively PapA and FimA), followed by a terminal flexible tip comprising various minor subunits (PapK, PapE, PapF in the P pilus and FimF, FimG in the type 1 pilus) (Fig. 3.1). The tip carries at its very top the critical adhesin subunit, PapG in P pili, that binds to the Galα1-4Gal of the kidney glycolipids, or FimH in type 1 pili that binds to mannosylated receptors at the surface of the bladder epithelium (Roberts *et al.*, 1994; Mulvey *et al.*, 1998). Recent works have unravelled an allosteric regulation of adhesin activity, whereby, once engaged within the tip, receptor-binding affinity becomes highly sensitive to the tensile strength applied onto the pilus (Le Trong *et al.*, 2010). The application of shear stress causes the adhesin switching from low to high affinity for the cognate sugar, presumably to prevent the bacteria from being washed away in the urinary tract. In contrast, F1 pili are composed of only one type of subunit (Caf1, also called polyadhesive subunit) that contains the receptor-binding site. Caf1 polymerizes to form a protective layer around the bacteria, preventing phagocytosis by macrophages or internalization by respiratory tract epithelial cells (Du *et al.*, 2002; Liu *et al.*, 2006).

In spite of morphological and functional differences between the P, type 1 and F1 pili, the biogenesis machinery of these organelles shares a common and remarkably well-conserved mechanism. Most pilus subunits are proteins consisting of a single 'pilin' domain. The exception to this rule is the adhesin subunit, which is generally made of two domains, a pilin domain and a receptor-binding or adhesin domain. Pilin domains

share the same structure, a C-terminally-truncated single immunoglobulin (Ig)-like fold preceded by an unstructured but essential N-terminal extension (Nte) of 10–20 residues (Fig. 3.2). Pilin domains indeed lack the seventh strand of a canonical Ig fold, resulting in a solvent exposed hydrophobic groove. Consequently, the newly synthesized subunit is not stable on its own and has to be stabilized *in trans* by a dedicated periplasmic chaperone (PapD, FimC or Caf1M respectively for P, type 1 and F1 pili), a 25 kDa protein consisting of two Ig-like domains arranged in a boomerang shape. The chaperone stabilizes the subunit by inserting one of its own strands (the G1 strand) into the subunit's exposed groove, an interaction termed 'donor-strand complementation' (DSC) (Choudhury *et al.*, 1999; Sauer *et al.*, 1999; Zavialov *et al.*, 2003). Importantly, DSC prevents non-productive aggregation or proteolytic degradation of the subunit in the periplasm. The chaperone G1 strand is characterized by a conserved motif of four alternating bulky and hydrophobic residues (P1 to P4 residues), which are inserted inside the corresponding P1 to P4 pockets of the receiving subunit's groove (Fig. 3.2). All subunits contain two cysteines involved in an invariant disulfide bond between β-strands A and B (Fig. 3.2), the formation of which depends on the periplasmic oxydoreductase DsbA (Jacob-Dubuisson *et al.*, 1994; Totsika *et al.*, 2009). Recent works have shown that the chaperone recognizes only the newly secreted subunit presenting a correctly formed disulfide bond, revealing another regulatory role of the chaperone as a control quality sensor (Crespo *et al.*, 2012). The combination of a large hydrophobic interface and main chain hydrogen bonding within the subunit groove enables the chaperone to bind rapidly to an early intermediate of the subunit and to catalyse its folding (Puorger *et al.*, 2011; Yu *et al.*, 2012). Finally, upon association with the subunit, a conformational change triggered by the rotation of two prolines at the beginning of the chaperone F1–G1 loop allows the chaperone:subunit complex to bind to the usher. By this recently described allosteric regulation, called 'two prolines lock' mechanism, the usher is able to discriminate between subunit-bound and free chaperone molecules (Di Yu *et al.*, 2012). These results illustrate more key functions of the chaperone in priming the subunit for the assembly process.

During pilus assembly, each pilus subunit becomes non-covalently linked by their Nte. Indeed, during assembly, the Nte of the subunit next in assembly replaces the chaperone G1 strand by a mechanism known as 'donor-strand exchange' (DSE) (Sauer *et al.*, 2002; Zavialov *et al.*, 2003) (Fig. 3.2). As observed for the chaperone's G1 strand, the Nte is also characterized by a conserved motif of alternating hydrophobic residues, termed P2 to P5 residues, interacting with the receiving subunit P2 to P5 pockets (Fig.

3.2). Importantly, the subunit P5 pocket is not occupied in the chaperone–subunit complex and this turns out to be mandatory for the initiation of the DSE mechanism (Verger *et al.*, 2006). As demonstrated by competition experiments, real-time native mass spectrometry and molecular dynamic simulation, the Nte progressively replaces the G1 strand in a concerted manner, comparable to a zippering mechanism, starting with the insertion of the incoming Nte's P5 residue into the receiving groove's P5 pocket, then followed by the progressive insertion (or zip-in) of the Nte into the receiving subunit's groove and the concerted withdrawal (or zip-out) of the chaperone's strand from that groove (Fig. 3.2) (Remaut *et al.*, 2006; Vetsch

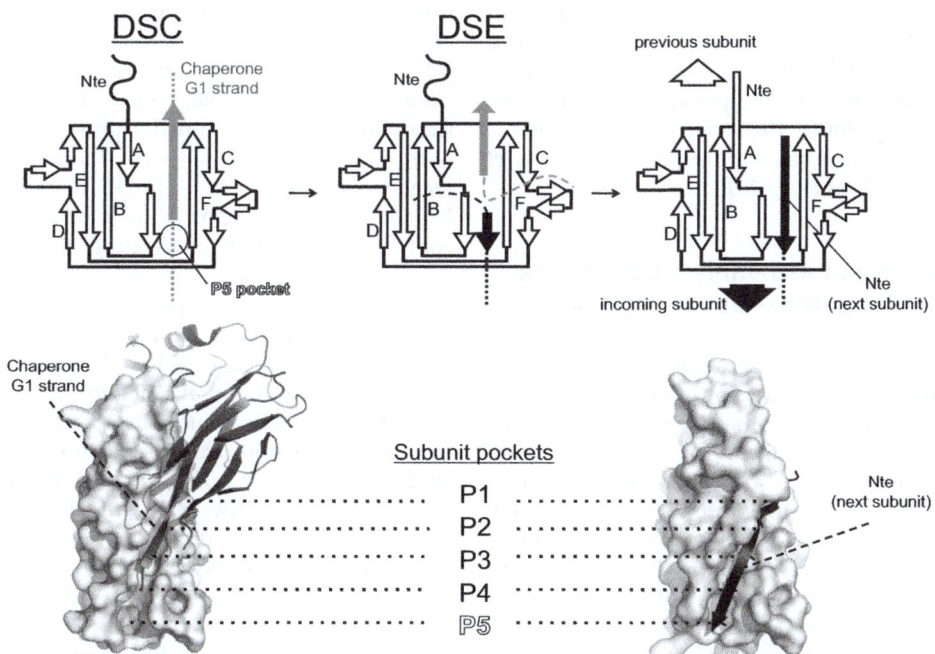

Fig. 3.2. Molecular details of the pilus polymerization mechanism. The donor strand complementation (DSC) and donor strand exchange (DSE) mechanisms are shown schematically (above) and through crystal structures (below, based on PDB entry 3JWN). The chaperone (in ribbon representation) donates one of its strands (the G1 strand) to complete the subunit Ig-like fold (in surface representation), occupying the subunit pockets P1 to P4 but leaving the P5 pocket free. During subunit polymerization, the N-terminal extension (Nte) of the next subunit in assembly (black) replaces progressively the chaperone's G1 strand in a zip-in–zip-out manner, starting by the insertion of the incoming subunit Nte's P5 residue into the receiving subunit's P5 pocket and progressively zippering in to occupy the P4 to P2 pockets of the receiving subunit.

et al., 2006; Vitagliano *et al.*, 2007; Rose *et al.*, 2008).

Pilus subunits polymerize in a defined order (Fig. 3.1) and the order of assembly within the pilus is dictated by the stereo-chemical fit between Ntes and grooves. Rose *et al.* (2008) examined the rate of DSE between subunits in the absence of the usher and found that DSE rates are higher between pairs of subunits known to associate naturally (termed 'cognate' subunits). Thus the order in which subunits come together in the pilus is kinetically determined. Leney *et al.* (2011) further examined the dependence of DSE rates on Nte residues and highlighted the importance of the P5 residues in determining DSE rates. Thus, subunit assembly *in vitro* and in the absence of the usher is ruled by a single interaction, that between the P5 residues in incoming subunits' Ntes and the P5 pocket of receiving subunits. While the same chaperone is able to accommodate a variety of subunits, the specific amino acid composition at the P5 pocket and its relative flexibility define the strict preference of the Nte for the cognate binding partner (Leney *et al.*, 2011; Ford *et al.*, 2012).

3.3 Pilus Assembly Model by the Chaperone–Usher Pathway

Pilus assembly is catalysed at the outer membrane by a conserved membrane transporter named the 'usher' (PapC, FimD and Caf1A respectively for the P, type 1 and F1 pili) (Allen *et al.*, 2012). In this review, FimD is used to illustrate the usher function since it is the best characterized in terms of structure and activity so far (Nishiyama *et al.*, 2005, 2008; Phan *et al.*, 2011). The usher consists of five distinct domains (Fig. 3.3A): an N-terminal domain (NTD), which is known to form the recruitment site for chaperone:subunit complexes (Nishiyama *et al.*, 2003, 2005; Eidam *et al.*, 2008; Henderson *et al.*, 2011), a 24 β-stranded pore domain embedded in the outer membrane and providing an exit portal for the nascent pilus (Remaut *et al.*, 2008), a plug domain sealing the lumen of the pore when the usher is not active (apo-usher) and preventing random

leakage across the outer membrane (Remaut *et al.*, 2008; Huang *et al.*, 2009), and two C-terminal domains (CTD1 and CTD2), which have been shown to form a second binding site for chaperone:subunit complexes (Dubnovitsky *et al.*, 2010; Ford *et al.*, 2010; Phan *et al.*, 2011).

3.3.1 Usher activating mechanism

It is known that recruitment of the chaperone:adhesin complex (FimC:FimH) is required for initiation of pilus biogenesis: without FimC:FimH, the usher is unable to make pili. Thus, recruitment of FimC:FimH to the usher must trigger some conformational changes within the usher, involving a structural transition to an active form. Crystal structures of the apo-usher pore domain and investigation of its gating mechanism by electrophysiology have shown a closed β-barrel with a dynamic plug domain (Remaut *et al.*, 2008; Mapingire *et al.*, 2009). Comparison of the apo-usher pore structure with that of the same pore domain in a ternary complex of FimD bound to FimC:FimH (Fig. 3.3A) shows a dramatic conformational change affecting the plug, which, from its initial position inside the barrel in the apo-usher pore, swings out of the barrel in the ternary complex to give way to the adhesin domain of FimH (Fig. 3.3 B). Another conformational change occurring upon FimC:FimH recruitment is a change in the shape of the pore, from elongated, kidney-shaped, to round. These conformational changes affecting the plug and the barrel must be part of the usher activating mechanism: indeed, the plug must swing out to let the nascent pilus through and the pore must adjust its shape to that of the nascent pilus. However, the full extent of the FimC:FimH-induced usher activating mechanism, notably the conformational changes affecting the usher NTD and CTDs relative to the barrel and the plug, is unclear, due to the lack of a full-length apo-usher structure.

The usher NTD has been identified as the recruitment site for all the chaperone:subunit substrates (Nishiyama *et al.*, 2005; Eidam *et al.*, 2008; Henderson *et al.*, 2011). It has

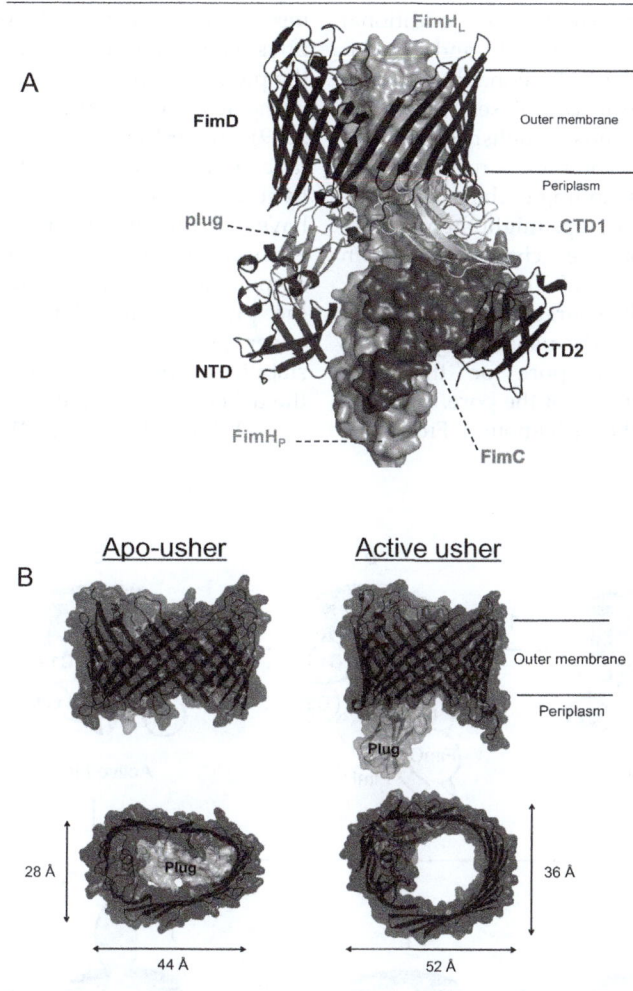

Fig. 3.3. Structural biology of the usher. (A) Crystal structure of the usher:chaperone:adhesin (FimD:FimC:FimH) complex from the type 1 pilus (PDB code 3RFZ). In the active conformation of the full-length usher FimD, the plug domain (plug) is displaced from the lumen of the β-barrel and interacts with the N-terminal domain (NTD) at the periplasmic side. The adhesin subunit (FimH) is inserted through the pore via its adhesin domain (FimH$_L$), whereas the pilin domain (FimHp) remains bound to the chaperone (FimC). The two C-terminal domains of the usher (CTD1 and CTD2) interact with the translocated FimC:FimH complex. (B) Crystal structures of the usher pore domain in two different conformations. The lengthwise and transverse views of the usher pore domain are represented for the apo-usher (left, PDB code 3OHN) and the active usher (right, PDB code 3RFZ). The Cα-Cα dimensions of the transversal view are indicated. The plug domain occludes the pore in the apo-usher, whereas it is displaced toward the periplasmic space in the active usher.

been demonstrated that the differential binding affinities for the different subunits reflect the order of assembly (Nishiyama and Glockshuber, 2010; Volkan *et al.*, 2012), with the chaperone:adhesin complex exhibiting the highest affinity for the usher NTD among all chaperone:subunit complexes. Interactions between the usher NTD and the chaperone:subunit complexes mostly involve the chaperone and the pilin domains of subunits (Nishiyama *et al.*, 2005; Eidam *et al.*, 2008; Di Yu *et al.*, 2012), but in the case of chaperone:adhesin complexes, additional interactions between the NTD and the adhesin domain explains the enhanced affinity of chaperone:adhesin complexes (Morrissey *et al.*, 2012). It is thus established that the chaperone:adhesin complex is recruited first to the usher via its binding to the usher NTD.

Two alternative steps might then operate after binding of the chaperone:adhesin complex to the usher NTD (Fig. 3.4A, step 1): either a yet-to-be-characterized allosteric mechanism unlocks the usher plug from its position inside the pore or the plug spontaneously flips out of the pore. Once the pore is opened, the NTD-bound FimH can

insert into it (Fig. 3.4A, view b). Electrophysiology experiments have suggested that the plug transiently flips out of the pore even in the absence of substrates (Mapingire *et al.*, 2009), suggesting that the plug might spontaneously give way to FimH. It is interesting to note that FimH is the only pilus subunit to have two domains and as a result, is extended enough to reach out to the pore when its pilin domain is still NTD-bound (Fig. 3.4A). Once the plug is out, interactions between the usher plug and the usher NTD probably stabilize the plug in its open form, maintaining the usher in its activated conformation (Phan *et al.*, 2011; Volkan *et al.*, 2012).

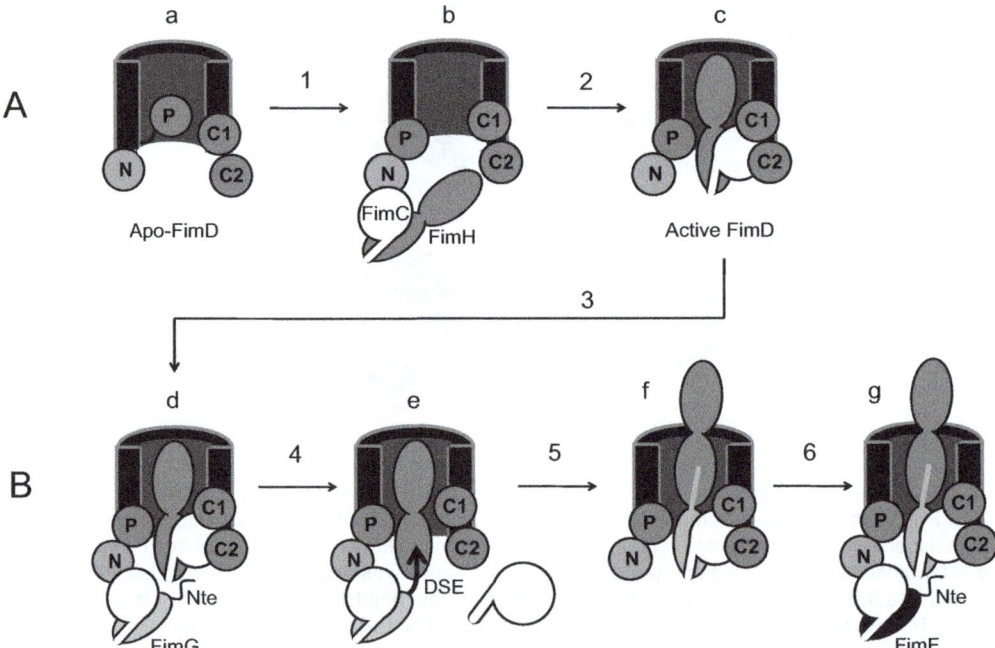

Fig. 3.4. Model for the subunit incorporation cycle by the chaperone–usher pathway. (A) Activation of the apo-usher. The apo-usher (view a) is occluded by its plug domain. In this view, only the structures of the usher pore and plug have been experimentally derived. The position of the usher NTD and CTDs is unknown. The recruitment of the FimC:FimH complex (step 1) at the usher NTD (view b) initiates usher activation. FimC:FimH subsequently inserts inside the usher pore domain (step 2) and the complex transfers to the usher CTDs (view c). (B) Subunits incorporation cycle. Once activated by the FimC:FimH complex (view c), the next subunit FimG is recruited at the usher NTD (step 3, view d). FimH undergoes DSE with FimG (step 4), releasing the chaperone of FimH (view e). FimG then transfers to the usher CTDs with concomitant translocation of FimH through the pore (step 5). The FimD:FimC:FimG:FimH complex (view f) is then ready for the next round of subunit recruitment (step 6) with FimF binding to the usher NTD (view g).

In the ternary complex of FimD bound to FimC:FimH, the chaperone:adhesin complex is seen bound to the CTDs (Fig. 3.4A, view c). Thus, after binding of FimC:FimH to the NTD (Fig. 3.4A, view b), displacement of the plug and subsequent insertion of FimH inside the pore, the chaperone:adhesin complex must be transferred to the CTDs. How this step occurs is still unclear. However, Volkan *et al.* (2012) have shown that CTD2 alone is able to displace chaperone:subunit complexes from the NTD. CTD2 was also shown to be able to bind chaperone:subunit complexes. These results suggest that CTD2 alone is able to operate the transfer of the chaperone:adhesin complex from the NTD to the CTDs, by actively promoting dissociation from the NTD and association with the CTDs.

3.3.2 The subunit incorporation cycle

The crystal structure of the FimD:FimC:FimH complex revealed a secondary binding site for chaperone:subunit complexes at the CTDs (Phan *et al.*, 2011), this was in addition to the already well-characterized site formed by the NTD (Nishiyama *et al.*, 2005). The NTD was shown to operate first (Phan *et al.*, 2011). Thus, the usher is endowed with two chaperone:subunit binding sites, a primary one (the NTD) where chaperone:subunit complexes are first recruited and a secondary one (the CTDs) where chaperone:subunit complexes appear to transfer subsequently. Modelling of the next chaperone:subunit complex in assembly (FimC:FimG) to the usher NTD in the FimD:FimC:FimH structure, using the known structure of a chaperone:subunit complex bound to an isolated usher NTD (Nishiyama *et al.*, 2005) (Fig. 3.4B, view d), shows FimG perfectly poised to undergo DSE with FimH: its Nte is not only being oriented toward the receiving groove of FimH, but also the FimG Nte's P5 residue is located right next to the FimH P5 pocket. Thus, the mechanism of pilus biogenesis by the usher follows a well-defined cycle of subunit incorporation into the nascent pilus: the next incoming chaperone:subunit complex is recruited to the usher NTD (Fig. 3.4, step 3); this binding

to the NTD positions the incoming subunit for DSE with the receiving subunit bound to the CTDs; DSE leads to the dissociation of the chaperone bound to the receiving subunit, thereby freeing the usher CTDs (Fig. 3.4B, step 4 and view e); the transfer of the incoming chaperone:subunit complex to the CTDs then occurs, likely catalysed by CTD2 (Fig. 3.4B, step 5); this frees the NTD site that can then be used for the recruitment of the next incoming chaperone:subunit complex, initiating a new round of subunit incorporation (Fig. 3.4B, step 6). Therefore, the usher's ability to catalyse pilus assembly resides in the optimal positioning of the subunits bound to its two chaperone:subunit binding sites prior to DSE and the alternating mobilization of these sites. The usher can thus be described as a nanomachine that cycles through a number of conformational and binding states to recruit, assemble and secrete pilus subunits.

The mechanism outlined above is thought to apply to any pili assembled by any chaperone:usher pathways. In this model, subunits are constantly assembled one at a time in a defined top to down order. For the P and type 1 pili class, pilus biogenesis starts with the adhesin and the following tip subunits to proceed with the subsequent incorporation of circa 1000 copies of the major subunit that constitutes the rod. As outlined above, the order in which subunits assemble at the usher is likely to be crucially dependent on the Nte-groove interaction and more specifically on the interaction between the incoming subunit's P5 residue and the receiving subunit's P5 pocket. However, it is also likely that the affinity of the primary recruitment site, the usher NTD, for chaperone:subunit complexes as well as the concentration of chaperone:subunit complexes within the periplasm play a role. These issues were investigated recently (Nishiyama *et al.*, 2008; Allen *et al.*, 2013), taking advantage of the remarkable demonstration by Nishiyama *et al.* (2008) that the usher is fully functional in detergents. Indeed, the usher FimD can be extracted from outer membranes in detergents, purified to homogeneity and is still capable of incorporating subunits when provided with pure and homogeneous preparations of chaperone:subunit complexes. This breakthrough paved the way for further biophysical investigations of the usher by Allen *et al.* (2013) which exploited a DSE monitoring systems set up by Phan *et al.* (2011), based on fluorescence labelling and polyacrylamide gel electrophoresis of subunit–subunit products. FimD usher-mediated DSE rates were then experimentally derived (Allen *et al.*, 2013). As for DSE rates measured in the absence of the usher, it was shown that usher-mediated DSE rates are the highest between cognate pairs, except for the FimH–FimG pair that exhibited DSE rates similar to those of the FimH–FimF pair. Crucially, the usher accelerates DSE rates about 100-fold. A kinetic model integrating experimentally derived usher-mediated DSE rates, chaperone:subunit complex affinities for the usher NTD and hypothetical concentrations of chaperone:subunit complexes in the periplasm have provided quantitative system-level insights into the assembly of the type 1 pilus.

Usher dimers have repeatedly been observed (Li *et al.*, 2004; Remaut *et al.*, 2008; Huang *et al.*, 2009) and an earlier model for the usher-mediated subunit incorporation cycle was proposed that invoked a functional usher dimer (Remaut *et al.*, 2008). At the time, only the usher NTD was known to bind chaperone:subunit complexes. Since there is a need of another chaperone:subunit binding site for the incorporation of the next subunit, two ushers would have been essential. However, the discovery of a second binding site within the usher monomer (the CTDs) made this assumption unnecessary. Experimental evidences have since weighed in favour of a functional monomeric usher. The FimD:FimC:FimH structure clearly demonstrates a 1:1:1 stoichiometry and AUC data by Allen *et al.* (2013) demonstrates that the same complex remains over a large concentration range and that within this concentration range, the complex is active. Thus monomeric FimD is fully functional. The dimeric usher model was also flawed since the second usher was hypothesized to only contribute its NTD to the dimeric usher mechanism, a considerable waste of energy: why would the cell produce a 830 residues protein to only

use 125 of them at the N-terminus? Thus, the reported observations of usher dimers are likely the results of the high concentration used experimentally.

During assembly, all subunits emerge from the usher pore linearly since its dimension can only accommodate one subunit at a time. However, after exiting the usher pore, the polymer is known to adopt a quaternary structure that is believed to provide the energy that pulls the polymer out of the pore towards the extracellular milieu. For example, the major subunit forming the rod in P and type 1 pili is known to form a right-handed superhelix of 3.3 subunits per turn. The formation of this superhelical structure at the exit of the usher potentially provides a driving force for pilus export. Finally, pilus biogenesis in P pili is known to end once a termination subunit (termed PapH) is incorporated into the pilus. PapH acts as a termination subunit because it lacks a P5 pocket, thus preventing other subunits from inserting into its groove. No such subunit has been identified for the type 1 pilus assembly termination, suggesting that a PapH-type termination mechanism is not universally employed.

3.4 Conclusion

The CU pathway is the best understood bacterial secretion system. Despite the diversity of CU pili structure and organization, from rigid rods to thin capsule organelles, the assembly and secretion mechanism of CU pili is conserved: a specialized periplasmic chaperone catalysing pilus subunit folding and escorting them to a multidomain membrane transporter that orchestrates the assembly of the pilus and its exposure to the bacterial cell surface. The 24 β-stranded usher is the largest known outer membrane β-barrel protein (Fairman *et al.*, 2011) and the crystal structures of the closed and active usher provide the first details of the structural rearrangements required for gating an outer membrane pore. Remarkably, no external energy is required to move the process forward: the chaperone fine-tunes the subunit's energy level so that it is the subunit itself that provides its own folding energy for its own assembly, the usher catalysing the assembly process by optimally positioning the various reaction partners.

Given the relevance of pili in bacterial pathogenesis, it is not surprising that the structural and molecular knowledge accumulated on the CU pathway has been exploited to design a new class of compounds, termed 'pilicides', that efficiently inhibit pilus biogenesis by disrupting a number of steps including chaperone:subunit complex formation, and recruitment of chaperone:subunit complexes to the usher NTD (Pinkner *et al.*, 2006; Chorell *et al.*, 2010). Such compounds are effective in not only preventing attachment to the bladder epithelium but also preventing biofilm formation (Cegelski *et al.*, 2009; Chorell *et al.*, 2012). The recent elucidation of previously uncharacterized steps in the usher-mediated subunit incorporation cycle has unravelled novel protein–protein interactions that can also be targeted for disruption. Uropathogenic infections are highly recurrent due to the capacity of bacterial strains to escape antibiotics treatment by forming so-called intracellular bacterial communities (IBCs) (Hannan *et al.*, 2012), bacterial reservoirs formed after infection. Thus, while antibiotics treatment appears to have cleared the infection, the pathogen lies dormant, waiting to be deployed again, an occurrence that does not require re-infection but just stress-induced release from the IBCs. Since IBCs formation is crucially dependent on the type 1 pili, pilicides would be useful tools to prevent recurrence. Thus, pilicides in conjunction with more classical antibiotics treatment might provide useful avenues for treatment of bacterial infections.

Acknowledgement

This work is supported by the Medical Research Council (MRC grant 85602).

References

Allen, W.J., Phan, G. and Waksman, G. (2012) Pilus biogenesis at the outer membrane of Gram-negative bacterial pathogens. *Current Opinion in Structural Biology* 22, 500–506.

Allen, W.J., Phan, G., Hultgren, S.J. and Waksman, G. (2013) Dissection of pilus tip assembly by the FimD usher monomer. *Journal of Molecular Biology* http://dx.doi.org/10.1016/j.jmb.2012.12.024.

Busch, A. and Waksman, G. (2012) Chaperone-usher pathways: diversity and pilus assembly mechanism. *Philosophical Transactions of the Royal Society B: Biological Sciences* 367, 1112–1122.

Cegelski, L., Pinkner, J.S., Hammer, N.D., Cusumano, C.K., Hung, C.S., Chorell, E., Aberg, V., Walker, J.N., Seed, P.C., Almqvist, F., Chapman, M.R. and Hultgren, S.J. (2009) Small-molecule inhibitors target *Escherichia coli* amyloid biogenesis and biofilm formation. *Nature Chemical Biology* 5, 913–919.

Chatzi, K.E., Sardis, M.F., Karamanou, S. and Economou, A. (2013) Breaking on through to the other side: protein export through the bacterial Sec system. *Biochemical Journal* 449, 25–37.

Chorell, E., Pinkner, J.S., Phan, G., Edvinsson, S., Buelens, F., Remaut, H., Waksman, G., Hultgren, S.J. and Almqvist, F. (2010) Design and synthesis of C-2 substituted thiazolo and dihydrothiazolo ring-fused 2-pyridones: pilicides with increased antivirulence activity. *Journal of Medicinal Chemistry* 53, 5690–5695.

Chorell, E., Pinkner, J.S., Bengtsson, C., Banchelin, T.S., Edvinsson, S., Linusson, A., Hultgren, S.J. and Almqvist, F. (2012) Mapping pilicide anti-virulence effect in *Escherichia coli*, a comprehensive structure-activity study. *Bioorganic and Medicinal Chemistry* 20, 3128–3142.

Choudhury, D., Thompson, A., Stojanoff, V., Langermann, S., Pinkner, J., Hultgren, S.J. and Knight, S.D. (1999) X-ray structure of the FimC-FimH chaperone-adhesin complex from uropathogenic *Escherichia coli*. *Science* 285, 1061–1066.

Crespo, M.D., Puorger, C., Scharer, M.A., Eidam, O., Grutter, M.G., Capitani, G. and Glockshuber, R. (2012) Quality control of disulfide bond formation in pilus subunits by the chaperone FimC. *Nature Chemical Biology* 8, 707–713.

Cusumano, C.K. and Hultgren, S.J. (2009) Bacterial adhesion – a source of alternate antibiotic targets. *IDrugs* 12, 699–705.

Di Yu, X., Dubnovitsky, A., Pudney, A.F., MacIntyre, S., Knight, S.D. and Zavialov, A.V. (2012) Allosteric mechanism controls traffic in the chaperone/usher pathway. *Structure* 20, 1861–1871.

Driessen, A.J. and Nouwen, N. (2008) Protein translocation across the bacterial cytoplasmic membrane. *Annual Review of Biochemistry* 77, 643–667.

Du, Y., Rosqvist, R. and Forsberg, A. (2002) Role of fraction 1 antigen of *Yersinia pestis* in inhibition of phagocytosis. *Infection and Immunity* 70, 1453–1460.

Dubnovitsky, A.P., Duck, Z., Kersley, J.E., Hard, T., MacIntyre, S. and Knight, S.D. (2010) Conserved hydrophobic clusters on the surface of the Caf1A usher C-terminal domain are important for F1 antigen assembly. *Journal of Molecular Biology* 403, 243–259.

Duguid, J.P., Smith, I.W., Dempster, G. and Edmunds, P.N. (1955) Non-flagellar filamentous appendages (fimbriae) and haemagglutinating activity in Bacterium coli. *Journal of Pathology and Bacteriology* 70, 335–348.

Durand, E., Verger, D., Rego, A.T., Chandran, V., Meng, G., Fronzes, R. and Waksman, G. (2009) Structural biology of bacterial secretion systems in gram-negative pathogens – potential for new drug targets. *Infectious Disorders Drug Targets* 9, 518–547.

Eidam, O., Dworkowski, F.S., Glockshuber, R., Grutter, M.G. and Capitani, G. (2008) Crystal structure of the ternary FimC-FimF(t)-FimD(N) complex indicates conserved pilus chaperone-subunit complex recognition by the usher FimD. *FEBS Letters* 582, 651–655.

Fairman, J.W., Noinaj, N. and Buchanan, S.K. (2011) The structural biology of beta-barrel membrane proteins: a summary of recent reports. *Current Opinion in Structural Biology* 21, 523–531.

Felek, S., Jeong, J.J., Runco, L.M., Murray, S., Thanassi, D.G. and Krukonis, E.S. (2011) Contributions of chaperone/usher systems to cell binding, biofilm formation and *Yersinia pestis* virulence. *Microbiology* 157, 805–818.

Ford, B., Rego, A.T., Ragan, T.J., Pinkner, J., Dodson, K., Driscoll, P.C., Hultgren, S. and Waksman, G. (2010) Structural homology between the C-terminal domain of the PapC usher and its plug. *Journal of Bacteriology* 192, 1824–1831.

Ford, B., Verger, D., Dodson, K., Volkan, E., Kostakioti, M., Elam, J., Pinkner, J., Waksman, G. and Hultgren, S. (2012) The structure of the PapD-PapGII pilin complex reveals an open and flexible P5 pocket. *Journal of Bacteriology* 194, 6390–6397.

Fronzes, R., Remaut, H. and Waksman, G. (2008) Architectures and biogenesis of non-flagellar protein appendages in Gram-negative bacteria. *EMBO Journal* 27, 2271–2280.

Geibel, S., Procko, E., Hultgren, S.J., Baker, D. and Waksman, G. (2013) Structural and energetic basis of folded-protein transport by FimD usher. *Nature* 496, 243–246.

Hahn, E., Wild, P., Hermanns, U., Sebbel, P., Glockshuber, R., Haner, M., Taschner, N., Burkhard, P., Aebi, U. and Muller, S.A. (2002) Exploring the 3D molecular architecture of *Escherichia coli* type 1 pili. *Journal of Molecular Biology* 323, 845–857.

Hannan, T.J., Totsika, M., Mansfield, K.J., Moore, K.H., Schembri, M.A. and Hultgren, S.J. (2012) Host–pathogen checkpoints and population bottlenecks in persistent and intracellular uropathogenic *Escherichia coli* bladder infection. *FEMS Microbiology Reviews* 36, 616–648.

Henderson, N.S., Ng, T.W., Talukder, I. and Thanassi, D.G. (2011) Function of the usher N-terminus in catalysing pilus assembly. *Molecular Microbiology* 79, 954–967.

Huang, Y., Smith, B.S., Chen, L.X., Baxter, R.H. and Deisenhofer, J. (2009) Insights into pilus assembly and secretion from the structure and functional characterization of usher PapC. *Proceedings of the National Academy of Sciences of the USA* 106, 7403–7407.

Jacob-Dubuisson, F., Pinkner, J., Xu, Z., Striker, R., Padmanhaban, A. and Hultgren, S.J. (1994) PapD chaperone function in pilus biogenesis depends on oxidant and chaperone-like activities of DsbA. *Proceedings of the National Academy of Sciences of the USA* 91, 11552–11556.

Korea, C.G., Badouraly, R., Prevost, M.C., Ghigo, J.M. and Beloin, C. (2010) *Escherichia coli* K-12 possesses multiple cryptic but functional chaperone-usher fimbriae with distinct surface specificities. *Environmental Microbiology* 12, 1957–1977.

Le Trong, I., Aprikian, P., Kidd, B.A., Forero-Shelton, M., Tchesnokova, V., Rajagopal, P., Rodriguez, V., Interlandi, G., Klevit, R., Vogel, V., Stenkamp, R.E., Sokurenko, E.V. and Thomas, W.E. (2010) Structural basis for mechanical force regulation of the adhesin FimH via finger trap-like beta sheet twisting. *Cell* 141, 645–655.

Leney, A.C., Phan, G., Allen, W., Verger, D., Waksman, G., Radford, S.E. and Ashcroft, A.E. (2011) Second order rate constants of donor-strand exchange reveal individual amino acid residues important in determining the subunit specificity of pilus biogenesis. *Journal of the American Society for Mass Spectrometry* 22, 1214–1223.

Li, H., Qian, L., Chen, Z., Thibault, D., Liu, G., Liu, T. and Thanassi, D.G. (2004) The outer membrane usher forms a twin-pore secretion complex. *Journal of Molecular Biology* 344, 1397–1407.

Liu, F., Chen, H., Galvan, E.M., Lasaro, M.A. and Schifferli, D.M. (2006) Effects of Psa and F1 on the adhesive and invasive interactions of *Yersinia pestis* with human respiratory tract epithelial cells. *Infection and Immunity* 74, 5636–5644.

Mapingire, O.S., Henderson, N.S., Duret, G., Thanassi, D.G. and Delcour, A.H. (2009) Modulating effects of the plug, helix, and N- and C-terminal domains on channel properties of the PapC usher. *Journal of Biological Chemistry* 284, 36324–36333.

Miller, J., Williamson, E.D., Lakey, J.H., Pearce, M.J., Jones, S.M. and Titball, R.W. (1998) Macromolecular organisation of recombinant *Yersinia pestis* F1 antigen and the effect of structure on immunogenicity. *FEMS Immunology and Medical Microbiology* 21, 213–221.

Morrissey, B., Leney, A.C., Toste Rego, A., Phan, G., Allen, W.J., Verger, D., Waksman, G., Ashcroft, A.E. and Radford, S.E. (2012) The role of chaperone-subunit usher domain interactions in the mechanism of bacterial pilus biogenesis revealed by ESI-MS. *Molecular and Cellular Proteomics* 11, M111 015289.

Mu, X.Q. and Bullitt, E. (2006) Structure and assembly of P-pili: a protruding hinge region used for assembly of a bacterial adhesion filament. *Proceedings of the National Academy of Sciences of the USA* 103, 9861–9866.

Mulvey, M.A., Lopez-Boado, Y.S., Wilson, C.L., Roth, R., Parks, W.C., Heuser, J. and Hultgren, S.J. (1998) Induction and evasion of host defenses by type 1-piliated uropathogenic *Escherichia coli*. *Science* 282, 1494–1497.

Nishiyama, M. and Glockshuber, R. (2010) The outer membrane usher guarantees the formation of functional pili by selectively catalyzing donor-strand exchange between subunits that are adjacent in the mature pilus. *Journal of Molecular Biology* 396, 1–8.

Nishiyama, M., Vetsch, M., Puorger, C., Jelesarov, I. and Glockshuber, R. (2003) Identification and characterization of the chaperone-subunit complex-binding domain from the type 1 pilus assembly platform FimD. *Journal of Molecular Biology* 330, 513–525.

Nishiyama, M., Horst, R., Eidam, O., Herrmann, T., Ignatov, O., Vetsch, M., Bettendorff, P., Jelesarov, I., Grutter, M.G., Wuthrich, K., Glockshuber, R. and Capitani, G. (2005) Structural basis of chaperone-subunit complex recognition by the type 1 pilus assembly platform FimD. *EMBO Journal* 24, 2075–2086.

Nishiyama, M., Ishikawa, T., Rechsteiner, H. and Glockshuber, R. (2008) Reconstitution of pilus assembly reveals a bacterial outer membrane catalyst. *Science* 320, 376–379.

Nuccio, S.P. and Baumler, A.J. (2007) Evolution of the chaperone/usher assembly pathway: fimbrial classification goes Greek. *Microbiology and Molecular Biology Reviews* 71, 551–575.

Phan, G., Remaut, H., Wang, T., Allen, W.J., Pirker, K.F., Lebedev, A., Henderson, N.S., Geibel, S., Volkan, E., Yan, J., Kunze, M.B., Pinkner, J.S., Ford, B., Kay, C.W., Li, H., Hultgren, S.J., Thanassi, D.G. and Waksman, G. (2011) Crystal structure of the FimD usher bound to its cognate FimC-FimH substrate. *Nature* 474, 49–53.

Pinkner, J.S., Remaut, H., Buelens, F., Miller, E., Aberg, V., Pemberton, N., Hedenstrom, M., Larsson, A., Seed, P., Waksman, G., Hultgren, S.J. and Almqvist, F. (2006) Rationally designed small compounds inhibit pilus biogenesis in uropathogenic bacteria. *Proceedings of the National Academy of Sciences of the USA* 103, 17897–17902.

Puorger, C., Vetsch, M., Wider, G. and Glockshuber, R. (2011) Structure, folding and stability of FimA, the main structural subunit of type 1 pili from uropathogenic *Escherichia coli* strains. *Journal of Molecular Biology* 412, 520–535.

Rego, A.T., Chandran, V. and Waksman, G. (2010) Two-step and one-step secretion mechanisms in Gram-negative bacteria: contrasting the type IV secretion system and the chaperone-usher pathway of pilus biogenesis. *Biochemical Journal* 425, 475–488.

Remaut, H., Rose, R.J., Hannan, T.J., Hultgren, S.J., Radford, S.E., Ashcroft, A.E. and Waksman, G. (2006) Donor-strand exchange in chaperone-assisted pilus assembly proceeds through a concerted beta strand displacement mechanism. *Molecular Cell* 22, 831–842.

Remaut, H., Tang, C., Henderson, N.S., Pinkner, J.S., Wang, T., Hultgren, S.J., Thanassi, D.G., Waksman, G. and Li, H. (2008) Fiber formation across the bacterial outer membrane by the chaperone/usher pathway. *Cell* 133, 640–652.

Roberts, J.A., Marklund, B.I., Ilver, D., Haslam, D., Kaack, M.B., Baskin, G., Louis, M., Mollby, R., Winberg, J. and Normark, S. (1994) The Gal(alpha 1-4)Gal-specific tip adhesin of *Escherichia coli* P-fimbriae is needed for pyelonephritis to occur in the normal urinary tract. *Proceedings of the National Academy of Sciences of the USA* 91, 11889–11893.

Rose, R.J., Verger, D., Daviter, T., Remaut, H., Paci, E., Waksman, G., Ashcroft, A.E. and Radford, S.E. (2008) Unraveling the molecular basis of subunit specificity in P pilus assembly by mass spectrometry. *Proceedings of the National Academy of Sciences of the USA* 105, 12873–12878.

Sauer, F.G., Futterer, K., Pinkner, J.S., Dodson, K.W., Hultgren, S.J. and Waksman, G. (1999) Structural basis of chaperone function and pilus biogenesis. *Science* 285, 1058–1061.

Sauer, F.G., Pinkner, J.S., Waksman, G. and Hultgren, S.J. (2002) Chaperone priming of pilus subunits facilitates a topological transition that drives fiber formation. *Cell* 111, 543–551.

Stathopoulos, C., Hendrixson, D.R., Thanassi, D.G., Hultgren, S.J., St Geme, J.W. 3rd and Curtiss, R. 3rd (2000) Secretion of virulence determinants by the general secretory pathway in gram-negative pathogens: an evolving story. *Microbes and Infection* 2, 1061–1072.

Thanassi, D.G., Bliska, J.B. and Christie, P.J. (2012) Surface organelles assembled by secretion systems of Gram-negative bacteria: diversity in structure and function. *FEMS Microbiology Reviews* 36, 1046–1082.

Totsika, M., Heras, B., Wurpel, D.J. and Schembri, M.A. (2009) Characterization of two homologous disulfide bond systems involved in virulence factor biogenesis in uropathogenic *Escherichia coli* CFT073. *Journal of Bacteriology* 191, 3901–3908.

Verger, D., Miller, E., Remaut, H., Waksman, G. and Hultgren, S. (2006) Molecular mechanism of P pilus termination in uropathogenic *Escherichia coli*. *EMBO Reports* 7, 1228–1232.

Vetsch, M., Erilov, D., Moliere, N., Nishiyama, M., Ignatov, O. and Glockshuber, R. (2006) Mechanism of fibre assembly through the chaperone-usher pathway. *EMBO Reports* 7, 734–738.

Vitagliano, L., Ruggiero, A., Pedone, C. and Berisio, R. (2007) A molecular dynamics study of pilus subunits: insights into pilus biogenesis. *Journal of Molecular Biology* 367, 935–941.

Volkan, E., Ford, B.A., Pinkner, J.S., Dodson, K.W., Henderson, N.S., Thanassi, D.G., Waksman, G. and Hultgren, S.J. (2012) Domain activities of PapC usher reveal the mechanism of action of an *Escherichia coli* molecular machine. *Proceedings of the National Academy of Sciences of the USA* 109, 9563–9568.

Waksman, G. and Hultgren, S.J. (2009) Structural biology of the chaperone-usher pathway of pilus biogenesis. *Nature Reviews Microbiology* 7, 765–774.

Yu, X.D., Fooks, L.J., Moslehi-Mohebi, E., Tischenko, V.M., Askarieh, G., Knight, S.D., MacIntyre, S. and Zavialov, A.V. (2012) Large is fast, small is tight: determinants of speed and affinity in subunit capture by a periplasmic chaperone. *Journal of Molecular Biology* 417, 294–308.

Zavialov, A.V., Berglund, J., Pudney, A.F., Fooks, L.J., Ibrahim, T.M., MacIntyre, S. and Knight, S.D. (2003) Structure and biogenesis of the capsular F1 antigen from *Yersinia pestis*: preserved folding energy drives fiber formation. *Cell* 113, 587–596.

Zav'yalov, V., Zavialov, A., Zav'yalova, G. and Korpela, T. (2010) Adhesive organelles of Gram-negative pathogens assembled with the classical chaperone/usher machinery: structure and function from a clinical standpoint. *FEMS Microbiology Reviews* 34, 317–378.

4 Type 1 and P Pili of Uropathogenic *Escherichia coli*

Colin Russell and Matthew Mulvey

Division of Microbiology and Immunology, University of Utah, Salt Lake City, USA

4.1 Introduction

Escherichia coli express a variety of proteinaceous structures known as pili or fimbriae that extend as hair-like fibres from the outer bacterial membrane. An adhesin protein located at the distal tip of each pilus fibre enables the bacteria to attach to various surfaces, facilitating colonization of environmental and host niches. In the case of uropathogenic *E. coli* (UPEC), type 1 pili (T1P) and P pili enable bacteria to bind to host epithelial cells within the urinary tract, which can lead to a urinary tract infection (UTI).

Several species of pathogenic bacteria can colonize and cause disease within the human urinary tract. However, UPEC are responsible for the overwhelming majority of UTIs (Ronald, 2002; Foxman, 2010). In general, UPEC colonize the urinary tract via an ascending route, moving from the urethra meatus through the urethra and into the bladder where they can cause cystitis (bladder infection) (Hannan *et al.*, 2012). UPEC can then climb the ureters and initiate pyelonephritis (kidney infection). Cystitis and pyelonephritis patients typically suffer from painful and frequent urination, while pyelonephritis patients can also experience more severe symptoms such as fever (Hooton, 2012). UTIs are one of the most common infections seen in the outpatient setting, accounting for well over 8 million doctor visits annually in the USA alone (Schappert and Rechtsteiner, 2011). Fifty percent of women will have a UTI by age 32 (Foxman and Brown, 2003), and 44% of patients will have another symptomatic episode within 1 year after initial infection (Ikaheimo *et al.*, 1996). The majority of UTI patients suffer from cystitis, whereas only 4% of patients develop pyelonephritis (Ikaheimo *et al.*, 1996). A small fraction of pyelonephritis infections will progress to urosepsis, in which UPEC disseminate systemically (Jolley *et al.*, 2012).

Colonization of the urinary tract by UPEC is expedited by T1P and P pili. UPEC attach to the bladder epithelium (or urothelium) via T1P, where they can subsequently trigger their internalization (Mulvey *et al.*, 1998; Martinez *et al.*, 2000). Within the cytosol of bladder superficial cells – which are binucleate and relatively large – the bacteria form T1P-dependent biofilm-like aggregates referred to as intracellular bacterial communities (IBC) (Eto *et al.*, 2006; Wright *et al.*, 2007; Jorgensen and Seed, 2012). As UPEC replicate within the host cell, individual bacterial cells can detach from the IBC, exit the host cell, and go on to bind, invade and colonize neighbouring superficial or underlying cells (Fig. 4.1) (Mulvey *et al.*, 2001; Justice *et al.*, 2004). In the case of pyelonephritis, P pili are important mediators of UPEC attachment to the kidney urothelium (O'Hanley et al., 1985). In addition to their adhesive properties, T1P and P pili also trigger host responses that result in the release of cytokines and the influx of neutrophils (Fischer et al., 2006; Godaly et al., 2007).

4.2 Genetics and Structure of Type 1 and P Pili

T1P and P pili are encoded by the *fim* and *pap* loci, respectively (Fig. 4.2A). Each set of loci encodes an adhesin, structural and assembly proteins, and regulators of pilus expression. *fimH* and *papG* encode the adhesins of T1P

and P pili, respectively, and are essential for pili function (Roberts et al., 1994; Connell et al., 1996). The structures of both pili are very similar (Fig. 4.2B), consisting of two parts: a helical rod and an adhesive fibrillar tip (Kuehn et al., 1992; Jones et al., 1995). The rods, which are comprised of 1000 or more FimA or PapA pilin subunits, are about 7 nm in diameter and vary from a few fractions of a micron to more than 5 μm in length. Each rod is anchored to the outer membrane through its interaction with the usher, a transmembrane pore-forming protein (Thanassi et al., 2012). Distally attached to each rod is a thinner (~3 nm), much shorter tip fibrillum made up of two to three adaptor proteins and an adhesin. During pilus biogenesis the structural subunits are translocated from the

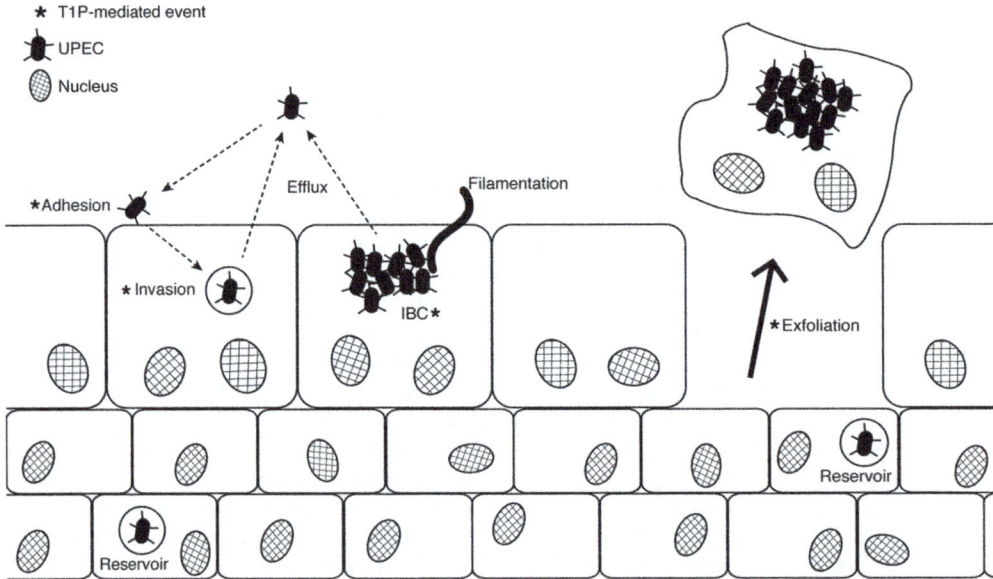

Fig. 4.1. Overview of bladder colonization by type 1-piliated UPEC. Pathogens enter the lumen of the bladder by an ascending route and subsequently bind and invade via T1P the large, binucleated superficial epithelial cells present on the apical surface of the urothelium. Some bacteria are quickly redirected back out into the bladder lumen, while others traffic to late endosome-like compartments where they can persist indefinitely in a quiescent state. Alternatively, UPEC can enter the host cytosol and rapidly multiply, forming IBCs. Infected superficial cells eventually die or exfoliate, taking with them large numbers of UPEC that can go on to infect other hosts or other sites within the urinary tract. Exfoliation, as well as the influx of immune cells like neutrophils, can compromise the barrier function of the urothelium. This allows UPEC to invade the smaller immature cells of the urothelium, where they can establish long-lived reservoirs that can persist in the presence of robust inflammatory responses and standard antibiotic treatments.

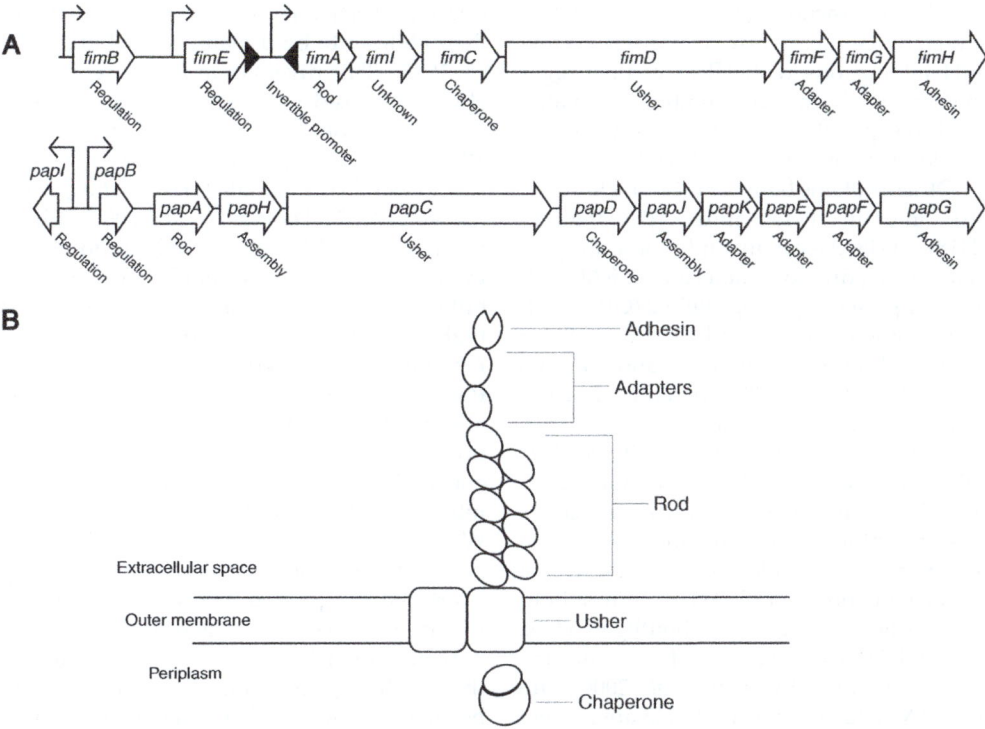

Fig. 4.2. (A) Organization of the *fim* and *pap* gene clusters, with (B) a basic model of pilus assembly by the chaperone–usher pathway.

bacterial cytosol into the periplasm. A chaperone then stabilizes and delivers the subunits to the usher protein in the outer membrane. Finally, the usher protein mediates addition of new subunits to the growing pilus (Allen *et al.*, 2012; Thanassi *et al.*, 2012).

4.3 Type 1 Pili

4.3.1 Adhesion

T1P-mediated adhesion of UPEC to superficial cells of the urothelium is one of the first steps during the infection process and can lead to host cell invasion, intracellular bacterial replication and persistence. FimH binds mannose on glycosylated host receptor proteins (Abraham *et al.*, 1985b; Krogfelt *et al.*, 1990). FimH binds mannose through a

binding pocket within an N-terminal lectin domain, which is joined by a linker to the C-terminal pilin domain, the portion of the protein that connects FimH to the pilus (Choudhury, 1999; Schembri *et al.*, 2000).

The receptors engaged during the course of a UTI are not yet completely detailed, although several cell-surface proteins that bind T1P have been identified *in vitro*. FimH can bind to the uroplakin protein UPIa (Zhou *et al.*, 2001), which – along with at least three other uroplakin proteins – is important for maintaining the permeability barrier of the urothelium (Wu *et al.*, 2009). Because of its abundance on the urothelial surface, UPIa may be the predominant receptor that is used by UPEC to initiate infection. High-resolution electron microscopy showing T1P engaging the uroplakin-covered urothelium within the bladder of a mouse infected with UPEC yielded indirect *in vivo* evidence for T1P-uroplakin interactions (Mulvey *et al.*, 1998). Cryo-electron microscopy demonstrated that FimH interactions with UP1a could trigger conformational changes within uroplakin complexes, possibly transducing signals into host bladder cells (Wang *et al.*, 2009).

Other proteins bound by T1P include the α3β1 integrin heterodimer, β2 integrin (CD18), CD48, the carcinoembryonic antigen-related cell adhesion molecule (CEACAM) family of proteins, glycoprotein 2 (GP2), Toll-like receptor 4 (TLR4), and the extracellular proteins fibronectin, collagen and laminin (Sauter *et al.*, 1991, 1993; Gbarah *et al.*, 1991; Sokurenko *et al.*, 1992; Kukkonen *et al.*, 1993; Baorto *et al.*, 1997; Pouttu *et al.*, 1999; Eto *et al.*, 2007; Mossman *et al.*, 2008; Carvalho *et al.*, 2009; Hase *et al.*, 2009; Yu and Lowe, 2009). These receptors exhibit distinct expression patterns and localization. α3β1 is located at cellular junctions and elsewhere throughout all layers of the urothelium (Southgate *et al.*, 1995). CEACAMs are also found on the urothelial surface (Kuespert *et al.*, 2006), and CEACAMs, β2 integrin and CD48 are present on the surface of immune cells such as neutrophils and macrophages (Gbarah *et al.*, 1991; Sauter *et al.*, 1991; Baorto *et al.*, 1997). Despite *in vitro* evidence for T1P interaction with these receptors, corroboration by *in vivo* experiments is still lacking.

Regardless of what host receptors T1P exploit, the interaction must be strong enough so as to withstand the shear forces produced by the bulk flow of urine during micturition. Interestingly, the receptor binding affinity of FimH rises as flow increases (Thomas *et al.*, 2002, 2004). Under low flow conditions the pilin domain of FimH inhibits the lectin domain in such a way that the mannose binding pocket is in a low affinity state (Le Trong *et al.*, 2010). As flow increases, the linker between the two domains stretches, releasing the inhibition on the lectin domain, allowing the binding pocket to adopt a high affinity conformation. Thus, as flow increases FimH binds more tightly to its receptor, ensuring that the bacteria are not dispelled into the urine.

Although *fimH* displays a high degree of conservation among various *E. coli* isolates (Abraham *et al.*, 1988; Weissman *et al.*, 2006),

different *fimH* alleles do exist that vary in their abilities to bind mannose (Sokurenko *et al.*, 1995). This diversity in mannose-binding ability is linked with sequence variations in the lectin domain and linker, as well as alterations in composition of the type 1 pilus rod (Schembri *et al.*, 2000; Hung *et al.*, 2002; Mulvey, 2002; Thomas *et al.*, 2002). FimH from most T1P+ *E. coli* – including those from the gut – are capable of binding trimannose, but UPEC-associated *fimH* alleles often have a higher binding affinity for monomannose residues (Sokurenko *et al.*, 1997, 1998). Phylogenetic evidence indicates that the *fimH* alleles carried by UPEC isolates evolved under positive selection from *fimH* alleles encoded by commensal strains within the intestinal microbiota (Sokurenko *et al.*, 1998; Weissman *et al.*, 2006). UPEC are believed to persist within the gut without eliciting any overt pathology, and many UPEC isolates found in urine closely match strains from the patient's colon (Russo *et al.*, 1995). Together, these data suggest a model in which faecal-borne UPEC strains that make their way into the urinary tract experience pressures that select *fimH* alleles that can mediate higher affinity interactions with receptors containing accessible monomannose residues. Since FimH makes up only a small portion of the gross T1P structure, it is theorized that *fimH* alleles in UPEC strains are not modified under selective pressure to evade antigenic recognition by the immune system (Weissman *et al.*, 2006). Rather, the composition and abundance of specific N-linked glycan receptors in the urinary tract probably help drive the selection process within individual hosts (Taganna *et al.*, 2011). Overall, these sorts of observations suggest that UPEC express *fimH* alleles that are uniquely evolved for life in the urinary tract.

4.3.2 Exfoliation

The presence of bacterial antigens within the bladder can stimulate the detachment of superficial cells from the urothelium (Mulvey *et al.*, 1998). This exfoliation process occurs via an apoptosis-like mechanism and provides the host a possible defence mechanism to dispose of infected host cells via the urine (Aronson *et al.*, 1988; Mulvey *et al.*, 1998). However, exfoliation temporarily exposes the deeper portions of the urothelium. This gives the bacteria the opportunity to bind and invade the underlying immature intermediate and basal epithelial cells of the urothelium where they may be able to evade the immune system and establish long-lived reservoirs (see below) (Mulvey *et al.*, 2001).

In mouse models of UTI, FimH is necessary to cause exfoliation, and the expression of FimH by non-pathogenic laboratory *E. coli* strains allows them to induce exfoliation *in vivo*, similar to UPEC (Mulvey *et al.*, 1998). Binding of FimH to bladder epithelial cells can inhibit the anti-apoptotic function of the host transcription factor NF-κB, contributing to the induction of apoptotic pathways (Klumpp *et al.*, 2001, 2006). Binding of FimH to UPIa may stimulate the exfoliation process by triggering pro-apoptotic signalling cascades via effects on the cytoplasmic tail of neighbouring UPIIIa proteins (Thumbikat *et al.*, 2009). Secreted bacterial toxins like α-hemolysin and cytotoxin necrotizing factor 1 can also promote the exfoliation process, acting synergistically with type 1 pili (Mills *et al.*, 2000; Wiles *et al.*, 2008; Dhakal and Mulvey, 2012). While UPEC and laboratory *E. coli* strains that lack FimH fail to induce appreciably high levels of exfoliation *in vivo*, these bacterial strains can effectively trigger exfoliation in *ex vivo* experiments using bladder tissue explants (unpublished observations). These results indicate that type 1 pili can accelerate bladder cell exfoliation, possibly by keeping bacteria in close contact with host cells against the flow of urine, but other as-yet-undefined factors and signalling events likely initiate the process.

4.3.3 Invasion

T1P-mediated adhesion of bacteria to urothelial cells can lead to internalization via an endocytic-like mechanism in which the host plasma membrane zippers around and envelopes bound bacteria (Mulvey *et al.*, 1998; Martinez *et al.*, 2000). Bacteria are then trafficked into acidic, late endosome-like compartments via a pathway that involves Rab27b, a regulator of vesicular traffic that has been implicated in the delivery of uroplakin complexes to the host cell surface (Chen *et al.*, 2003; Eto *et al.*, 2006, 2008; Mysorekar and Hultgren, 2006; Bishop *et al.*, 2007). Within immature bladder epithelial cells, UPEC typically remains bound in the endosomal compartments, which are often enmeshed within a matrix of actin filaments (Eto *et al.*, 2006). These bacteria are highly resistant to antibiotics, many of which cannot cross host membranes and are ineffective against non-replicating microbes (Blango and Mulvey, 2010). UPEC can persist in a quiescent state within endosomal compartments for many days to weeks, serving as potential reservoirs for recurrent, relapsing and chronic UTIs that afflict many individuals (Mulvey *et al.*, 1998, 2001; Hvidberg *et al.*, 2000; Kerrn *et al.*, 2005; Mysorekar and Hultgren, 2006; Blango and Mulvey, 2010).

The FimH adhesin within T1P stimulates bacterial internalization by causing host receptors and signalling complexes to cluster and activate at sites of attachment, but the specific mechanisms remain incompletely defined (Martinez *et al.*, 2000; Dhakal *et al.*, 2008). Receptors implicated in the invasion process include UP1a and α3β1 integrin, though other receptors may also be involved (Eto *et al.*, 2007; Wang *et al.*, 2009). For example, complement and the host receptor for opsonized bacteria, CD46, can synergize with T1P to enhance bacterial internalization (Li *et al.*, 2009). The recruitment of host receptors and subsequent activation of signalling cascades that lead to UPEC internalization is apparently modulated by cholesterol-rich membrane domains known as lipid rafts, which serve as staging sites for many signalling events in the host plasma membrane (Duncan *et al.*, 2004; Eto *et al.*, 2008). T1P-mediated entry into host cells also requires actin rearrangements and, not surprisingly, many different host regulators of actin dynamics (Martinez *et al.*, 2000; Martinez and Hultgren, 2002; Duncan *et al.*, 2004; Eto *et al.*, 2007; Visvikis *et al.*, 2011). These include Src and focal adhesin kinases, phosphoinositides 3-kinase, Rho-family

GTPases like Cdc42 and Rac1, and actin stabilizing factors such as α-actinin and vinculin. The recruitment of clathrin and associated adaptor and accessory proteins like AP-2 also promote the invasion process (Eto *et al.*, 2008). Of note, clathrin itself does not actually encase the invading bacteria in classic clathrin-coated pits, but may be recruited primarily to help drive localized actin rearrangements.

In addition to actin, UPEC entry into host cells is reliant on functional microtubules, the microtubule-associated motor protein kinesin-1 and histone deacetylase 6 (HDAC6), a cytosolic enzyme that alters microtubule stability by deacetylating α-tubulin (Dhakal and Mulvey, 2009). Ongoing work suggests that kinesin-1 delivers via microtubules one or more factors needed to control actin cytoskeletal dynamics at sites of bacterial entry, with the net effect being the envelopment and internalization of bound UPEC. Scission of the host membrane and delivery of UPEC into a membrane-bound vacuole within the host cytosol is accomplished by the guanosine triphosphatase dynamin 2 (DNM2) (Eto *et al.*, 2008; Wang *et al.*, 2011). While these findings offer insight into the mechanism of host cell invasion mediated by T1P, additional work is needed to define how these various host factors interact and function as a system within the immature and terminally differentiated cells of the urothelium.

4.3.4 Biofilm formation

Disruption of host actin filaments, or perturbation of endosomal membranes, enables internalized UPEC to enter the host cytosol where it can rapidly multiply, forming large biofilm-like aggregates in close association with host cytoskeletal intermediate filaments (Eto *et al.*, 2006). These aggregates, known as intracellular bacterial communities (IBCs), can contain several thousand bacteria (Mulvey *et al.*, 2001; Anderson *et al.*, 2003; Justice *et al.*, 2004). *In vivo*, IBCs are typically seen only within terminally differentiated superficial bladder epithelial cells, in which the actin cytoskeleton is sparsely distributed (Eto *et al.*, 2006). It is

feasible that the redistribution of actin filaments during the terminal stages of bladder cell differentiation provides a trigger for the resurgence of UPEC reservoirs that become sequestered within immature bladder cells during the acute phase of a UTI. IBCs provide a means for UPEC to build up high numbers within a protected host environment. UPEC can be released from IBCs into the bladder lumen, and sometimes being extruded as long filamentous forms that are resistant to killing by host phagocytes (Mulvey *et al.*, 2001; Justice *et al.*, 2004, 2006). However, IBCs do not represent long-term bacterial reservoirs within the urinary tract, as IBC-containing superficial cells are eventually destroyed from within or shed (see Fig. 4.1) (Mulvey *et al.*, 2001; Blango and Mulvey, 2010).

The development of IBCs is dependent upon expression of T1P within bladder epithelial cells. Immunohistochemistry of IBCs showed that the bacteria within these structures express T1P (Anderson *et al.*, 2003; Wright *et al.*, 2007). Using a tetracycline-inducible expression system to temporally control the expression of T1P, Wright *et al.* (2007) showed that T1P were functionally important within IBCs. Bacteria grown with tetracycline (and expressing T1P) were able to bind and invade urothelial cells. Once within the host cells, T1P expression was discontinued due to the absence of tetracycline. Without continued T1P expression the bacteria were unable to form structured IBCs and they dispersed throughout the cytoplasm. The inability to form IBCs correlated with decreased bacterial persistence within the bladders of infected mice. These observations correlate with *in vitro* results showing that T1P also facilitate formation of biofilm-like communities on abiotic surfaces – such as catheters – in static growth conditions (Pratt and Kolter, 1998) and in the presence of hydrodynamic flow (Schembri and Klemm, 2001). Thus, T1P may contribute to the establishment and progression of UTI in multiple ways in both the community at large and in catheterized hospitalized patients, a group that is exceptionally prone to UTI arising from catheter-associated biofilms (Stickler, 2008).

4.3.5 Regulation of type 1 pili expression

T1P undergo phase variation, in which the expression of the *fim* operon can be turned on or off by an epigenetic mechanism (Abraham *et al.*, 1985a). *In vitro*, individual bacterial cells in a population will switch expression of T1P on and off at a low rate, resulting in a mixed population of T1P+ and T1P- cells (Freitag *et al.*, 1985). Phase variation has also been observed in human infection (Kisielius *et al.*, 1989). The switch from one state to another is accomplished by a DNA recombination event in which the *fim* promoter is inverted by the recombinases FimB and FimE (Fig. 4.2) (Abraham *et al.*, 1985a). FimE primarily changes the orientation from on to off, and FimB changes the orientation equally from off to on and from on to off (Klemm, 1986; McClain *et al.*, 1991; Gally *et al.*, 1996).

The expression of T1P is controlled by several transcription factors (Table 4.1). T1P synthesis is modulated via transcriptional regulation of the recombinases, and by alteration of the conformation of DNA to facilitate recombination. The conditions that induce expression of T1P are not well elucidated, but studies of T1P regulators suggest that nutrient availability and environmental stresses play key roles.

T1P phase variation is also driven by crosstalk with P pili. The P pili regulator PapB dampens the expression of T1P by up-regulating expression of FimE and inhibiting the synthesis of FimB (Xia *et al.*, 2000; Holden *et al.*, 2006). Conversely, T1P expression is capable of inhibiting P pili expression (Snyder *et al.*, 2005). The complexity of these regulatory networks is further exasperated if one considers potential crosstalk with other systems, such as those that control bacterial motility and chemotaxis (Simms and Mobley, 2008; Cooper *et al.*, 2012).

4.3.6 Relevance of type 1 pili in pathogenesis

There is overwhelming evidence that T1P are important in animal models of UTI. UPEC strains that express T1P are able to infect mice better than non-piliated strains (Hultgren *et al.*, 1985; Schaeffer *et al.*, 1987). In competitive assays in which a specific mutant bacterial strain is mixed one-to-one with the wild-type parent strain and injected into mice, Keith *et al.* (1986) found that mutants lacking functional T1P are outcompeted by wild-type bacteria within the bladder. In non-competitive assays, *fimH* null mutants have reduced survival within the mouse urinary tract and are significantly less immunogenic (Connell *et al.*, 1996). Blocking T1P attachment to the murine urothelium using soluble FimH receptor analogues like α-D-mannosides inhibits UPEC colonization (Klein *et al.*, 2010; Cusumano *et al.*, 2011; Jiang *et al.*, 2012), as does passive immunization with antibodies directed against T1P (Abraham *et al.*, 1985). Immunization of animals with purified pili or, even better, the purified FimH adhesin or antigenically modified variants of FimH can provide significant protection against experimental UTI (Silverblatt and Cohen, 1979; Silverblatt *et al.*, 1982; Langermann, 1997; Poggio *et al.*, 2006; Karam *et al.*, 2012). These observations are underscored by results from an unbiased genetic screen in mice that showed T1P to be important for UPEC survival within the urinary tract (Bahrani-Mougeot *et al.*, 2002), and microarray analysis of bacteria taken straight from the urine of infected mice indicating that T1P are upregulated during UTI (Snyder *et al.*, 2004).

Evidence that T1P are crucial for UPEC colonization of the urinary tract in humans is less straightforward. Despite the fact that most UPEC isolates encode T1P (Norinder *et al.*, 2012), several studies have found that T1P are often absent on the surface of *E. coli* strains present in urine samples collected from patients with UTI (Ofek *et al.*, 1981; Pere *et al.*, 1987; Lim *et al.*, 1998; Hagan *et al.*, 2010). Such observations, coupled with the fact that T1P are also encoded by most commensal non-pathogenic *E. coli* isolates within the gut (Hagberg *et al.*, 1983), brings into question the role of T1P as virulence factors within the human host. The existence of a group of *E. coli* strains that cause asymptomatic bacteriuria (ABU) further questions the utility of T1P as mediators of bacterial colonization in the human urinary tract. ABU isolates can efficiently colonize humans, growing to very

Table 4.1 Regulators of T1P expression.

Regulator	Description	Relevant environmental signals	T1P +/–	Mechanism of T1P regulation	Reference
RpoS	Regulates genes important for response to stress and stationary phase	Expression of *fim* is inhibited during stationary phase	–	Inhibits transcription of *fimB*, inhibiting T1P expression	Dove *et al.* (1997)
LrhA	Inhibition of flagella, motility, chemotaxis	Reciprocal regulation of T1P and flagella. Inhibits adherence when motility genes activated, and vice-versa	–	Activates transcription of *fimE*	Blumer *et al.* (2005), Simms and Mobley (2008)
ppGpp	Regulates the stringent response (during nutrient deprivation or environmental stress)	Amino acid shortage, etc.	+	Interacts with RNA polymerase to activate transcription of *fimB*	Blomfield *et al.* (1993), Aberg *et al.* (2008)
DksA	Regulates the stringent response (during nutrient deprivation or environmental stress)	Augments ppGpp effects	+	Interacts with RNA polymerase to activate transcription of *fimB*	Aberg *et al.* (2008)
leuX	tRNA that recognizes rare codons for leucine to facilitate translation		+	Facilitates the translation of *fimB* by recognizing rare leucine-encoding codons in *fimB* mRNA	Newman *et al.* (1994)
Lrp	Regulates expression of genes important for survival during nutrient deprivation; plays a role in DNA packaging	Recombination is stimulated by presence of aliphatic amino acids	+/–	Facilitates recombination by bending DNA. It is biased towards the ON phase	Gally *et al.* (1993)
IHF	Integration host factor, modulates transcription of genes by bending DNA		+	Alters DNA conformation to facilitate recombination towards the ON phase	Corcoran and Dorman (2009)
H-NS	Supercoils DNA to silence genes with high A+T content		–	Supercoils DNA, facilitates recombination towards the OFF phase	O'Gara and Dorman (2000)
CRP	Regulates genes for catabolism of secondary carbon sources	Growth in presence of glucose increases T1P	–	Stimulates DNA gyrase activity which supercoils DNA and inhibits transcription	Muller *et al.* (2009)

high numbers within the lumen of the bladder without causing any overt inflammatory responses and without binding to or invading urothelial cells (Hull *et al.*, 1999; Bergsten *et al.*, 2005; Klemm *et al.*, 2007). These strains typically lack functional T1P and P pili, as well as many other factors that are often associated with UPEC isolates (Klemm *et al.*, 2006). ABU strains can outcompete more pathogenic UPEC isolates, and are being developed as therapeutic tools to interfere with UPEC colonization of the host (Falagas *et al.*, 2008). The engineered expression of recombinant T1P by the prototypical ABU strain 83972 causes increased host inflammatory responses in a mouse UTI model, but this effect was not observed in human volunteers (Bergsten *et al.*, 2007). Furthermore, an older study concluded that the 83972 strain expressing T1P was rapidly outcompeted by the non-adherent wild-type 83972 strain in the human urinary tract (Andersson *et al.*, 1991). While informative, these results are difficult to interpret because the patient volunteers had complicated histories of recurrent or chronic UTIs that were unresponsive to antibiotic therapy.

Overall, these results indicate that T1P are not an absolute requirement for bacterial colonization of the human urinary tract. For example, ABU isolates can apparently employ alternate T1P-independent strategies to persist within the host, probably taking advantage of more robust metabolic pathways present within their streamlined genomes to grow more efficiently within the extracellular urine-saturated milieu of the bladder. In the case of *bona fide* UPEC isolates, the necessity of T1P remains debatable. While UPEC strains present in the urine of patients with UTI do not always have T1P present, the expression of these fibres can almost always be induced when the strains are grown in broth culture (Ofek *et al.*, 1981; Pere *et al.*, 1987; Lim *et al.*, 1998; Hagan *et al.*, 2010). It can therefore be argued that the free bacteria present in the urine simply have their T1P in the off phase, while those in the on phase remain associated with the urothelium and allow UPEC to persist within the urinary tract and ultimately cause disease (Hultgren *et al.*, 1985; Lim *et al.*, 1998). Indeed, evidence

that T1P-mediated adherence, host cell invasion and IBC development are critical to the progression of UTIs in humans is mounting, as exemplified by data showing the remnants of IBCs within urothelial cells that are shed into the urine of women suffering from UTI (Rosen *et al.*, 2007).

4.4 P Pili

4.4.1 Adhesion

P pili play a part in colonization of the urinary tract by mediating adhesion to the Galα1-4Gal disaccharide moiety found on globo-series glycosphingolipids located on the surface of host cell membranes (Leffler and Svanborg-Eden, 1981; O'Hanley *et al.*, 1985; Wold *et al.*, 1988). Because there is a strong correlation between P pili and pyelonephritis, P pili have mostly been studied in the context of kidney infection (Kallenius *et al.*, 1981; Plos *et al.*, 1990). However, Galα1-4Gal glycolipids are found throughout the entire urothelium (O'Hanley *et al.*, 1985), suggesting that P pili may play a role in cystitis in some instances. The binding of P pili to glycosphingolipids in the urothelium is currently the only known function of P pili. They are dispensable for other events such as host cell invasion (Martinez *et al.*, 2000).

Like T1P, P pili may enable UPEC to better withstand the shear flow of urine. P pili can unwind *in vitro* to five times their length in a reversible manner (Gong and Makowski, 1992; Bullitt and Makowski, 1995; Thanassi *et al.*, 1998; Fallman *et al.*, 2005). It is hypothesized that this unwinding gives the pilus more flexibility during urine flow, preventing the pilus from breaking or detaching from the urothelium.

The adhesive tip subunit of P pili, encoded by *papG*, is highly polymorphic. Distinct alleles of *papG* correlate with UPEC strains isolated from different host species, as well as specific host-associated niches. These alleles have differing affinities for Galα1-4Gal, depending on the topography of the glycolipid receptors present (Stromberg *et al.*, 1990). Class II *papG* alleles are found in human pyelonephritis and urosepsis isolates

(Otto *et al.*, 1993; Johanson *et al.*, 1993; Johnson, 1998; Johnson *et al.*, 1998a), while class III alleles are encoded by human cystitis isolates as well as canine UTI strains (Johanson *et al.*, 1993; Johnson *et al.*, 1998b, 2000). Other *papG* allele classes are less common and little is known about their binding propensities.

4.4.2 Regulation of P pili expression

Similar to T1P and most other fimbrial adhesins, P pili are subject to phase variation, switching between an on and off state (Blyn *et al.*, 1989; Kisielius *et al.*, 1989). The frequency of phase switching can vary greatly dependent upon strain background and culture conditions (Holden *et al.*, 2007). Unlike T1P, phase variation in P pili is driven by differential methylation of the *pap* promoter and not by DNA recombination (Blyn *et al.*, 1990). Methylation occurs at two GATC sequences within the promoter. GATC sites are found throughout the entire genome and are typically recognized and methylated by

the enzyme DNA adenine methyltransferase (Dam). When the site proximal to the *papBA* operon (GATC^{prox}) is methylated, *pap* genes are transcribed and P pili are produced (Fig. 4.3). Conversely, when GATC^{prox} is unmethylated and the distal GATC site (GATC^{dist}) is methylated, the main *pap* promoter P_{BA} is not activated and P pili are not made.

Binding of the Leucine-responsive protein (Lrp) modulates the differential methylation of the *pap* promoter in an autoregulatory fashion by competing with Dam for binding sites within the *pap* (P_{BA}) promoter (Braaten *et al.*, 1992; Kawamura *et al.*, 2011). There are six binding sites for Lrp within the *pap* promoter (Fig. 4.3), two of which overlap the GATC^{prox} and GATC^{dist} methylation sites. When GATC^{prox} is methylated, Lrp binds more readily to GATC^{dist} where it can stimulate expression of PapI (Goransson *et al.*, 1989). PapI promotes interactions between Lrp and distal Lrp binding sites within the promoter, allowing Dam and RNA polymerase access to proximal

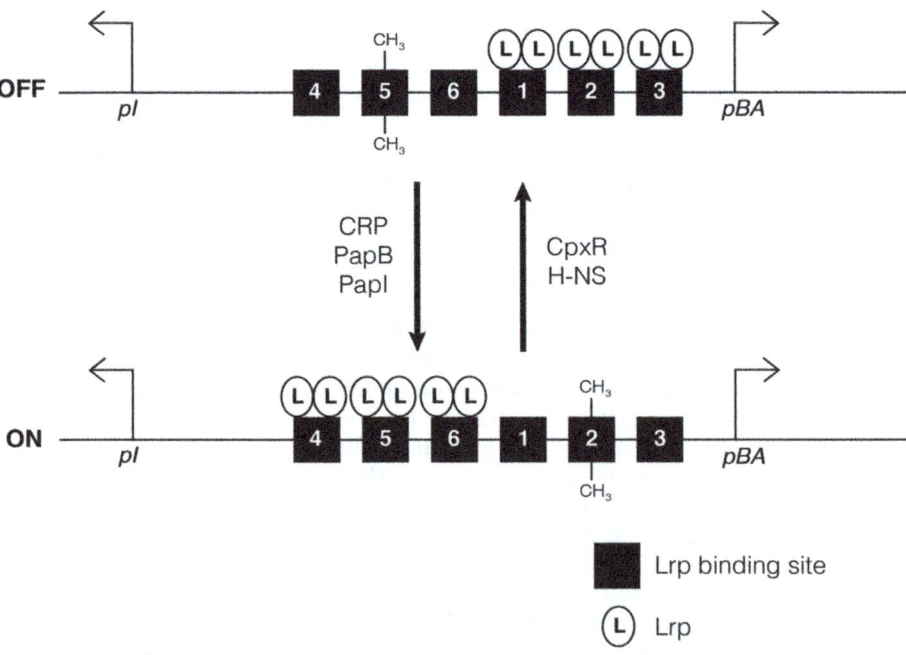

Fig. 4.3. Model of phase switching of *pap* genes by multiple regulators. Methylation (CH$_3$) sites and promoters for *papI* and *papBA* are indicated.

sites upstream of *papBA* and thereby activating P pili expression (Nou *et al.*, 1993; Kaltenbach *et al.*, 1995; Kawamura *et al.*, 2011). Increased expression of PapB can result in re-occupation of the proximal, higher affinity Lrp binding sites, resulting in the inhibition of P pili expression. Although Lrp typically activates genes important for amino acid metabolism and nutrient transport (Newman and Lin, 1995), the activity of Lrp within the *pap* gene cluster is not responsive to leucine (Braaten *et al.*, 1992).

Transcription of *papBA* is inhibited by the response regulator CpxR, part of the Cpx two-component system that senses and responds to membrane stress and unfolded periplasmic proteins (Hung *et al.*, 2001; Hernday *et al.*, 2004). When misfolded pilin subunits are sensed, CpxR downregulates the expression of P pili by blocking Lrp-binding (Hernday *et al.*, 2004). The *papBA* operon is also repressed by the histone-like nucleoid-structuring protein H-NS (Goransson *et al.*, 1990; White-Ziegler *et al.*, 1998), a global regulator that can preferentially silence the transcription of A+T-rich DNA sequences by inducing DNA supercoiling (Fang and Rimsky, 2008). Repression by H-NS is relieved by PapB and PapB-mediated effects on PapI expression, as well as Catabolite Responsive Protein (CRP) (Baga *et al.*, 1985; Forsman *et al.*, 1989, 1992; Goransson *et al.*, 1989). CRP enables UPEC and other bacteria to respond to changing carbon sources and may act to enhance P pili expression when glucose availability is low (Baga *et al.*, 1985; Goransson *et al.*, 1989; Forsman *et al.*, 1992; Weyand *et al.*, 2001).

DNA replication itself can also affect phase variable expression of the *pap* operon. After DNA replication, Dam methylates the new, unmethylated DNA strand at most of the GATC sites in the genome. However, GATC sites within the *pap* promoter are not optimal targets for Dam (Peterson and Reich, 2006). Slower methylation of sites within the *pap* promoter enables Lrp to competitively bind to the GATC sequences, blocking their methylation by Dam and repressing P pili expression (Hernday *et al.*, 2003; Peterson and Reich, 2008). These regulatory effects are dependent upon the concentrations of Lrp,

and are likely complicated further by myriad regulatory factors that come into play during the course of a UTI.

4.4.3 Relevance of P pili in pathogenesis

The importance of P pili in UTI has been shown in rodents, monkeys and humans. In mice, the delivery of P pili-specific antibodies provides passive protection against kidney colonization by UPEC (O'Hanley *et al.*, 1985). UPEC strains, and even faecal isolates, that express recombinant P pili can outcompete P pili-negative counterparts within the kidneys of mice (Hagberg *et al.*, 1983), but in other work P pili expression in mice was inconsequential to the ability of a prototypic UPEC isolate to cause pyelonephritis (Mobley *et al.*, 1993). In monkeys, immunization with purified P pili can interfere with the ability of UPEC to cause pyelonephritis (Roberts *et al.*, 1984), and in at least one instance disruption of the *papG* allele was shown to attenuate UPEC colonization of monkey kidneys (Roberts *et al.*, 1994). Corroborating these primate studies is work showing that human pyelonephritic strains are more likely to have the *pap* genes and express P pili than cystitis, ABU and faecal isolates (Kallenius *et al.*, 1981; Plos *et al.*, 1990). In some studies with human volunteers, transformation of the ABU strain 83972 with *pap* genes enhanced bacterial growth, survival and inflammatory effects within the urinary tract (Wullt *et al.*, 2000, 2001; Wullt, 2003), but no effects were seen in a similar, older study (Andersson *et al.*, 1991). These varying results may be attributable to differences in the expression constructs used, the specific *pap* alleles examined and the medical histories of the patient volunteers.

In total, these results from animal and human studies indicate that P pili expression can impact the establishment and progression of kidney infections, but this may vary dependent on both strain and host background. The precise mechanisms by which P pili promote bacterial colonization of the kidneys is not entirely clear, though recent work employing intravital imaging of infected rat kidneys suggest that P pili may act in a synergistic fashion with T1P within the proximal tubules of the kidneys (Melican et al., 2011). Specifically, P pili seem to promote early colonization of the kidney tubules, allowing UPEC to resist flow generated by filtration processes within the kidneys. Later, T1P expression may help adherent bacteria within the tubules assemble into biofilm-like aggregates that are even more resistant to elimination.

4.5 Type 1 Pili, P Pili and the Immune Response

The immune response to UPEC is mostly mediated by innate defences, with T and B cells and other components of adaptive immunity playing a smaller and less well-studied role (Ragnarsdottir and Svanborg, 2012). The innate response involves activation of pattern recognition receptors such as TLR4 and subsequent upregulation of pro-inflammatory responses, including increased expression of cytokines like IL-6 and IL-8 (Frendeus et al., 2001; Ragnarsdottir et al., 2011). IL-6 stimulates fever, increases the production of C-reactive protein to activate complement and stimulates B cells to secrete mucosal IgG (Ragnarsdottir and Svanborg, 2012). IL-8 is a chemokine important for recruiting neutrophils to sites of infection where they can target and kill bacteria (Haraoka et al., 1999). These inflammatory responses can be initiated by a panoply of pathogen-associated molecules, including lipopolysaccharide (LPS). LPS binds and activates TLR4 via association with the TLR4 co-receptor CD14 (Palsson-Mcdermott and O'Neill, 2004). Interestingly, human urothelial cells in both the kidneys and bladder lack membrane-bound CD14, but are still responsive to UPEC (Samuelsson et al., 2004). In vitro studies suggest that FimH, by binding TLR4 directly, may inadvertently present LPS to TLR4 and thereby bypasses the need for CD14 (Hedlund et al., 2001; Samuelsson et al., 2004; Fischer et al., 2006; Mossman et al., 2008). P pili can also stimulate TLR4 signalling, though through a more circuitous pathway (Fischer et al., 2007, 2010). Rather than interacting with TLR4 directly, the engagement of glycosphingolipid receptors

by P pili leads to the release of ceramide from the host cell and this in turn binds and activates TLR4 even in the absence of LPS (Hedlund et al., 1999; Fischer et al., 2007, 2010). These processes may enable the urothelium to modulate its sensitivity to LPS, avoiding the activation of potentially damaging inflammatory responses unless truly virulent T1P- or P pili-positive pathogens are present (Samuelsson et al., 2004).

Activation of TLR4 via T1P and P pili may also benefit UPEC by eliciting tissue damage that could facilitate bacterial dissemination within the host. UPEC strains display an ability to survive longer than non-pathogenic E. coli strains within macrophages, neutrophils and mast cells, and could potentially hijack these immune effector cells as an additional means to spread within the urinary tract (Baorto et al., 1997; Shin, 2000; Nazareth et al., 2007; Bokil et al., 2011). The ability of some UPEC isolates to dampen inflammatory responses under some conditions may enable the pathogens to fine-tune the host environment over the course of an infection (Hunstad et al., 2005; Billips et al., 2007; Wiles et al., 2008; Hilbert et al., 2008; Loughman and Hunstad, 2011, 2012; Dhakal and Mulvey, 2012). This is, however, a potentially dangerous game for UPEC, as T1P interactions with receptors on neutrophils, macrophages and mast cells can stimulate oxidative burst and other antibacterial responses (Sauter et al., 1991; Tewari et al., 1993; Malaviya et al., 1999; Ashkar et al., 2008). These can include the release of antimicrobial peptides, which can work in conjunction with soluble proteins like Tamm-Horsfall Protein (THP) and secretory IgA (sIgA) that can bind FimH and block UPEC interactions with host cells (Wold et al., 1990; Bates et al., 2004; Mo et al., 2004).

4.6 Conclusion

Data accumulated over the past few decades indicate that T1P and P pili are preeminent virulence and fitness factors utilized by many UPEC isolates to effectively colonize host tissues within the urinary tract. Ongoing work demonstrates that these adhesive organelles can function in a complex array of biological activities, ranging from epithelial cell attachment and invasion to biofilm formation and the modulation of host inflammatory responses. The central roles that these pili have as mediators of UPEC pathogenesis within the urinary tract has made them prime targets for the development of vaccines and other antimicrobial therapeutics. Future studies will undoubtedly highlight new ways in which these organelles can affect the onset and progression of UTIs. It is likewise probable that the coming years will reveal effects that T1P and P pili have on the fitness of both pathogens and commensal strains at sites outside of the urinary tract.

References

Aberg, A., Shingler, V. and Balsalobre, C. (2008) Regulation of the *fimB* promoter: a case of differential regulation by ppGpp and DksA *in vivo. Molecular Microbiology* 67, 1223–1241.

Abraham, J.M., Freitag, C.S., Clements, J.R. and Eisenstein, B.I. (1985a) An invertible element of DNA controls phase variation of type 1 fimbriae of *Escherichia coli. Proceedings of the National Academy of Sciences of the United States of America* 82, 5724–5727.

Abraham, S.N., Babu, J.P., Giampapa, C.S., Hasty, D.L., Simpson, W.A. and Beachey, E.H. (1985b) Protection against *Escherichia coli*-induced urinary tract infections with hybridoma antibodies directed against type 1 fimbriae or complementary D-mannose receptors. *Infection and Immunity* 48, 625–628.

Abraham, S.N., Sun, D., Dale, J.B. and Beachey, E.H. (1988) Conservation of the D-mannose-adhesion protein among type 1 fimbriated members of the family Enterobacteriaceae. *Nature* 336, 682–684.

Allen, W.J., Phan, G. and Waksman, G. (2012) Pilus biogenesis at the outer membrane of Gram-negative bacterial pathogens. *Current Opinion in Structural Biology* 22, 500–506.

Anderson, G.G., Palermo, J.J., Schilling, J.D., Roth, R., Heuser, J. and Hultgren, S.J. (2003) Intracellular bacterial biofilm-like pods in urinary tract infections. *Science* 301, 105–107.

Andersson, P., Engberg, I., Lidin-Janson, G., Lincoln, K., Hull, R., Hull, S. and Svanborg, C. (1991) Persistence of *Escherichia coli* bacteriuria is not determined by bacterial adherence. *Infection and Immunity* 59, 2915–2921.

Aronson, M., Medalia, O., Amichay, D. and Nativ, O. (1988) Endotoxin-induced shedding of viable uroepithelial cells is an antimicrobial defense mechanism. *Infection and Immunity* 56, 1615–1617.

Ashkar, A.A., Mossman, K.L., Coombes, B.K., Gyles, C.L. and Mackenzie, R. (2008) FimH adhesin of type 1 fimbriae is a potent inducer of innate antimicrobial responses which requires TLR4 and type 1 interferon signalling. *PLoS Pathogens* 4, e1000233.

Baga, M., Goransson, M., Normark, S. and Uhlin, B.E. (1985) Transcriptional activation of a pap pilus virulence operon from uropathogenic *Escherichia coli. EMBO Journal* 4, 3887–3893.

Bahrani-Mougeot, F.K., Buckles, E.L., Lockatell, C.V., Hebel, J.R., Johnson, D.E., Tang, C.M. and Donnenberg, M.S. (2002) Type 1 fimbriae and extracellular polysaccharides are preeminent uropathogenic *Escherichia coli* virulence determinants in the murine urinary tract. *Molecular Microbiology* 45, 1079–1093.

Baorto, D.M., Gao, Z., Malaviya, R., Dustin, M.L., Van Der Merwe, A., Lublin, D.M. and Abraham, S.N. (1997) Survival of FimH-expressing enterobacteria in macrophages relies on glycolipid traffic. *Nature* 389, 636–639.

Bates, J.M., Raffi, H.M., Prasadan, K., Mascarenhas, R., Laszik, Z., Maeda, N., Hultgren, S.J. and Kumar, S. (2004) Tamm-Horsfall protein knockout mice are more prone to urinary tract infection: rapid communication. *Kidney International* 65, 791–797.

Bergsten, G., Wullt, B. and Svanborg, C. (2005) *Escherichia coli*, fimbriae, bacterial persistence and host response induction in the human urinary tract. *International Journal of Medical Microbiology: IJMM* 295, 487–502.

Bergsten, G., Wullt, B., Schembri, M.A., Leijonhufvud, I. and Svanborg, C. (2007) Do type 1 fimbriae promote inflammation in the human urinary tract? *Cellular Microbiology* 9, 1766–1781.

Billips, B.K., Forrestal, S.G., Rycyk, M.T., Johnson, J.R., Klumpp, D.J. and Schaeffer, A.J. (2007) Modulation of host innate immune response in the bladder by uropathogenic *Escherichia coli. Infection and Immunity* 75, 5353–5360.

Bishop, B.L., Duncan, M.J., Song, J., Li, G., Zaas, D. and Abraham, S.N. (2007) Cyclic AMP-regulated exocytosis of *Escherichia coli* from infected bladder epithelial cells. *Nature Medicine* 13, 625–630.

Blango, M.G. and Mulvey, M.A. (2010) Persistence of uropathogenic *Escherichia coli* in the face of multiple antibiotics. *Antimicrobial Agents and Chemotherapy* 54, 1855–1863.

Blomfield, I.C., Calie, P.J., Eberhardt, K.J., McClain, M.S. and Eisenstein, B.I. (1993) Lrp stimulates phase variation of type 1 fimbriation in *Escherichia coli* K-12. *Journal of Bacteriology* 175, 27–36.

Blumer, C., Kleefeld, A., Lehnen, D., Heintz, M., Dobrindt, U., Nagy, G., Michaelis, K., Emody, L., Polen, T., Rachel, R., Wendisch, V.F. and Unden, G. (2005) Regulation of type 1 fimbriae synthesis and biofilm formation by the transcriptional regulator LrhA of *Escherichia coli*. *Microbiology* 151, 3287–3298.

Blyn, L.B., Braaten, B.A., White-Ziegler, C.A., Rolfson, D.H. and Low, D.A. (1989) Phase-variation of pyelonephritis-associated pili in *Escherichia coli*: evidence for transcriptional regulation. *EMBO Journal* 8, 613–620.

Blyn, L.B., Braaten, B.A. and Low, D.A. (1990) Regulation of pap pilin phase variation by a mechanism involving differential dam methylation states. *EMBO Journal* 9, 4045–4054.

Bokil, N.J., Totsika, M., Carey, A.J., Stacey, K.J., Hancock, V., Saunders, B.M., Ravasi, T., Ulett, G.C., Schembri, M.A. and Sweet, M.J. (2011) Intramacrophage survival of uropathogenic *Escherichia coli*: differences between diverse clinical isolates and between mouse and human macrophages. *Immunobiology* 216, 1164–1171.

Braaten, B.A., Platko, J.V., Van Der Woude, M.W., Simons, B.H., De Graaf, F.K., Calvo, J.M. and Low, D.A. (1992) Leucine-responsive regulatory protein controls the expression of both the pap and fan pili operons in *Escherichia coli*. *Proceedings of the National Academy of Sciences of the United States of America* 89, 4250–4254.

Bullitt, E. and Makowski, L. (1995) Structural polymorphism of bacterial adhesion pili. *Nature* 373, 164–167.

Carvalho, F.A., Barnich, N., Sivignon, A., Darcha, C., Chan, C.H., Stanners, C.P. and Darfeuille-Michaud, A. (2009) Crohn's disease adherent-invasive *Escherichia coli* colonize and induce strong gut inflammation in transgenic mice expressing human CEACAM. *Journal of Experimental Medicine* 206, 2179–2189.

Chen, Y., Guo, X., Deng, F.M., Liang, F.X., Sun, W., Ren, M., Izumi, T., Sabatini, D.D., Sun, T.T. and Kreibich, G. (2003) Rab27b is associated with fusiform vesicles and may be involved in targeting uroplakins to urothelial apical membranes. *Proceedings of the National Academy of Sciences of the United States of America* 100, 14012–14017.

Choudhury, D. (1999) X-ray structure of the FimC-FimH chaperone-adhesin complex from uropathogenic *Escherichia coli*. *Science* 285, 1061–1066.

Connell, I., Agace, W., Klemm, P., Schembri, M., Marild, S. and Svanborg, C. (1996) Type 1 fimbrial expression enhances *Escherichia coli* virulence for the urinary tract. *Proceedings of the National Academy of Sciences of the United States of America* 93, 9827–9832.

Cooper, L.A., Simmons, L.A. and Mobley, H.L. (2012) Involvement of mismatch repair in the reciprocal control of motility and adherence of uropathogenic *Escherichia coli*. *Infection and Immunity* 80, 1969–1979.

Corcoran, C.P. and Dorman, C.J. (2009) DNA relaxation-dependent phase biasing of the fim genetic switch in *Escherichia coli* depends on the interplay of H-NS, IHF and LRP. *Molecular Microbiology* 74, 1071–1082.

Cusumano, C.K., Pinkner, J.S., Han, Z., Greene, S.E., Ford, B.A., Crowley, J.R., Henderson, J.P., Janetka, J.W. and Hultgren, S.J. (2011) Treatment and prevention of urinary tract infection with orally active FimH inhibitors. *Science Translational Medicine* 3, 109ra115.

Dhakal, B.K. and Mulvey, M.A. (2009) Uropathogenic *Escherichia coli* invades host cells via an HDAC6-modulated microtubule-dependent pathway. *Journal of Biological Chemistry* 284, 446–454.

Dhakal, B.K. and Mulvey, M.A. (2012) The UPEC pore-forming toxin alpha-hemolysin triggers proteolysis of host proteins to disrupt cell adhesion, inflammatory, and survival pathways. *Cell Host & Microbe* 11, 58–69.

Dhakal, B.K., Kulesus, R.R. and Mulvey, M.A. (2008) Mechanisms and consequences of bladder cell invasion by uropathogenic *Escherichia coli*. *European Journal of Clinical Investigation* 38 Supplement 2, 2–11.

Dove, S.L., Smith, S.G. and Dorman, C.J. (1997) Control of *Escherichia coli* type 1 fimbrial gene expression in stationary phase: a negative role for RpoS. *Molecular & General Genetics* 254, 13–20.

Duncan, M.J., Li, G., Shin, J.S., Carson, J.L. and Abraham, S.N. (2004) Bacterial penetration of bladder epithelium through lipid rafts. *Journal of Biological Chemistry* 279, 18944–18951.

Eto, D.S., Sundsbak, J.L. and Mulvey, M.A. (2006) Actin-gated intracellular growth and resurgence of uropathogenic *Escherichia coli*. *Cellular Microbiology* 8, 704–717.

Eto, D.S., Jones, T.A., Sundsbak, J.L. and Mulvey, M.A. (2007) Integrin-mediated host cell invasion by type 1-piliated uropathogenic *Escherichia coli*. *PLoS Pathogens* 3, e100.

Eto, D.S., Gordon, H.B., Dhakal, B.K., Jones, T.A. and Mulvey, M.A. (2008) Clathrin, AP-2, and the NPXY-binding subset of alternate endocytic adaptors facilitate FimH-mediated bacterial invasion of host cells. *Cellular Microbiology* 10, 2553–2567.

Falagas, M.E., Rafailidis, P.I. and Makris, G.C. (2008) Bacterial interference for the prevention and treatment of infections. *International Journal of Antimicrobial Agents* 31, 518–522.

Fallman, E., Schedin, S., Jass, J., Uhlin, B.E. and Axner, O. (2005) The unfolding of the P pili quaternary structure by stretching is reversible, not plastic. *EMBO Reports* 6, 52–56.

Fang, F.C. and Rimsky, S. (2008) New insights into transcriptional regulation by H-NS. *Current Opinion in Microbiology* 11, 113–120.

Fischer, H., Yamamoto, M., Akira, S., Beutler, B. and Svanborg, C. (2006) Mechanism of pathogen-specific TLR4 activation in the mucosa: fimbriae, recognition receptors and adaptor protein selection. *European Journal of Immunology* 36, 267–277.

Fischer, H., Ellstrom, P., Ekstrom, K., Gustafsson, L., Gustafsson, M. and Svanborg, C. (2007) Ceramide as a TLR4 agonist; a putative signalling intermediate between sphingolipid receptors for microbial ligands and TLR4. *Cellular Microbiology* 9, 1239–1251.

Fischer, H., Lutay, N., Ragnarsdottir, B., Yadav, M., Jonsson, K., Urbano, A., Al Hadad, A., Ramisch, S., Storm, P., Dobrindt, U., Salvador, E., Karpman, D., Jodal, U. and Svanborg, C. (2010) Pathogen specific, IRF3-dependent signaling and innate resistance to human kidney infection. *PLoS Pathogens*, 6, e1001109.

Forsman, K., Goransson, M. and Uhlin, B.E. (1989) Autoregulation and multiple DNA interactions by a transcriptional regulatory protein in *E. coli* pili biogenesis. *EMBO Journal* 8, 1271–1277.

Forsman, K., Sonden, B., Goransson, M. and Uhlin, B.E. (1992) Antirepression function in *Escherichia coli* for the cAMP-cAMP receptor protein transcriptional activator. *Proceedings of the National Academy of Sciences of the United States of America* 89, 9880–9884.

Foxman, B. (2010) The epidemiology of urinary tract infection. *Nature Reviews Urology* 7, 653–660.

Foxman, B. and Brown, P. (2003) Epidemiology of urinary tract infections: transmission and risk factors, incidence, and costs. *Infectious Disease Clinics of North America* 17, 227–241.

Freitag, C.S., Abraham, J.M., Clements, J.R. and Eisenstein, B.I. (1985) Genetic analysis of the phase variation control of expression of type 1 fimbriae in *Escherichia coli. Journal of Bacteriology* 162, 668–675.

Frendeus, B., Wachtler, C., Hedlund, M., Fischer, H., Samuelsson, P., Svensson, M. and Svanborg, C. (2001) *Escherichia coli* P fimbriae utilize the Toll-like receptor 4 pathway for cell activation. *Molecular Microbiology* 40, 37–51.

Gally, D.L., Bogan, J.A., Eisenstein, B.I. and Blomfield, I.C. (1993) Environmental regulation of the fim switch controlling type 1 fimbrial phase variation in *Escherichia coli* K-12: effects of temperature and media. *Journal of Bacteriology* 175, 6186–6193.

Gally, D.L., Leathart, J. and Blomfield, I.C. (1996) Interaction of FimB and FimE with the fim switch that controls the phase variation of type 1 fimbriae in *Escherichia coli* K-12. *Molecular Microbiology* 21, 725–738.

Gbarah, A., Gahmberg, C.G., Ofek, I., Jacobi, U. and Sharon, N. (1991) Identification of the leukocyte adhesion molecules CD11 and CD18 as receptors for type 1-fimbriated (mannose-specific) *Escherichia coli. Infection and Immunity* 59, 4524–4530.

Godaly, G., Otto, G., Burdick, M.D., Strieter, R.M. and Svanborg, C. (2007) Fimbrial lectins influence the chemokine repertoire in the urinary tract mucosa. *Kidney International* 71, 778–786.

Gong, M. and Makowski, L. (1992) Helical structure of P pili from *Escherichia coli*. Evidence from X-ray fiber diffraction and scanning transmission electron microscopy. *Journal of Molecular Biology* 228, 735–742.

Goransson, M., Forsman, P., Nilsson, P. and Uhlin, B.E. (1989) Upstream activating sequences that are shared by two divergently transcribed operons mediate cAMP-CRP regulation of pilus-adhesin in *Escherichia coli. Molecular Microbiology* 3, 1557–1565.

Goransson, M., Sonden, B., Nilsson, P., Dagberg, B., Forsman, K., Emanuelsson, K. and Uhlin, B.E. (1990) Transcriptional silencing and thermoregulation of gene expression in *Escherichia coli. Nature* 344, 682–685.

Hagan, E.C., Lloyd, A.L., Rasko, D.A., Faerber, G.J. and Mobley, H.L. (2010) *Escherichia coli* global gene expression in urine from women with urinary tract infection. *PLoS Pathogens* 6, e1001187.

Hagberg, L., Hull, R., Hull, S., Falkow, S., Freter, R. and Svanborg Eden, C. (1983) Contribution of adhesion to bacterial persistence in the mouse urinary tract. *Infection and Immunity* 40, 265–272.

Hannan, T.J., Totsika, M., Mansfield, K.J., Moore,

K.H., Schembri, M.A. and Hultgren, S.J. (2012) Host–pathogen checkpoints and population bottlenecks in persistent and intracellular uropathogenic *Escherichia coli* bladder infection. *FEMS Microbiology Reviews* 36, 616–648.

Haraoka, M., Hang, L., Frendeus, B., Godaly, G., Burdick, M., Strieter, R. and Svanborg, C. (1999) Neutrophil recruitment and resistance to urinary tract infection. *Journal of Infectious Diseases* 180, 1220–1229.

Hase, K., Kawano, K., Nochi, T., Pontes, G.S., Fukuda, S., Ebisawa, M., Kadokura, K., Tobe, T., Fujimura, Y., Kawano, S., Yabashi, A., Waguri, S., Nakato, G., Kimura, S., Murakami, T., Iimura, M., Hamura, K., Fukuoka, S., Lowe, A.W., Itoh, K., Kiyono, H. and Ohno, H. (2009) Uptake through glycoprotein 2 of FimH(+) bacteria by M cells initiates mucosal immune response. *Nature* 462, 226–230.

Hedlund, M., Wachtler, C., Johansson, E., Hang, L., Somerville, J.E., Darveau, R.P. and Svanborg, C. (1999) P fimbriae-dependent, lipopolysaccharide-independent activation of epithelial cytokine responses. *Molecular Microbiology* 33, 693–703.

Hedlund, M., Frendeus, B., Wachtler, C., Hang, L., Fischer, H. and Svanborg, C. (2001) Type 1 fimbriae deliver an LPS- and TLR4-dependent activation signal to CD14-negative cells. *Molecular Microbiology* 39, 542–552.

Hernday, A.D., Braaten, B.A. and Low, D.A. (2003) The mechanism by which DNA adenine methylase and PapI activate the pap epigenetic switch. *Molecular Cell* 12, 947–957.

Hernday, A.D., Braaten, B.A., Broitman-Maduro, G., Engelberts, P. and Low, D.A. (2004) Regulation of the pap epigenetic switch by CpxAR: phosphorylated CpxR inhibits transition to the phase ON state by competition with Lrp. *Molecular Cell* 16, 537–547.

Hilbert, D.W., Pascal, K.E., Libby, E.K., Mordechai, E., Adelson, M.E. and Trama, J.P. (2008) Uropathogenic *Escherichia coli* dominantly suppress the innate immune response of bladder epithelial cells by a lipopolysaccharide- and Toll-like receptor 4-independent pathway. *Microbes and Infection* 10, 114–121.

Holden, N.J., Totsika, M., Mahler, E., Roe, A.J., Catherwood, K., Lindner, K., Dobrindt, U. and Gally, D.L. (2006) Demonstration of regulatory cross-talk between P fimbriae and type 1 fimbriae in uropathogenic *Escherichia coli*. *Microbiology* 152, 1143–1153.

Holden, N., Totsika, M., Dixon, L., Catherwood, K. and Gally, D.L. (2007) Regulation of P-fimbrial phase variation frequencies in *Escherichia coli* CFT073. *Infection and Immunity* 75, 3325–3334.

Hooton, T.M. (2012) Clinical practice. Uncomplicated urinary tract infection. *New England Journal of Medicine* 366, 1028–1037.

Hull, R.A., Rudy, D.C., Donovan, W.H., Wieser, I.E., Stewart, C. and Darouiche, R.O. (1999) Virulence properties of *Escherichia coli* 83972, a prototype strain associated with asymptomatic bacteriuria. *Infection and Immunity* 67, 429–432.

Hultgren, S.J., Porter, T.N., Schaeffer, A.J. and Duncan, J.L. (1985) Role of type 1 pili and effects of phase variation on lower urinary tract infections produced by *Escherichia coli*. *Infection and Immunity* 50, 370–377.

Hung, C.S., Bouckaert, J., Hung, D., Pinkner, J., Widberg, C., Defusco, A., Auguste, C.G., Strouse, R., Langermann, S., Waksman, G. and Hultgren, S.J. (2002) Structural basis of tropism of *Escherichia coli* to the bladder during urinary tract infection. *Molecular Microbiology* 44, 903–915.

Hung, D.L., Raivio, T.L., Jones, C.H., Silhavy, T.J. and Hultgren, S.J. (2001) Cpx signaling pathway monitors biogenesis and affects assembly and expression of P pili. *EMBO Journal* 20, 1508–1518.

Hunstad, D.A., Justice, S.S., Hung, C.S., Lauer, S.R. and Hultgren, S.J. (2005) Suppression of bladder epithelial cytokine responses by uropathogenic *Escherichia coli*. *Infection and Immunity* 73, 3999–4006.

Hvidberg, H., Struve, C., Krogfelt, K.A., Christensen, N., Rasmussen, S.N. and Frimodt-Moller, N. (2000) Development of a long-term ascending urinary tract infection mouse model for antibiotic treatment studies. *Antimicrobial Agents and Chemotherapy* 44, 156–163.

Ikaheimo, R., Siitonen, A., Heiskanen, T., Karkkainen, U., Kuosmanen, P., Lipponen, P. and Makela, P.H. (1996) Recurrence of urinary tract infection in a primary care setting: analysis of a 1-year follow-up of 179 women. *Clinical Infectious Diseases* 22, 91–99.

Jiang, X., Abgottspon, D., Kleeb, S., Rabbani, S., Scharenberg, M., Wittwer, M., Haug, M., Schwardt, O. and Ernst, B. (2012) Antiadhesion therapy for urinary tract infections – a balanced PK/PD profile proved to be key for success. *Journal of Medicinal Chemistry* 55, 4700–4713.

Johanson, I.M., Plos, K., Marklund, B.I. and Svanborg, C. (1993) Pap, papG and prsG DNA sequences in *Escherichia coli* from the fecal flora and the urinary tract. *Microbial Pathogenesis* 15, 121–129.

Johnson, J.R. (1998) papG alleles among

Escherichia coli strains causing urosepsis: associations with other bacterial characteristics and host compromise. *Infection and Immunity* 66, 4568–4571.

Johnson, J.R., Brown, J.J. and Maslow, J.N. (1998a) Clonal distribution of the three alleles of the Gal(alpha1-4)Gal-specific adhesin gene papG among *Escherichia coli* strains from patients with bacteremia. *Journal of Infectious Diseases* 177, 651–661.

Johnson, J.R., Russo, T.A., Brown, J.J. and Stapleton, A. (1998b) papG alleles of *Escherichia coli* strains causing first-episode or recurrent acute cystitis in adult women. *Journal of Infectious Diseases* 177, 97–101.

Johnson, J.R., O'Bryan, T.T., Low, D.A., Ling, G., Delavari, P., Fasching, C., Russo, T.A., Carlino, U. and Stell, A.L. (2000) Evidence of commonality between canine and human extraintestinal pathogenic *Escherichia coli* strains that express papG allele III. *Infection and Immunity* 68, 3327–3336.

Jolley, J.A., Kim, S. and Wing, D.A. (2012) Acute pyelonephritis and associated complications during pregnancy in 2006 in US hospitals. *Journal of Maternal-Fetal and Neonatal Medicine* 25, 2494–2498.

Jones, C.H., Pinkner, J.S., Roth, R., Heuser, J., Nicholes, A.V., Abraham, S.N. and Hultgren, S.J. (1995) FimH adhesin of type 1 pili is assembled into a fibrillar tip structure in the Enterobacteriaceae. *Proceedings of the National Academy of Sciences of the United States of America* 92, 2081–2085.

Jorgensen, I. and Seed, P.C. (2012) How to make it in the urinary tract: a tutorial by *Escherichia coli*. *PLoS Pathogens* 8, e1002907.

Justice, S.S., Hung, C., Theriot, J.A., Fletcher, D.A., Anderson, G.G., Footer, M.J. and Hultgren, S.J. (2004) Differentiation and developmental pathways of uropathogenic *Escherichia coli* in urinary tract pathogenesis. *Proceedings of the National Academy of Sciences of the United States of America* 101, 1333–1338.

Justice, S.S., Hunstad, D.A., Seed, P.C. and Hultgren, S.J. (2006) Filamentation by *Escherichia coli* subverts innate defenses during urinary tract infection. *Proceedings of the National Academy of Sciences of the United States of America* 103, 19884–19889.

Kallenius, G., Mollby, R., Svenson, S.B., Helin, I., Hultberg, H., Cedergren, B. and Winberg, J. (1981) Occurrence of P-fimbriated *Escherichia coli* in urinary tract infections. *Lancet* 2, 1369–1372.

Kaltenbach, L.S., Braaten, B.A. and Low, D.A. (1995) Specific binding of PapI to Lrp-pap DNA complexes. *Journal of Bacteriology* 177, 6449–

6455.

Karam, M.R., Oloomi, M., Mahdavi, M., Habibi, M. and Bouzari, S. (2012) Assessment of immune responses of the flagellin (FliC) fused to FimH adhesin of uropathogenic *Escherichia coli*. *Molecular Immunology* 54, 32–39.

Kawamura, T., Vartanian, A.S., Zhou, H. and Dahlquist, F.W. (2011) The design involved in PapI and Lrp regulation of the pap operon. *Journal of Molecular Biology* 409, 311–332.

Keith, B.R., Maurer, L., Spears, P.A. and Orndorff, P.E. (1986) Receptor-binding function of type 1 pili effects bladder colonization by a clinical isolate of *Escherichia coli*. *Infection and Immunity* 53, 693–696.

Kerrn, M.B., Struve, C., Blom, J., Frimodt-Moller, N. and Krogfelt, K.A. (2005) Intracellular persistence of *Escherichia coli* in urinary bladders from mecillinam-treated mice. *Journal of Antimicrobial Chemotherapy* 55, 383–386.

Kisielius, P.V., Schwan, W.R., Amundsen, S.K., Duncan, J.L. and Schaeffer, A.J. (1989) *In vivo* expression and variation of *Escherichia coli* type 1 and P pili in the urine of adults with acute urinary tract infections. *Infection and Immunity* 57, 1656–1662.

Klein, T., Abgottspon, D., Wittwer, M., Rabbani, S., Herold, J., Jiang, X., Kleeb, S., Luthi, C., Scharenberg, M., Bezencon, J., Gubler, E., Pang, L., Smiesko, M., Cutting, B., Schwardt, O. and Ernst, B. (2010) FimH antagonists for the oral treatment of urinary tract infections: from design and synthesis to *in vitro* and *in vivo* evaluation. *Journal of Medicinal Chemistry* 53, 8627–8641.

Klemm, P. (1986) Two regulatory fim genes, fimB and fimE, control the phase variation of type 1 fimbriae in *Escherichia coli*. *EMBO Journal* 5, 1389–1393.

Klemm, P., Roos, V., Ulett, G.C., Svanborg, C. and Schembri, M.A. (2006) Molecular characterization of the *Escherichia coli* asymptomatic bacteriuria strain 83972: the taming of a pathogen. *Infection and Immunity* 74, 781–785.

Klemm, P., Hancock, V. and Schembri, M.A. (2007) Mellowing out: adaptation to commensalism by *Escherichia coli* asymptomatic bacteriuria strain 83972. *Infection and Immunity* 75, 3688–3695.

Klumpp, D.J., Weiser, A.C., Sengupta, S., Forrestal, S.G., Batler, R.A. and Schaeffer, A.J. (2001) Uropathogenic *Escherichia coli* potentiates type 1 pilus-induced apoptosis by suppressing NF-kappaB. *Infection and Immunity* 69, 6689–6695.

Klumpp, D.J., Rycyk, M.T., Chen, M.C., Thumbikat, P., Sengupta, S. and Schaeffer, A.J. (2006)

Uropathogenic *Escherichia coli* induces extrinsic and intrinsic cascades to initiate urothelial apoptosis. *Infection and Immunity* 74, 5106–5113.

Krogfelt, K.A., Bergmans, H. and Klemm, P. (1990) Direct evidence that the FimH protein is the mannose-specific adhesin of *Escherichia coli* type 1 fimbriae. *Infection and Immunity* 58, 1995–1998.

Kuehn, M.J., Heuser, J., Normark, S. and Hultgren, S.J. (1992) P pili in uropathogenic *E. coli* are composite fibres with distinct fibrillar adhesive tips. *Nature* 356, 252–255.

Kuespert, K., Pils, S. and Hauck, C.R. (2006) CEACAMs: their role in physiology and pathophysiology. *Current Opinion in Cell Biology* 18, 565–571.

Kukkonen, M., Raunio, T., Virkola, R., Lahteenmaki, K., Makela, P.H., Klemm, P., Clegg, S. and Korhonen, T.K. (1993) Basement membrane carbohydrate as a target for bacterial adhesion: binding of type I fimbriae of *Salmonella enterica* and *Escherichia coli* to laminin. *Molecular Microbiology* 7, 229–237.

Langermann, S. (1997) Prevention of mucosal *Escherichia coli* infection by FimH-adhesin-based systemic vaccination. *Science* 276, 607–611.

Le Trong, I., Aprikian, P., Kidd, B.A., Forero-Shelton, M., Tchesnokova, V., Rajagopal, P., Rodriguez, V., Interlandi, G., Klevit, R., Vogel, V., Stenkamp, R.E., Sokurenko, E.V. and Thomas, W.E. (2010) Structural basis for mechanical force regulation of the adhesin FimH via finger trap-like beta sheet twisting. *Cell* 141, 645–655.

Leffler, H. and Svanborg-Eden, C. (1981) Glycolipid receptors for uropathogenic *Escherichia coli* on human erythrocytes and uroepithelial cells. *Infection and Immunity* 34, 920–929.

Li, K., Zhou, W., Hong, Y., Sacks, S.H. and Sheerin, N.S. (2009) Synergy between type 1 fimbriae expression and C3 opsonisation increases internalisation of *E. coli* by human tubular epithelial cells. *BMC Microbiology* 9, 64.

Lim, J.K., Gunther, N.W.T., Zhao, H., Johnson, D.E., Keay, S.K. and Mobley, H.L. (1998) *In vivo* phase variation of *Escherichia coli* type 1 fimbrial genes in women with urinary tract infection. *Infection and Immunity* 66, 3303–3310.

Loughman, J.A. and Hunstad, D.A. (2011) Attenuation of human neutrophil migration and function by uropathogenic bacteria. *Microbes and Infection* 13, 555–565.

Loughman, J.A. and Hunstad, D.A. (2012) Induction of indoleamine 2,3-dioxygenase by uro-pathogenic bacteria attenuates innate responses to epithelial infection. *Journal of Infectious Diseases* 205, 1830–1839.

Malaviya, R., Gao, Z., Thankavel, K., Van Der Merwe, P.A. and Abraham, S.N. (1999) The mast cell tumor necrosis factor alpha response to FimH-expressing *Escherichia coli* is mediated by the glycosylphosphatidylinositol-anchored molecule CD48. *Proceedings of the National Academy of Sciences of the United States of America* 96, 8110–8115.

Martinez, J.J. and Hultgren, S.J. (2002) Requirement of Rho-family GTPases in the invasion of Type 1-piliated uropathogenic *Escherichia coli*. *Cellular Microbiology* 4, 19–28.

Martinez, J.J., Mulvey, M.A., Schilling, J.D., Pinkner, J.S. and Hultgren, S.J. (2000) Type 1 pilus-mediated bacterial invasion of bladder epithelial cells. *EMBO Journal* 19, 2803–2812.

McClain, M.S., Blomfield, I.C. and Eisenstein, B.I. (1991) Roles of fimB and fimE in site-specific DNA inversion associated with phase variation of type 1 fimbriae in *Escherichia coli*. *Journal of Bacteriology* 173, 5308–5314.

Melican, K., Sandoval, R.M., Kader, A., Josefsson, L., Tanner, G.A., Molitoris, B.A. and Richter-Dahlfors, A. (2011) Uropathogenic *Escherichia coli* P and Type 1 fimbriae act in synergy in a living host to facilitate renal colonization leading to nephron obstruction. *PLoS Pathogens* 7, e1001298.

Mills, M., Meysick, K.C. and O'Brien, A.D. (2000) Cytotoxic necrotizing factor type 1 of uropathogenic *Escherichia coli* kills cultured human uroepithelial 5637 cells by an apoptotic mechanism. *Infection and Immunity* 68, 5869–5880.

Mo, L., Zhu, X.H., Huang, H.Y., Shapiro, E., Hasty, D.L. and Wu, X.R. (2004) Ablation of the Tamm-Horsfall protein gene increases susceptibility of mice to bladder colonization by type 1-fimbriated *Escherichia coli*. *American Journal of Physiology. Renal Physiology* 286, F795–802.

Mobley, H.L., Jarvis, K.G., Elwood, J.P., Whittle, D.I., Lockatell, C.V., Russell, R.G., Johnson, D.E., Donnenberg, M.S. and Warren, J.W. (1993) Isogenic P-fimbrial deletion mutants of pyelonephritogenic *Escherichia coli*: the role of alpha Gal(1-4) beta Gal binding in virulence of a wild-type strain. *Molecular Microbiology* 10, 143–155.

Mossman, K.L., Mian, M.F., Lauzon, N.M., Gyles, C.L., Lichty, B., Mackenzie, R., Gill, N. and Ashkar, A.A. (2008) Cutting edge: FimH adhesin of type 1 fimbriae is a novel TLR4 ligand. *Journal of Immunology* 181, 6702–6706.

Muller, C.M., Aberg, A., Straseviciene, J., Emody, L., Uhlin, B.E. and Balsalobre, C. (2009) Type 1 fimbriae, a colonization factor of uropathogenic *Escherichia coli*, are controlled by the metabolic sensor CRP-cAMP. *PLoS Pathogens* 5, e1000303.

Mulvey, M.A. (2002) Adhesion and entry of uropathogenic *Escherichia coli*. *Cellular Microbiology* 4, 257–271.

Mulvey, M.A., Lopez-Boado, Y.S., Wilson, C.L., Roth, R., Parks, W.C., Heuser, J. and Hultgren, S.J. (1998) Induction and evasion of host defenses by type 1-piliated uropathogenic *Escherichia coli*. *Science* 282, 1494–1497.

Mulvey, M.A., Schilling, J.D. and Hultgren, S.J. (2001) Establishment of a persistent *Escherichia coli* reservoir during the acute phase of a bladder infection. *Infection and Immunity* 69, 4572–4579.

Mysorekar, I.U. and Hultgren, S.J. (2006) Mechanisms of uropathogenic *Escherichia coli* persistence and eradication from the urinary tract. *Proceedings of the National Academy of Sciences of the United States of America* 103, 14170–14175.

Nazareth, H., Genagon, S.A. and Russo, T.A. (2007) Extraintestinal pathogenic *Escherichia coli* survives within neutrophils. *Infection and Immunity* 75, 2776–2785.

Newman, E.B. and Lin, R. (1995) Leucine-responsive regulatory protein: a global regulator of gene expression in *E. coli*. *Annual Review of Microbiology* 49, 747–775.

Newman, J.V., Burghoff, R.L., Pallesen, L., Krogfelt, K.A., Kristensen, C.S., Laux, D.C. and Cohen, P.S. (1994) Stimulation of *Escherichia coli* F-18Col- type-1 fimbriae synthesis by leuX. *FEMS Microbiology Letters* 122, 281–287.

Norinder, B.S., Koves, B., Yadav, M., Brauner, A. and Svanborg, C. (2012) Do *Escherichia coli* strains causing acute cystitis have a distinct virulence repertoire? *Microbial Pathogenesis* 52, 10–16.

Nou, X., Skinner, B., Braaten, B., Blyn, L., Hirsch, D. and Low, D. (1993) Regulation of pyelonephritis-associated pili phase-variation in *Escherichia coli*: binding of the PapI and the Lrp regulatory proteins is controlled by DNA methylation. *Molecular Microbiology* 7, 545–553.

Ofek, I., Mosek, A. and Sharon, N. (1981) Mannose-specific adherence of *Escherichia coli* freshly excreted in the urine of patients with urinary tract infections, and of isolates subcultured from the infected urine. *Infection and Immunity* 34, 708–711.

O'Gara, J.P. and Dorman, C.J. (2000) Effects of local transcription and H-NS on inversion of the fim switch of *Escherichia coli*. *Molecular Microbiology* 36, 457–466.

O'Hanley, P., Lark, D., Falkow, S. and Schoolnik, G. (1985) Molecular basis of *Escherichia coli* colonization of the upper urinary tract in BALB/c mice. Gal-Gal pili immunization prevents *Escherichia coli* pyelonephritis in the BALB/c mouse model of human pyelonephritis. *Journal of Clinical Investigation* 75, 347–360.

Otto, G., Sandberg, T., Marklund, B.I., Ulleryd, P. and Svanborg, C. (1993) Virulence factors and pap genotype in *Escherichia coli* isolates from women with acute pyelonephritis, with or without bacteremia. *Clinical Infectious Diseases* 17, 448–456.

Palsson-McDermott, E.M. and O'Neill, L.A. (2004) Signal transduction by the lipopolysaccharide receptor, Toll-like receptor-4. *Immunology* 113, 153–162.

Pere, A., Nowicki, B., Saxen, H., Siitonen, A. and Korhonen, T.K. (1987) Expression of P, type-1, and type-1C fimbriae of *Escherichia coli* in the urine of patients with acute urinary tract infection. *Journal of Infectious Diseases* 156, 567–574.

Peterson, S.N. and Reich, N.O. (2006) GATC flanking sequences regulate Dam activity: evidence for how Dam specificity may influence pap expression. *Journal of Molecular Biology* 355, 459–472.

Peterson, S.N. and Reich, N.O. (2008) Competitive Lrp and Dam assembly at the *pap* regulatory region: implications for mechanisms of epigenetic regulation. *Journal of Molecular Biology* 383, 92–105.

Plos, K., Carter, T., Hull, S., Hull, R. and Svanborg Eden, C. (1990) Frequency and organization of pap homologous DNA in relation to clinical origin of uropathogenic *Escherichia coli*. *Journal of Infectious Diseases* 161, 518–524.

Poggio, T.V., La Torre, J.L. and Scodeller, E.A. (2006) Intranasal immunization with a recombinant truncated FimH adhesin adjuvanted with CpG oligodeoxynucleotides protects mice against uropathogenic *Escherichia coli* challenge. *Canadian Journal of Microbiology* 52, 1093–1102.

Pouttu, R., Puustinen, T., Virkola, R., Hacker, J., Klemm, P. and Korhonen, T.K. (1999) Amino acid residue Ala-62 in the FimH fimbrial adhesin is critical for the adhesiveness of meningitis-associated *Escherichia coli* to collagens. *Molecular Microbiology* 31, 1747–1757.

Pratt, L.A. and Kolter, R. (1998) Genetic analysis of *Escherichia coli* biofilm formation: roles of flagella, motility, chemotaxis and type I pili. *Molecular Microbiology* 30, 285–293.

Ragnarsdottir, B. and Svanborg, C. (2012) Susceptibility to acute pyelonephritis or asymptomatic bacteriuria: host–pathogen interaction in urinary tract infections. *Pediatric Nephrology* 27, 2017–2029.

Ragnarsdottir, B., Lutay, N., Gronberg-Hernandez, J., Koves, B. and Svanborg, C. (2011) Genetics of innate immunity and UTI susceptibility. *Nature Reviews. Urology* 8, 449–468.

Roberts, J.A., Hardaway, K., Kaack, B., Fussell, E.N. and Baskin, G. (1984) Prevention of pyelonephritis by immunization with P-fimbriae. *Journal of Urology* 131, 602–607.

Roberts, J.A., Marklund, B.I., Ilver, D., Haslam, D., Kaack, M.B., Baskin, G., Louis, M., Mollby, R., Winberg, J. and Normark, S. (1994) The Gal(alpha 1-4)Gal-specific tip adhesin of *Escherichia coli* P-fimbriae is needed for pyelonephritis to occur in the normal urinary tract. *Proceedings of the National Academy of Sciences of the United States of America* 91, 11889–11893.

Ronald, A. (2002) The etiology of urinary tract infection: traditional and emerging pathogens. *American Journal of Medicine* 113 Suppl 1A, 14S–19S.

Rosen, D.A., Hooton, T.M., Stamm, W.E., Humphrey, P.A. and Hultgren, S.J. (2007) Detection of intracellular bacterial communities in human urinary tract infection. *PLoS Medicine* 4, e329.

Russo, T.A., Stapleton, A., Wenderoth, S., Hooton, T.M. and Stamm, W.E. (1995) Chromosomal restriction fragment length polymorphism analysis of *Escherichia coli* strains causing recurrent urinary tract infections in young women. *Journal of Infectious Diseases* 172, 440–445.

Samuelsson, P., Hang, L., Wullt, B., Irjala, H. and Svanborg, C. (2004) Toll-like receptor 4 expression and cytokine responses in the human urinary tract mucosa. *Infection and Immunity* 72, 3179–3186.

Sauter, S.L., Rutherfurd, S.M., Wagener, C., Shively, J.E. and Hefta, S.A. (1991) Binding of non-specific cross-reacting antigen, a granulocyte membrane glycoprotein, to *Escherichia coli* expressing type 1 fimbriae. *Infection and Immunity* 59, 2485–2493.

Sauter, S.L., Rutherfurd, S.M., Wagener, C., Shively, J.E. and Hefta, S.A. (1993) Identification of the specific oligosaccharide sites recognized by type 1 fimbriae from *Escherichia coli* on nonspecific cross-reacting antigen, a CD66 cluster granulocyte glycoprotein. *Journal of Biological Chemistry* 268, 15510–15516.

Schaeffer, A.J., Schwan, W.R., Hultgren, S.J. and Duncan, J.L. (1987) Relationship of type 1 pilus expression in *Escherichia coli* to ascending urinary tract infections in mice. *Infection and Immunity* 55, 373–380.

Schappert, S.M. and Rechtsteiner, E.A. (2011) Ambulatory medical care utilization estimates for 2007. *Vital and Health Statistics. Series 13, Data from the National Health Survey*, 1–38.

Schembri, M.A. and Klemm, P. (2001) Biofilm formation in a hydrodynamic environment by novel fimh variants and ramifications for virulence. *Infection and Immunity* 69, 1322–1328.

Schembri, M.A., Sokurenko, E.V. and Klemm, P. (2000) Functional flexibility of the FimH adhesin: insights from a random mutant library. *Infection and Immunity* 68, 2638–2646.

Shin, J.S. (2000) Involvement of cellular caveolae in bacterial entry into mast cells. *Science* 289, 785–788.

Silverblatt, F.J. and Cohen, L.S. (1979) Antipili antibody affords protection against experimental ascending pyelonephritis. *Journal of Clinical Investigation* 64, 333–336.

Silverblatt, F.J., Weinstein, R. and Rene, P. (1982) Protection against experimental pyelonephritis by antibodies to pili. *Scandinavian Journal of Infectious Diseases. Supplementum* 33, 79–82.

Simms, A.N. and Mobley, H.L. (2008) Multiple genes repress motility in uropathogenic *Escherichia coli* constitutively expressing type 1 fimbriae. *Journal of Bacteriology* 190, 3747–3756.

Snyder, J.A., Haugen, B.J., Buckles, E.L., Lockatell, C.V., Johnson, D.E., Donnenberg, M.S., Welch, R.A. and Mobley, H.L. (2004) Transcriptome of uropathogenic *Escherichia coli* during urinary tract infection. *Infection and Immunity* 72, 6373–6381.

Snyder, J.A., Haugen, B.J., Lockatell, C.V., Maroncle, N., Hagan, E.C., Johnson, D.E., Welch, R.A. and Mobley, H.L. (2005) Coordinate expression of fimbriae in uropathogenic *Escherichia coli*. *Infection and Immunity* 73, 7588–7596.

Sokurenko, E.V., Courtney, H.S., Abraham, S.N., Klemm, P. and Hasty, D.L. (1992) Functional heterogeneity of type 1 fimbriae of *Escherichia coli*. *Infection and Immunity* 60, 4709–4719.

Sokurenko, E.V., Courtney, H.S., Maslow, J., Siitonen, A. and Hasty, D.L. (1995) Quantitative differences in adhesiveness of type 1 fimbriated *Escherichia coli* due to structural differences in fimH genes. *Journal of Bacteriology* 177, 3680–3686.

Sokurenko, E.V., Chesnokova, V., Doyle, R.J. and Hasty, D.L. (1997) Diversity of the *Escherichia coli* type 1 fimbrial lectin. Differential binding to mannosides and uroepithelial cells. *Journal of Biological Chemistry* 272, 17880–17886.

Sokurenko, E.V., Chesnokova, V., Dykhuizen, D.E., Ofek, I., Wu, X.R., Krogfelt, K.A., Struve, C., Schembri, M.A. and Hasty, D.L. (1998) Pathogenic adaptation of *Escherichia coli* by natural variation of the FimH adhesin. *Proceedings of the National Academy of Sciences of the United States of America* 95, 8922–8926.

Southgate, J., Kennedy, W., Hutton, K.A. and Trejdosiewicz, L.K. (1995) Expression and *in vitro* regulation of integrins by normal human urothelial cells. *Cell Adhesion and Communication* 3, 231–242.

Stickler, D.J. (2008) Bacterial biofilms in patients with indwelling urinary catheters. *Nature Clinical Practice. Urology* 5, 598–608.

Stromberg, N., Marklund, B.I., Lund, B., Ilver, D., Hamers, A., Gaastra, W., Karlsson, K.A. and Normark, S. (1990) Host-specificity of uropathogenic *Escherichia coli* depends on differences in binding specificity to Gal alpha 1-4Gal-containing isoreceptors. *EMBO Journal* 9, 2001–2010.

Taganna, J., De Boer, A.R., Wuhrer, M. and Bouckaert, J. (2011) Glycosylation changes as important factors for the susceptibility to urinary tract infection. *Biochemical Society Transactions* 39, 349–354.

Tewari, R., MacGregor, J.I., Ikeda, T., Little, J.R., Hultgren, S.J. and Abraham, S.N. (1993) Neutrophil activation by nascent FimH subunits of type 1 fimbriae purified from the periplasm of *Escherichia coli. Journal of Biological Chemistry* 268, 3009–3015.

Thanassi, D.G., Saulino, E.T., Lombardo, M.J., Roth, R., Heuser, J. and Hultgren, S.J. (1998) The PapC usher forms an oligomeric channel: implications for pilus biogenesis across the outer membrane. *Proceedings of the National Academy of Sciences of the United States of America* 95, 3146–3151.

Thanassi, D.G., Bliska, J.B. and Christie, P.J. (2012) Surface organelles assembled by secretion systems of Gram-negative bacteria: diversity in structure and function. *FEMS Microbiology Reviews* 36, 1046–1082.

Thomas, W.E., Trintchina, E., Forero, M., Vogel, V. and Sokurenko, E.V. (2002) Bacterial adhesion to target cells enhanced by shear force. *Cell* 109, 913–923.

Thomas, W.E., Nilsson, L.M., Forero, M., Sokurenko, E.V. and Vogel, V. (2004) Shear-dependent 'stick-and-roll' adhesion of type 1

fimbriated *Escherichia coli. Molecular Microbiology* 53, 1545–1557.

Thumbikat, P., Berry, R.E., Zhou, G., Billips, B.K., Yaggie, R.E., Zaichuk, T., Sun, T.T., Schaeffer, A.J. and Klumpp, D.J. (2009) Bacteria-induced uroplakin signaling mediates bladder response to infection. *PLoS Pathogens* 5, e1000415.

Visvikis, O., Boyer, L., Torrino, S., Doye, A., Lemonnier, M., Lores, P., Rolando, M., Flatau, G., Mettouchi, A., Bouvard, D., Veiga, E., Gacon, G., Cossart, P. and Lemichez, E. (2011) *Escherichia coli* producing CNF1 toxin hijacks Tollip to trigger Rac1-dependent cell invasion. *Traffic* 12, 579–590.

Wang, H., Min, G., Glockshuber, R., Sun, T.T. and Kong, X.P. (2009) Uropathogenic *E. coli* adhesin-induced host cell receptor conformational changes: implications in transmembrane signaling transduction. *Journal of Molecular Biology* 392, 352–361.

Wang, Z., Humphrey, C., Frilot, N., Wang, G., Nie, Z., Moniri, N.H. and Daaka, Y. (2011) Dynamin2- and endothelial nitric oxide synthase-regulated invasion of bladder epithelial cells by uropathogenic *Escherichia coli. Journal of Cell Biology* 192, 101–110.

Weissman, S.J., Chattopadhyay, S., Aprikian, P., Obata-Yasuoka, M., Yarova-Yarovaya, Y., Stapleton, A., Ba-Thein, W., Dykhuizen, D., Johnson, J.R. and Sokurenko, E.V. (2006) Clonal analysis reveals high rate of structural mutations in fimbrial adhesins of extraintestinal pathogenic *Escherichia coli. Molecular Microbiology* 59, 975–988.

Weyand, N.J., Braaten, B.A., Van Der Woude, M., Tucker, J. and Low, D.A. (2001) The essential role of the promoter-proximal subunit of CAP in pap phase variation: Lrp- and helical phase-dependent activation of *papBA* transcription by CAP from -215. *Molecular Microbiology* 39, 1504–1522.

White-Ziegler, C.A., Angus Hill, M.L., Braaten, B.A., Van Der Woude, M.W. and Low, D.A. (1998) Thermoregulation of *Escherichia coli pap* transcription: H-NS is a temperature-dependent DNA methylation blocking factor. *Molecular Microbiology* 28, 1121–1137.

Wiles, T.J., Dhakal, B.K., Eto, D.S. and Mulvey, M.A. (2008) Inactivation of host Akt/protein kinase B signaling by bacterial pore-forming toxins. *Molecular Biology of the Cell* 19, 1427–1438.

Wold, A.E., Thorssen, M., Hull, S. and Eden, C.S. (1988) Attachment of *Escherichia coli* via mannose- or Gal alpha 1–4Gal beta-containing receptors to human colonic epithelial cells. *Infection and Immunity* 56, 2531–2537.

Wold, A.E., Mestecky, J., Tomana, M., Kobata, A., Ohbayashi, H., Endo, T. and Eden, C.S. (1990) Secretory immunoglobulin A carries oligosaccharide receptors for *Escherichia coli* type 1 fimbrial lectin. *Infection and Immunity* 58, 3073–3077.

Wright, K.J., Seed, P.C. and Hultgren, S.J. (2007) Development of intracellular bacterial communities of uropathogenic *Escherichia coli* depends on type 1 pili. *Cellular Microbiology* 9, 2230–2241.

Wu, X.R., Kong, X.P., Pellicer, A., Kreibich, G. and Sun, T.T. (2009) Uroplakins in urothelial biology, function, and disease. *Kidney International* 75, 1153–1165.

Wullt, B. (2003) The role of P fimbriae for *Escherichia coli* establishment and mucosal inflammation in the human urinary tract. *International Journal of Antimicrobial Agents* 21, 605–621.

Wullt, B., Bergsten, G., Connell, H., Rollano, P., Gebretsadik, N., Hull, R. and Svanborg, C. (2000) P fimbriae enhance the early establishment of *Escherichia coli* in the human urinary tract. *Molecular Microbiology* 38, 456–464.

Wullt, B., Bergsten, G., Connell, H., Rollano, P., Gebratsedik, N., Hang, L. and Svanborg, C. (2001) P-fimbriae trigger mucosal responses to *Escherichia coli* in the human urinary tract. *Cellular Microbiology* 3, 255–264.

Xia, Y., Gally, D., Forsman-Semb, K. and Uhlin, B.E. (2000) Regulatory cross-talk between adhesin operons in *Escherichia coli*: inhibition of type 1 fimbriae expression by the PapB protein. *EMBO Journal* 19, 1450–1457.

Yu, S. and Lowe, A.W. (2009) The pancreatic zymogen granule membrane protein, GP2, binds *Escherichia coli* Type 1 fimbriae. *BMC Gastroenterology* 9, 58.

Zhou, G., Mo, W.J., Sebbel, P., Min, G., Neubert, T.A., Glockshuber, R., Wu, X.R., Sun, T.T. and Kong, X.P. (2001) Uroplakin Ia is the urothelial receptor for uropathogenic *Escherichia coli*: evidence from *in vitro* FimH binding. *Journal of Cell Science* 114, 4095–4103.

5 Type IV Pili: Functions and Biogenesis

Michaella Georgiadou and Vladimir Pelicic

MRC Centre for Molecular Bacteriology and Infection, Section of Microbiology, Imperial College London, London, UK

5.1 Introduction

Adhesion to host surfaces is a key virulence attribute of pathogenic bacteria, instrumental in their ability to cause disease. Although diverse species can use a variety of mechanisms to adhere firmly to host cells and/or extracellular matrix, the most common strategy is to use pili (Proft and Baker, 2009). Pili (or fimbriae) are hair-like appendages that extend from the surface of many bacteria, and are polymers of primarily one protein generically named pilin. Out of the many types of pili that have been identified (many of which are reviewed in this book), and classified according to their morphological and/or molecular characteristics, type IV pili (Tfp) are undoubtedly the most widespread (Pelicic, 2008). Tfp are likely to be present in hundreds of different species based either on direct observation of filaments and/or twitching motility (a form of surface translocation exclusively mediated by Tfp), or the identification of genes involved in Tfp biology in a myriad of genome sequencing projects (Mattick, 2002; Pelicic, 2008). Indeed, these organelles are the only pili found both in Gram-negative and Gram-positive species, spanning at least 14/30 phyla in the Bacteria domain (Acidobacteria, Actinobacteria, Aquificae, Caldiserica, Cyanobacteria, Defferibacteres, Deinococcus-Thermus, Dictyoglomi, Fibrobacteres, Firmicutes, Gemmatimon-adetes, Nitrospira, Proteobacteria and Thermodesulfobacteria). Strikingly, similar organelles recently named archaella (Jarrell and Albers, 2012) are also found in another domain of life (Archaea) where they mediate swimming. In addition, numerous bacterial species use machineries extremely similar to the one involved in Tfp biogenesis either to secrete proteins through a process known as type II secretion (T2S) (Douzi et al., 2012), or to mediate uptake of free DNA to use as a source of food, for repairing DNA damage or for generating genetic diversity (Chen and Dubnau, 2004). These different biological systems are therefore evolutionarily related, and represent variations on the theme of transport of macromolecules across membranes in prokaryotes. It is tempting to speculate that pil genes were already present in a common ancestor to Bacteria and Archaea in which they likely encoded a rudimentary macromolecule transport machinery.

5.1.1 Morphological and molecular characteristics defining Tfp

Tfp have a typical morphology, as seen by electron microscopy, that was originally the basis of their classification into a specific pilus type (type IV), which gave them their name (Fig. 5.1a). Tfp are thin (5–8 nm in width), long (up to several microns in length) and

flexible filaments. Most often, several filaments will interact laterally to form highly
distinctive bundles. There is now ample
molecular and structural evidence confirming
that Tfp are a homogeneous class of filaments.
The pilin subunits have been extensively
studied in multiple systems, which led to
common principles about Tfp structure and
assembly (Craig *et al.*, 2004). Although they
are extremely variable in sequence and in
length, all type IV pilins have a conserved
N-terminal domain of about 20–25 aa (Fig.
5.1b), the first of which is methylated, and
that are highly hydrophobic except for a
glutamate in fifth position (PS00409 domain)
(Dalrymple and Mattick, 1987). Moreover,
pilins are synthesized as precursors with a
leader peptide of hydrophilic residues ending
with a glycine, which needs to be cleaved by
a dedicated prepilin peptidase prior to
assembly (Nunn and Lory, 1991). This motif
has recently been named class III signal
peptide (Szabo *et al.*, 2007).

Lengths of leader peptides and mature
pilins have been used to define two Tfp
subtypes, type IVa (Tfpa) and type IVb (Tfpb),
which is consistent with clear differences
between the machineries involved in their
biogenesis (see below) (Pelicic, 2008). Tfpa
are formed of pilins typically 150–160 aa long,
with short leader peptides of less than 10
residues. In contrast, Tfpb pilins display
longer leader peptides (about 15–30 aa), and
are either long (180–200 aa) or very short (40–
50 aa) in the case of Tad (tight adherence) pili
(Craig *et al.*, 2004; Tomich *et al.*, 2007). Much
of the widespread nature of Tfp noted above
is due to Tfpa (Pelicic, 2008), which have been
most extensively studied in three model
species: the human pathogens *Neisseria
gonorrhoeae* and *Neisseria meningitidis*, and
the opportunistic pathogen *Pseudomonas
aeruginosa*. In contrast, Tfpb are found in a
small number of genera, mainly enteric
Proteobacteria (although Tad pili might be
present in distant phyla such as Actinobateria
and Chlorobi) (Tomich *et al.*, 2007) and they
have been most extensively studied in *Vibrio
cholerae* (toxin co-regulated pilus, or Tcp),
enteropathogenic *Escherichia coli* (bundle-
forming pilus, or Bfp), and *E. coli* (thin pilus
of plasmid R64, or R64 Pil).

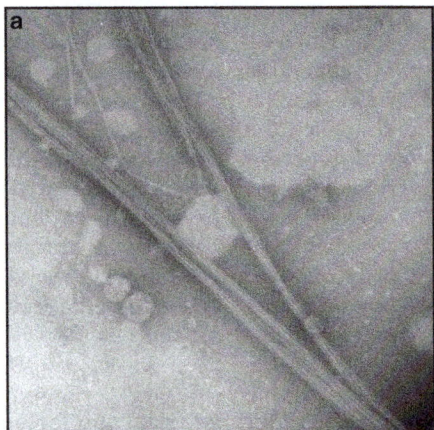

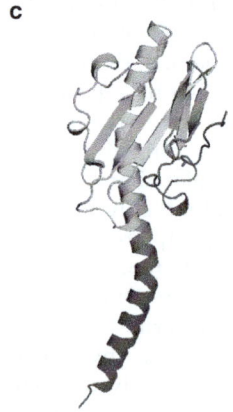

b

[KRHEQSTAG]-G-[FYLIVM]-[ST]-[LT]-[LIVP]-E-[LIVMFWSTAG]$_{14}$

Fig. 5.1. Distinctive features of Tfp. (a) Morphology. Transmission electron microscopy picture of
N. meningitidis filaments. (b) Sequence. Conserved N-terminal motifs found in all type IV prepilins. (c)
Structure. 3D structure of *N. gonorrhoeae* pilin (PDB entry 1AY2).

Despite these differences, atomic-resolution 3D structures of several Tfpa and Tfpb pilins have revealed a shared 'lollipop' architecture (Fig. 5.1c). Tfp pilins consist of a 'stick' formed by an extended N-terminal α-helix (α1), whose C-terminal half (α1-C) is hydrophobically packed against 4–5 anti-parallel β-sheet forming a globular head (Craig *et al.*, 2004). The N-terminal half of the long α-helix (α1-N), which corresponds to the highly conserved N-terminal hydrophobic domain, protrudes from the protein as seen with the full-length pilin structures (from *N. gonorrhoeae*, *P. aeruginosa* and *Dichelobacter nodosus*) (Parge *et al.*, 1995; Craig *et al.*, 2003; Hartung *et al.*, 2011). Finally, flanking the anti-parallel β-sheet of the globular domain on opposite sides, there are two regions that show extensive variation in sequence, length and structure. These are the α/β loop, which connects the 'stick' and the globular head, and the D-region that is delimited by a disulfide bond between two conserved C-terminal cysteines (Craig *et al.*, 2004).

Consistent with the similar structures of different pilins, models predicting how they are arranged within pili all agree that Tfp are helical polymers (Craig *et al.*, 2004). However, the rise between one subunit and the next, the rotation between subunits and the number of subunits per turn can vary substantially. The α1-N domains provide the principal assembly interface between the pilins, and are thus buried within the core of the filament, while the α/β loop and the D-region are exposed on the surface of the pilus and modulate many of its associated functions.

5.1.2 Tfp mediate an astonishing array of different functions

As illustrated in this book, pathogenic bacteria very often use pili to facilitate adhesion to host surfaces. Tfp are thus key virulence factors in many important human pathogens, including meningococci, gonococci and *V. cholerae*. In addition, owing to their wide distribution, Tfp have been associated with adhesion to a wide variety of biotic and abiotic surfaces. Usually, Tfp-facilitated adhesion to host cells is a two-step process (Pujol *et al.*, 1999). In the first step, sometimes called localized adherence, adhering bacteria form 3D microcolonies (or aggregates) on cells, which are the counterpart of the aggregates that are seen in liquid culture. Adhesion and aggregation are thus entangled Tfp-mediated properties, since a reduced propensity to form bacterial aggregates is invariably accompanied by reduced adherence/colonization (Helaine *et al.*, 2005). Therefore, it is not always clear whether Tfp promote adhesion by directly interacting with the cells or by promoting the formation of aggregates on the cells. In the second step of Tfp-facilitated adhesion, sometimes called diffuse adherence, aggregates disappear, and the bacteria adhere intimately to the cell surface.

Another reason for Tfp's celebrity status in the field of prokaryotic organelles is that they are not only adhesive filaments, but they mediate a vast array of remarkably diverse functions (Mattick, 2002). Besides sometimes contributing to biofilm formation, being receptors for bacteriophages, and even acting as nanowires involved in transfer of electrons, Tfp have been mainly studied for two other more broadly distributed properties, namely twitching motility (Mattick, 2002) and DNA uptake during natural transformation (Chen and Dubnau, 2004), which are directly powered by Tfp's uncommon ability to retract (Merz *et al.*, 2000). However, it should be noted that pilus retraction has so far been formally demonstrated only in Tfpa-expressing bacteria, and it remains an open question whether Tfpb can retract since many of the bacteria that express them lack the gene encoding the ATPase that powers this process (PilT). Twitching motility is a form of bacterial translocation over surfaces in the presence of humidity, which can be directly involved in the rapid colonisation of new surfaces (Merz *et al.*, 2000). If the pilus is tethered to a surface, its retraction generates remarkable mechanical force, in the nanonewton range for bundled filaments (Biais *et al.*, 2008), which pulls the bacteria toward the site of attachment like a 'grappling hook'. Significant tension can thus be exerted on the surface to which bacteria adhere, and in the case of *N. gonorrheae* adhesion to human cells results in dramatic changes in cell cortex (Higashi *et al.*, 2009). The other Tfp-mediated property that has been widely studied is their role in DNA uptake during natural transformation. However, not all the bacteria that harbour Tfp are naturally competent, and in many competent species Tfp cannot be seen, suggesting that elusive short structures termed competence pseudopili are used instead (Chen and Dubnau, 2004). Competence is a multi-stage process and Tfp are always involved in the first step during which bacteria bind free DNA and transport it from the extracellular milieu through the outer membrane in Gram-negative or the thick layer of peptidoglycan in Gram-positive species. The finding that PilT, which powers Tfp retraction, is necessary for DNA uptake (Wolfgang *et al.*, 1998a) suggests that Tfp/pseudopili pull DNA through the outer membrane/peptidoglycan upon retraction. Transformation directly contributes to virulence as it allows competent bacteria to evolve extremely rapidly through horizontal gene transfer, with possibly devastating consequences such as the acquisition of antibiotic resistance, or the emergence of highly virulent isolates.

5.2 Tfp Biogenesis

Although we are still not able to answer the fundamental question, 'how do bacteria assemble Tfp', much progress has been made recently.

5.2.1 Complex and conserved multi-protein machineries are involved in Tfp biogenesis

Systematic genetic studies in several model organisms have defined the complete sets of proteins dedicated to the biogenesis of both Tfpa (in *P. aeruginosa* and *N. meningitidis*) and Tfpb (Bfp, Tcp and R64 Pil) (Table 5.1), and have led to important conclusions. It is unfortunate however that the nomenclatures used in different systems are extremely diverse, with the prefix usually indicating the system (Bfp, Tcp, Pil) and the letter being specific for each protein. Even in highly homologous systems orthologues have different name in different species, which makes the field difficult for Tfp non-experts. Unless expressly stated, the *N. meningitidis* nomenclature will be used here.

The first finding was that although Tfp are polymers of primarily a single protein, the pilin, their biogenesis requires extremely complex machineries composed of between 10 (Tcp) and 18 (*P. aeruginosa*) proteins (Pelicic, 2008). In *N. meningitidis*, there are 15 proteins: PilC1/2, PilD, PilE, PilF, PilG, PilH, PilI, PilJ, PilK, PilM, PilN, PilO, PilP, PilQ and PilW (Carbonnelle *et al.*, 2006). However, to make things even more complex, it should be noted that in each system there are additional proteins that show signature motifs of Pil proteins, which are dispensable for Tfp biogenesis but are key players in Tfp biology since they modulate the various functions mediated by Tfp. In *N. meningitidis*, there are three minor pilins (ComP, PilV and PilX), three traffic ATPases (PilT, PilT2 and PilU) and PilZ (Brown *et al.*, 2010).

The second important finding was that Tfp are a homogeneous class of filaments since their biogenesis requires a universally conserved set of proteins (Pelicic, 2008). These consist of (i) several proteins, besides the pilin, with class III signal peptides; (ii) a dedicated prepilin peptidase; (iii) a traffic ATPase; (iv) an inner membrane polytopic protein from the GspF family; and (iv) an integral outer membrane protein named secretin (obviously absent in piliated Gram-positive species) (Table 5.1). These conserved proteins are often described as 'core', in opposition to the non-conserved (non-core) proteins, but recent structural findings suggest that this distinction is not meaningful and should probably be abandoned. Interestingly, clear homologues of these conserved proteins are present in the machineries mediating T2S (Table 5.1), and DNA uptake. This strengthens the notion that these distinct biological processes are evolutionary related, and often allows important parallels to be drawn.

The third important finding were differences consistent with the Tfpa and Tfp subdivision (Pelicic, 2008). Genes involved in

Table 5.1. Components of the machinery involved in Tfpa biogenesis, and their known counterparts in Tfp biogenesis and T2S.

| Function | Characteristic features | Tfp biogenesis | | | | T2S |
| | | Tfpa | | Tfpb | | |
		Neisseria spp.	P. aeruginosa	Bfp	Tcp	
Pilin	Class III signal peptide	PilE	PilA	BfpA	TcpA	GspG
Traffic ATPase	AAA+ ATPase	PilF	PilB	BfpD	TcpT	GspE
Prepilin peptidase		PilD	PilD	BfpP	TcpJ	GspO
Pilus assembly	Cytoplasmic protein	PilM	PilM	BfpC		GspL
	Inner membrane protein	PilN	PilN			GspL
	Inner membrane protein	PilO	PilO			GspM
	Outer membrane protein	PilP	PilP			
Pilus stabilization	Mediates adhesion	PilC	PilY1			
	'Core' integral membrane protein	PilG	PilC	BfpE	TcpE	GspF
	Class III signal peptide	PilH	FimU			GspH
	Class III signal peptide	PilI	PilV			GspI
	Class III signal peptide	PilJ	PilW			GspJ
	Class III signal peptide	PilK	PilK			GspK
	Outer membrane lipoprotein	PilW	PilF			
Pilus emergence on the surface	Secretin	PilQ	PilQ	BfpB	TcpC	GspD

Tfpa biogenesis are scattered throughout the genome, but the same genes or clusters of gene are always in the same genomic locations even in distant bacteria. Moreover, all these genes are extremely conserved in all Tfpa-expressing bacteria, indicating that Tfpa are a very homogeneous group. In contrast, genes involved in Tfpb biogenesis are fewer (10–12), are always clustered as operons and there is no sequence conservation of the proteins besides the 'core' proteins (Tad pili are an exception). This suggested that Tfpb are less homogeneous than Tfpa, and even led to questioning whether a single mechanism was used for the assembly of Tfpa and Tfpb (Pelicic, 2008). However, recent structural data showing that very different non-core proteins have similar 3D structures suggest that this is very likely to be the case. For example, despite no obvious sequence similarity, the structure of the N-terminal part of BfpC is similar to PilM and to GspL that is involved in T2S (Yamagata et al., 2012), which further extend the mechanistic similarities between Tfpa and Tfpb biogenesis, and T2S.

5.2.2 Molecular mechanisms of Tfp biogenesis

Although the molecular mechanisms of Tfp assembly are still mysterious, genetic studies have now defined clearly distinct steps, and several recent studies have illuminated the topology of the different protein subcomplexes involved (Fig. 5.2).

Prepilin transport and processing

As shown in T2S, prepilins are cotranslationally targeted to the Sec machinery that translocates them across the cytoplasmic membrane (Arts et al., 2007; Francetic et al., 2007). Prepilins are henceforth bitopic proteins with their charged leader peptides in the cytoplasm, the α1-N helices in the membrane, and their globular heads in the periplasm. Mutational analyses have revealed that the only prepilin residue key for efficient processing is the last and invariant glycine residue of the leader peptide, while the invariant glutamate at +5 is dispensable for processing but essential

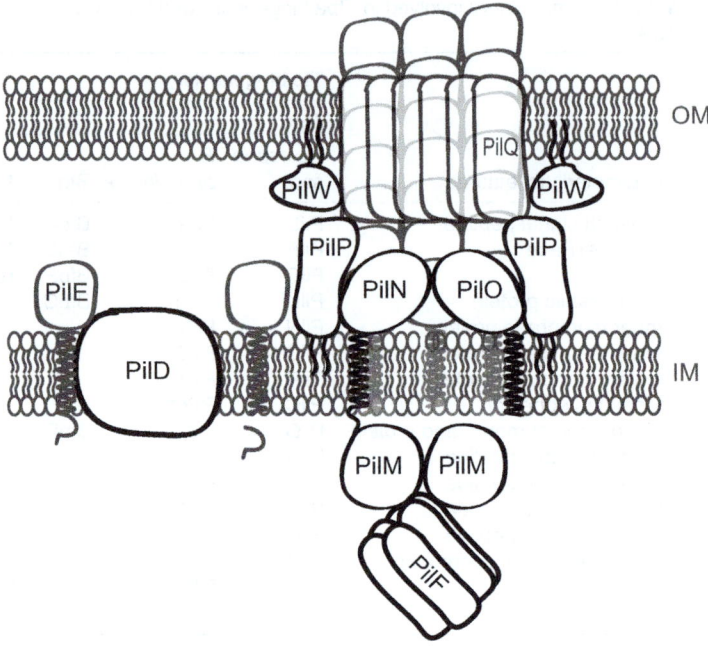

Fig. 5.2. Schematic representation of the machinery involved in Tfpa biogenesis. The interactions between the different components are discussed in the text. For the sake of clarity, the proteins in this cartoon are not drawn to scale.

for pilus assembly (Strom and Lory, 1991). In the R64 Pil system, several other residues distributed throughout the prepilin sequence are also important for efficient processing (Horiuchi and Komano, 1998). Although the molecular mechanism of processing remains to be understood, it is clear that the bitopic topology of the prepilin is key for processing by PilD, which occurs in the periplasm unlike what is seen with other leader peptidases (Fig. 5.2). Two conserved aspartate residues in the C-terminal cytoplasmic loop of PilD are critical for prepilin processing, while the N-terminal cytoplasmic domain is not, and prepilin peptidases therefore represent a novel class of aspartate proteases (LaPointe and Taylor, 2000). Concomitantly with processing, the first residue of the mature pilin is methylated by PilD (Strom *et al.*, 1993). However, while cleavage of the leader peptide is critical for the assembly of the pilus, methylation is dispensable and its functional role is still to be elucidated.

Pilus assembly

In each system there are proteins dispensable for Tfp biogenesis that play a key role in Tfp biology, and the most notorious one is PilT. This traffic ATPase, found in all Tfpa systems but almost never in Tfpb (Bfp might be an exception) powers the retraction (depolymerization) of the filaments, with pilins being dispersed in the cytoplasmic membrane and forming a pool ready to be polymerized again (Morand *et al.*, 2004). PilT is thus playing an opposite role to the 'core' ATPase PilF that powers pilus assembly, and the degree of piliation is therefore the net equilibrium between the relative activities of these two proteins. It was therefore not inconceivable that some Pil proteins could play a role in Tfp biogenesis not by affecting pilus assembly itself, but rather by hindering the retraction promoted by PilT. Therefore, the lack of piliation in the corresponding mutants would be due to uncontrolled pilus retraction, which was expected to be suppressed upon the introduction of a second mutation in *pilT*. This is precisely what has been reported in *N. gonorrhoeae* for the *pilC/T* mutant, which was found to be piliated, showing that PilC is not involved in pilus assembly per se (Wolfgang *et al.*, 1998b). Surprisingly, a systematic analysis in *N. meningitidis* in which a *pilT* mutation was introduced in each of the 15 non-piliated mutants showed that piliation could be restored in a majority of them, demonstrating that the corresponding Pil proteins (PilC1/2, PilG, PilH, PilI, PilJ, PilK, PilQ and PilW) are not involved in pilus assembly (Carbonnelle *et al.*, 2006). As a corollary, a surprisingly small number of both 'core' and non-core proteins (PilE, PilD, PilF, PilM, PilN, PilO and PilP) are necessary for Tfp assembly. Unexpectedly, it was found that the universally conserved membrane protein PilG is dispensable for pilus assembly. This is at odds with the general belief that PilG might be the platform of the pilus assembly machinery. However, experimental evidence for such a role has never been produced.

The roles of three of the proteins involved in Tfp assembly were known. PilE is the main pilus subunit, PilD is necessary for its maturation and PilF is the traffic ATPase that provides the energy necessary for pilus assembly. Based on biochemical and structural analysis of several traffic ATPases, most notably PilT (Satyshur *et al.*, 2007; Misic *et al.*, 2010), PilF forms homohexamers that generate mechanical force by undergoing major conformational changes upon ATP hydrolysis. This mechanical energy is thought to promote extrusion of the pilin subunits from the cytoplasmic membrane through a 'push and pull' mechanism (Misic *et al.*, 2010). Therefore, the remaining four non-core proteins PilM, PilN, PilO and PilP, encoded by an operon universal among bacteria expressing Tfpa, most likely form the actual machinery involved in pilus assembly (Fig. 5.2). Indirect evidence that these proteins form a complex came from protein stability assays showing that deletion of one of these proteins often has a negative impact on the stability of the others (Ayers *et al.*, 2009; Georgiadou *et al.*, 2012). Much progress has been made recently in their characterization,

demonstrating that they indeed form a sub-complex at the cytoplasmic membrane (Fig. 5.2).

The structure of PilM form *Thermus thermophilus* was recently solved (Karuppiah and Derrick, 2011). This cytoplasmic protein has a structure most similar to actin-like protein FtsA with which it shares the ability to bind ATP. However, PilM has no ATPase activity, but ATP binding might be modulating its interaction with PilN. Consistently, high-quality crystals of PilM could be obtained only in the presence of a short peptide corresponding to the N-terminus of PilN, which is the most highly conserved part of this latter protein (Karuppiah and Derrick, 2011). A recent large-scale analysis of protein–protein interactions between 11 Pil proteins of *N. meningitidis* using bacterial two-hybrid (BACTH) (Georgiadou *et al.*, 2012) shed light on the PilM–PilN interaction. PilM was found to interact with itself, and with PilN. Since it was shown in the same study that PilN is a bitopic protein in the cytoplasmic membrane with its highly conserved N-terminus in the cytoplasm, and the bulk of the protein in the periplasm, the PilM–PilN interaction was predicted to depend on that conserved INLLPY motif. This was demonstrated by showing that a truncated version of PilN consisting mainly of the short cytoplasmic domain and the transmembrane helix was capable of interacting with PilM as well as full-length PilN (Georgiadou *et al.*, 2012). The functional significance of the INLLPY motif was further demonstrated by site-directed mutagenesis, since point mutations abolished both the PilM–PilN interaction, and piliation in *N. meningitidis* (Georgiadou *et al.*, 2012). Interestingly, a systematic effort for determining the structure of all the components of the T2S machinery showed that the cytoplasmic domain of GspL is similar to PilM (Abendroth *et al.*, 2004a), while its periplasmic portion is similar to PilN (see below), showing that in this system the two proteins are fused.

As shown by BACTH, PilN, which is very likely to adopt the same circular permutation of the ferredoxin fold as the periplasmic domain of GspL (Abendroth *et al.*, 2009), interacts with itself, and with PilO, which has a similar bitopic topology

(Georgiadou *et al.*, 2012). The PilN–PilO interaction has been confirmed biochemically by showing that the periplasmic domains of these two proteins purify as a heterodimer when co-expressed in *E. coli* (Sampaleanu *et al.*, 2009). Interestingly, the structure of PilO from *P. aeruginosa* was determined by crystallography and shown to consist of a circular permutation of the ferredoxin fold (Sampaleanu *et al.*, 2009), similar to its homolog GspM in the T2S machinery (Abendroth *et al.*, 2004b). PilO was also shown to interact with itself by BACTH (Georgiadou *et al.*, 2012). Again, this is similar to what has been observed in the T2S system, where GspL and GspM interact with themselves and each other (Korotkov *et al.*, 2012).

Finally PilP, which is a cytoplasmic membrane lipoprotein whose structure consists of a modified β-sandwich fold (Golovanov *et al.*, 2006; Tammam *et al.*, 2011), was found to be an integral part of this complex. Co-immunoprecipitation studies in *N. meningitidis* showed that PilP co-immunoprecipitates with PilM, PilN and PilO (Georgiadou *et al.*, 2012). When co-expressed in *E. coli*, a soluble version of PilP and the periplasmic domains of PilN and PilO were found to purify as a stable hetero-trimer (Tammam *et al.*, 2011). When only PilP and PilO were co-expressed, they could not be co-purified, indicating that PilP either interacts with PilN, which could not be tested because PilN is not soluble in the absence of PilO, or that it interacts with the PilN–PilO hetero-dimer. Critically, when the entire *pilMNOP* operon from *N. meningitidis* was expressed in *E. coli*, in the absence of any other *pil* gene, PilP again co-immunoprecipitated with PilM, PilN and PilO, which suggests that the pilus assembly machinery self-assembles (Georgiadou *et al.*, 2012).

The BACTH analysis (Georgiadou *et al.*, 2012) also revealed that PilN and PilO both interact with PilE (Fig. 5.2), which is consistent with their role in pilus assembly. What was intriguing, however, was that this was observed in the absence of PilD and hence with a prepilin. Although it is admitted that the prepilin is first cleaved by PilD and then 'loaded' on the pilus assembly machinery, the

reverse order is not difficult to imagine and would agree with the above finding. Therefore, investigating the chain of events during Tfp assembly would be extremely interesting. Although no interaction between PilF and the other proteins involved in assembly could be detected, the fact that in T2S the traffic ATPase GspE interacts with the cytoplasmic domain of GspL (that is homologous to PilM) which stimulates its ATPase activity (Camberg *et al.*, 2007), PilF might similarly be powering pilus retraction through an interaction with PilM (Fig. 5.2). Interestingly, the crystal structure of *T. thermophilus* PilM showed a stretch of conserved residues on the opposite site of the PilN binding groove (Karuppiah and Derrick, 2011), which could be implicated in the binding of PilF. Taken together, these findings show that the machineries involved in Tfp biogenesis and T2S are far more similar than previously anticipated (Korotkov *et al.*, 2011), which suggests that advances in one field are more likely than ever to lead to advances in the other.

Pilus stabilization and functional maturation

This remains a poorly defined step. As mentioned above, it was found that piliation could be restored in a majority of the non-piliated *N. meningitidis* mutants by a concurrent mutation in *pilT*. This showed that seven 7/15 proteins essential for Tfp biogenesis (PilC1/2, PilG, PilH, PilI, PilJ, PilK and PilW) act after pilus assembly to prevent uncontrolled retraction of newly polymerized filaments (Carbonnelle *et al.*, 2006). These proteins were thus proposed to hinder, or counteract, pilus retraction. The phenotypic characterization of the double mutants for all the functions not abolished in the absence of retraction (aggregation and adhesion to human cells) showed that although their fibres were morphologically normal and produced to wild-type levels, they were mainly non-functional. This suggested that these proteins are important for the 'functional maturation' of the Tfp (Carbonnelle *et al.*, 2006).

In order to understand this late step of Tfp biogenesis, one must understand better the process it antagonizes, i.e. pilus retraction. It is clear that PilT is a hexameric traffic ATPase that directly powers pilus retraction (Satyshur *et al.*, 2007; Misic *et al.*, 2010). However, although much progress has been made, the molecular mechanism of PilT-mediated retraction remains speculative. The structural characterization of this protein led to two possible mechanisms being proposed (Misic *et al.*, 2010). The first one is referred to as the 'direct interaction model', in which PilT engages the bottom-most pilin subunit of a pilus filament via its N-terminal α-helix and subsequently pulls and transfers it back into the cytoplasmic membrane. The second model involves interaction of PilT with a conserved cytoplasmic membrane protein that would in turn, upon a conformational change, pull back into the cytoplasmic membrane the last pilin subunit in the filament. However, the picture is likely to be far more complex since in every system there are PilT paralogues, the most notable being PilU, which is encoded by a gene in operon with *pilT* (Pelicic, 2008). In *N. meningitidis*, there is even a second paralogue named PilT2. A systematic phenotypic analysis in *N. meningitidis* revealed that these proteins are involved in fine-tuning of the Tfp-linked functions (Brown *et al.*, 2010). Furthermore, BACTH provided evidence that the meningococcal traffic ATPases (PilF, PilT, PilT2 and PilU) interact with themselves and each other, suggesting the intriguing possibility that they might form hetero-hexamers that would modulate retraction and alter pilus dynamics (Georgiadou *et al.*, 2012). Nevertheless, regardless of the actual molecular mechanism of PilT-mediated retraction, it could be speculated that the above proteins interact directly with the pilus and hinder retraction by acting like a 'pincer'.

Several important studies concerning these proteins acting after pilus assembly have been published recently, giving a better view of this late step in Tfp biogenesis. Due to space limitations, the following three examples have been chosen. PilC has been extensively studied in pathogenic *Neisseria* species because in addition to its role in Tfp biogenesis, it is involved in mediating adhesion to human cells (Nassif *et al.*, 1994;

Rudel *et al.*, 1995). A series of elegant studies characterizing the PilY1 orthologue of PilC in *P. aeruginosa* recently shed light on this protein by showing that its N-terminal domain is associated with adhesion, while its C-terminal domain is involved in Tfp biogenesis. First, it was confirmed that while a *pilY1* mutant is non-piliated, piliation was restored in a double *pilY1/T* mutant, confirming in a different species that this protein although essential for Tfp biogenesis is dispensable for pilus assembly (Heiniger *et al.*, 2010). As in *N. meningitidis*, the *pilY1/T* mutant was unable to adhere to human cells, suggesting that PilY1 also ensures Tfp's 'functional maturation'. The crystal structure of the C-terminal domain of PilY1, which is conserved in all PilC (while the N-terminal domain is species-specific), revealed a modified β-propeller fold with a distinct EF-hand-like calcium-binding site (Orans *et al.*, 2010). Interestingly, calcium binding was found to be important for Tfp biogenesis. In a calcium-bound state, PilY1 inhibited PilT-mediated pilus retraction, which is consistent with its role in pilus stabilization. In a calcium-free state, PilY1 was unable to hinder PilT-mediated pilus retraction thus resulting in a non-piliated phenotype (Orans *et al.*, 2010). Since this calcium-binding site is conserved, this is expected to be a common feature of PilC orthologues. Finally, it was shown that the N-terminal domain of PilY1 is directly involved in adhesion to host epithelial cells. An integrin-binding RGD motif was identified in this domain, and the purified PilY1 was found to bind integrin *in vitro*, in an RGD-dependent manner (Johnson *et al.*, 2011). Therefore, since the N-terminal domain of PilC is directly involved in adhesion, it is now clear why this domain is different in species that adhere to diverse surfaces.

PilW is an outer-membrane lipoprotein that, in addition to its role in pilus stabilization and 'functional maturation' of the filaments, is required for the stability or assembly of PilQ multimers through which Tfp emerge on the bacterial surface. In the absence of PilW, only monomeric PilQ is present (Carbonnelle *et al.*, 2005). PilQ multimers could be detected in the absence of any other component involved in Tfp biogenesis (except, of course, PilQ) (Carbonnelle et al., 2006). The structure of PilW, which was solved by crystallography both in *P. aeruginosa* and *N. meningitidis*, revealed 13 anti-parallel α-helices that fold into six TPR motifs (Kim *et al.*, 2006; Koo *et al.*, 2008; Trindade *et al.*, 2008). The overall structure is organized as a super-helix of which the two halves define a deep groove. This structure was used as a guide to perform a structure/function analysis of PilW in the meningococcus (Szeto *et al.*, 2011). This showed that PilW was functional even in the absence of N-terminal lipidation, which was also shown in *P. aeruginosa* (Koo *et al.*, 2008), and is a clear evidence that PilW is not a pilot protein whose role would be to 'pilot' PilQ to its final destination in the outer membrane. Another important finding was that con-served residues in the deep groove of PilW were important for Tfp-mediated phenotypes without affecting PilQ multimerization (Szeto *et al.*, 2011). This showed that role of PilW in the 'functional maturation' of Tfp can be genetically uncoupled from its role in the assembly/stabilization of the secretin chan-nels, suggesting that it is a bi-functional protein.

The last proteins to be showcased here are the 'core' pilin-like proteins PilH, PilI, PilJ and PilK (Table 5.1). In an elegant study of the *K. oxytoca* T2S system, it was proposed that two of these pseudopilins (PulI and PulJ) form a complex in the cytoplasmic membrane, to which a third pseudopilin (PulK) binds, which results in its partial extraction from the membrane (Cisneros *et al.*, 2012). This was speculated to initiate filament assembly, with the resulting complex capping the filaments. However, it is unclear whether this model could be extended to Tfp biogenesis. First, these proteins were found to be distributed throughout the fibres in *P. aeruginosa* (Giltner *et al.*, 2010). Second, piliation can be restored in *pilH*, *pilI*, *pilJ* and *pilK* mutants in the presence of a concurrent mutation in *pilT* in *N. meningitidis*, *N. gonorrhoeae* and *P. aeruginosa* (Winther-Larsen *et al.*, 2005; Carbonnelle *et al.*, 2006; Giltner *et al.*, 2010). Actually, in *N. gonorrhoeae*, piliation was seen in the concurrent absence of the four proteins (Winther-Larsen *et al.*, 2005), provided that

pilus retraction was abolished, which seems incompatible with a role in initiating pilus assembly. However, it remains formally pos-sible that this initiation might be dispensable in the absence of pilus retraction, which awaits further characterization.

Emergence of the filaments on the bacterial surface

Upon assembly and stabilization, Tfp emerge on the bacterial surface through a channel formed by the secretin PilQ (Fig. 5.2). As determined by electron microscopy studies, PilQ channels consist of 12 subunits adopting a ring-like cylindrical structure with a large funnel-shaped central cavity sealed at the bottom (Collins *et al.*, 2004). The size of this cavity is large enough to accommodate a pilus, which has been experimentally confirmed in *N. meningitidis* (Collins *et al.*, 2005). Confirmation that PilQ channels are the only passage through which Tfp emerge on the bacterial surface came from reports in *N. gonorrhoeae* and *N. meningitidis* showing that piliation was restored in a double *pilQ/T* mutant, but that the filaments remained trapped within the periplasm (Wolfgang *et al.*, 2000; Carbonnelle *et al.*, 2006). Recently, the 3D structure of the periplasmic N-terminal portion of the GspD secretin revealed three subdomains forming two lobes connected by a flexible linker (Korotkov *et al.*, 2009). The first subdomain adopts a fold seen in TonB-dependent receptors, while the second and third subdomains exhibit a nuclear ribo-nucleoprotein K homology fold. Although the structure of the C-terminal portion of secretins is still to be determined, it is thought to form a β-strand assembly embedded in the outer membrane.

It is likely that the PilW lipoprotein forms a sub-complex with PilQ in the outer membrane (Fig. 5.2), although this remains to be formally demonstrated. This sub-complex might be linked to the sub-complex in the cytoplasmic membrane through the lipo-protein PilP that interacts with PilQ (Balasingham *et al.*, 2007). Secretins sometimes rely on the interaction with small outer membrane lipoproteins named pilotins to accompany them to the outer membrane, and thus prevent their mis-localization and premature multimerization in the cytoplasmic membrane (Guilvout *et al.*, 2006). However, PilP and PilW are unlikely to be pilotins, although they are often ascribed such a role in the literature. Unlike *bona fide* pilot proteins, when PilW is mis-targeted to the periplasm both in *P. aeruginosa* and *N. meningitidis*, PilQ multimers were still formed, piliation was not abolished and Tfp were partly functional (Koo *et al.*, 2008; Szeto *et al.*, 2011).

5.3 Conclusion

Despite considerable recent progress, there are still important gaps to be filled in our understanding of the molecular mechanisms of Tfp biogenesis and Tfp-mediated functions. Further structural, biochemical and genetic characterization of the different Pil proteins including in new models (such as a genetically tractable Gram-positive species), further unravelling the way they interact to form different parts of the Tfp biogenesis machinery, findings in evolutionarily related systems, are all expected to provide new and exciting insights in the years to come. Such advances are likely to have a major practical impact, such as providing the clues for the rational design of compounds that could inhibit the assembly of Tfp, which will be broad-spectrum due to the amazing distribution of these filaments in prokaryotes.

References

Abendroth, J., Bagdasarian, M., Sandkvist, M. and Hol, W.G. (2004a) The structure of the cytoplasmic domain of EpsL, an inner membrane component of the type II secretion system of *Vibrio cholerae*: an unusual member of the actin-like ATPase superfamily. *Journal of Molecular Biology* 344, 619–633.

Abendroth, J., Rice, A.E., McLuskey, K., Bagdasarian, M. and Hol, W.G. (2004b) The crystal structure of the periplasmic domain of the type II secretion system protein EpsM from *Vibrio cholerae*: the simplest version of the ferredoxin fold. *Journal of Molecular Biology* 338, 585–596.

Abendroth, J., Kreger, A.C. and Hol, W.G. (2009) The dimer formed by the periplasmic domain of EpsL from the type II secretion system of *Vibrio parahaemolyticus*. *Journal of Structural Biology* 168, 313–322.

Arts, J., van Boxtel, R., Filloux, A., Tommassen, J. and Koster, M. (2007) Export of the pseudopilin XcpT of the *Pseudomonas aeruginosa* type II secretion system via the signal recognition particle-Sec pathway. *Journal of Bacteriology* 189, 2069–2076.

Ayers, M., Sampaleanu, L.M., Tammam, S., Koo, J., Harvey, H., Howell, P.L. and Burrows, L.L. (2009) PilM/N/O/P proteins form an inner membrane complex that affects the stability of the *Pseudomonas aeruginosa* type IV pilus secretin. *Journal of Molecular Biology* 394, 128–142.

Balasingham, S.V., Collins, R.F., Assalkhou, R., Homberset, H., Frye, S.A., Derrick, J.P. and Tønjum, T. (2007) Interactions between the lipoprotein PilP and the secretin PilQ in *Neisseria meningitidis*. *Journal of Bacteriology* 189, 5716–5727.

Biais, N., Ladoux, B., Higashi, D., So, M. and Sheetz, M. (2008) Cooperative retraction of bundled type IV pili enables nanonewton force generation. *PLoS Biology* 6, e87.

Brown, D., Helaine, S., Carbonnelle, E. and Pelicic, V. (2010) Systematic functional analysis reveals that a set of 7 genes is involved in fine tuning of the multiple functions mediated by type IV pili in *Neisseria meningitidis*. *Infection and Immunity* 78, 3053–3063.

Camberg, J.L., Johnson, T.L., Patrick, M., Abendroth, J., Hol, W.G. and Sandkvist, M. (2007) Synergistic stimulation of EpsE ATP hydrolysis by EpsL and acidic phospholipids. *EMBO Journal* 26, 19–27.

Carbonnelle, E., Helaine, S., Prouvensier, L., Nassif, X. and Pelicic, V. (2005) Type IV pilus biogenesis in *Neisseria meningitidis*: PilW is involved in a step occuring after pilus assembly, essential for fiber stability and function. *Molecular Microbiology* 55, 54–64.

Carbonnelle, E., Helaine, S., Nassif, X. and Pelicic, V. (2006) A systematic genetic analysis in *Neisseria meningitidis* defines the Pil proteins required for assembly, functionality, stabilization and export of type IV pili. *Molecular Microbiology* 61, 1510–1522.

Chen, I. and Dubnau, D. (2004) DNA uptake during bacterial transformation. *Nature Reviews Microbiology* 2, 241–249.

Cisneros, D.A., Bond, P.J., Pugsley, A.P., Campos, M. and Francetic, O. (2012) Minor pseudopilin self-assembly primes type II secretion pseudopilus elongation. *EMBO Journal* 31, 1041–1053.

Collins, R.F., Frye, S.A., Kitmitto, A., Ford, R.C., Tønjum, T. and Derrick, J.P. (2004) Structure of the *Neisseria meningitidis* outer membrane PilQ secretin complex at 12 Å resolution. *Journal of Biological Chemistry* 279, 39750–39756.

Collins, R.F., Frye, S.A., Balasingham, S., Ford, R.C., Tønjum, T. and Derrick, J.P. (2005) Interaction with type IV pili induces structural changes in the bacterial outer membrane secretin PilQ. *Journal of Biological Chemistry* 280, 18923–18930.

Craig, L., Taylor, R.K., Pique, M.E., Adair, B.D., Arvai, A.S., Singh, M., Lloyd, S.J., Shin, D.S., Getzoff, E.D., Yeager, M., Forest, K.T. and Tainer, J.A. (2003) Type IV pilin structure and assembly. X-ray and EM analyses of *Vibrio cholerae* toxin-coregulated pilus and *Pseudomonas aeruginosa* PAK pilin. *Molecular Cell* 11, 1139–1150.

Craig, L., Pique, M.E. and Tainer, J.A. (2004) Type IV pilus structure and bacterial pathogenicity. *Nature Reviews Microbiology* 2, 363–378.

Dalrymple, B. and Mattick, J.S. (1987) An analysis of the organization and evolution of type 4 fimbrial (MePhe) subunit proteins. *Journal of Molecular Evolution* 25, 261–269.

Douzi, B., Filloux, A. and Voulhoux, R. (2012) On the path to uncover the bacterial type II secretion system. *Philosophical Transactions of the Royal Society B: Biological Sciences* 367, 1059–1072.

Francetic, O., Buddelmeijer, N., Lewenza, S., Kumamoto, C.A. and Pugsley, A.P. (2007) Signal recognition particle-dependent inner membrane targeting of the PulG pseudopilin component of a type II secretion system. *Journal of Bacteriology* 189, 1783–1793.

Georgiadou, M., Castagnini, M., Karimova, G., Ladant, D. and Pelicic, V. (2012) Large-scale study of the interactions between proteins involved in type IV pilus biology in *Neisseria meningitidis*: characterization of a subcomplex involved in pilus assembly. *Molecular Microbiology* 84, 857–873.

Giltner, C.L., Habash, M. and Burrows, L.L. (2010) *Pseudomonas aeruginosa* minor pilins are incorporated into type IV pili. *Journal of Molecular Biology* 398, 444–461.

Golovanov, A.P., Balasingham, S., Tzitzilonis, C., Goult, B.T., Lian, L.Y., Homberset, H., Tønjum, T. and Derrick, J.P. (2006) The solution structure of a domain from the *Neisseria meningitidis* lipoprotein PilP reveals a new β-sandwich fold. *Journal of Molecular Biology* 364, 186–195.

Guilvout, I., Chami, M., Engel, A., Pugsley, A.P. and Bayan, N. (2006) Bacterial outer membrane secretin PulD assembles into and inserts into the inner membrane in the absence of its pilotin. *EMBO Journal* 25, 5241–5249.

Hartung, S., Arvai, A.S., Wood, T., Kolappan, S., Shin, D.S., Craig, L. and Tainer, J.A. (2011) Ultra-high resolution and full-length pilin structures with insights for filament assembly, pathogenic functions, and vaccine potential. *Journal of Biological Chemistry* 286, 44254–44265.

Heiniger, R.W., Winther-Larsen, H.C., Pickles, R.J., Koomey, M. and Wolfgang, M.C. (2010) Infection of human mucosal tissue by *Pseudomonas aeruginosa* requires sequential and mutually dependent virulence factors and a novel pilus-associated adhesin. *Cellular Microbiology* 12, 1158–1173.

Helaine, S., Carbonnelle, E., Prouvensier, L., Beretti, J.-L., Nassif, X. and Pelicic, V. (2005) PilX, a pilus-associated protein essential for bacterial aggregation, is a key to pilus-facilitated attachment of *Neisseria meningitidis* to human cells. *Molecular Microbiology* 55, 65–77.

Higashi, D.L., Zhang, G.H., Biais, N., Myers, L.R., Weyand, N.J., Elliott, D.A. and So, M. (2009) Influence of type IV pilus retraction on the architecture of the *Neisseria gonorrhoeae*-infected cell cortex. *Microbiology* 155, 4084–4092.

Horiuchi, T. and Komano, T. (1998) Mutational analysis of plasmid R64 thin pilus prepilin: the entire prepilin sequence is required for processing by type IV prepilin peptidase. *Journal of Bacteriology* 180, 4613–4620.

Jarrell, K.F. and Albers, S.V. (2012) The archaellum: an old motility structure with a new name. *Trends in Microbiology* 20, 307–312.

Johnson, M.D., Garrett, C.K., Bond, J.E., Coggan, K.A., Wolfgang, M.C. and Redinbo, M.R. (2011) *Pseudomonas aeruginosa* PilY1 binds integrin in an RGD- and calcium-dependent manner. *PLoS One* 6, e29629.

Karuppiah, V. and Derrick, J.P. (2011) Structure of the PilM-PilN inner membrane type IV pilus biogenesis complex from *Thermus thermophilus*. *Journal of Biological Chemistry* 27, 24434–24442.

Kim, K., Oh, J., Han, D., Kim, E.E., Lee, B. and Kim, Y. (2006) Crystal structure of PilF: functional implication in the type 4 pilus biogenesis in *Pseudomonas aeruginosa*. *Biochemistry and Biophysics Research Communications* 340, 1028–1038.

Koo, J., Tammam, S., Ku, S.Y., Sampaleanu, L.M.,

Burrows, L.L. and Howell, P.L. (2008) PilF is an outer membrane lipoprotein required for multimerization and localization of the *Pseudomonas aeruginosa* Type IV pilus secretin. *Journal of Bacteriology* 190, 6961–6969.

Korotkov, K.V., Pardon, E., Steyaert, J. and Hol, W.G. (2009) Crystal structure of the N-terminal domain of the secretin GspD from ETEC determined with the assistance of a nanobody. *Structure* 17, 255–265.

Korotkov, K.V., Gonen, T. and Hol, W.G. (2011) Secretins: dynamic channels for protein transport across membranes. *Trends in Biochemical Sciences* 36, 433–443.

Korotkov, K.V., Sandkvist, M. and Hol, W.G. (2012) The type II secretion system: biogenesis, molecular architecture and mechanism. *Nature Reviews Microbiology* 10, 336–351.

LaPointe, C.F. and Taylor, R.K. (2000) The type 4 prepilin peptidases comprise a novel family of aspartic acid proteases. *Journal of Biological Chemistry* 275, 1502–1510.

Mattick, J.S. (2002) Type IV pili and twitching motility. *Annual Review of Microbiology* 56, 289–314.

Merz, A.J., So, M. and Sheetz, M.P. (2000) Pilus retraction powers bacterial twitching motility. *Nature* 407, 98–102.

Misic, A.M., Satyshur, K.A. and Forest, K.T. (2010) *P. aeruginosa* PilT structures with and without nucleotide reveal a dynamic type IV pilus retraction motor. *Journal of Molecular Biology* 400, 1011–1021.

Morand, P.C., Bille, E., Morelle, S., Eugène, E., Beretti, J.L., Wolfgang, M., Meyer, T.F., Koomey, M. and Nassif, X. (2004) Type IV pilus retraction in pathogenic *Neisseria* is regulated by the PilC proteins. *EMBO Journal* 23, 2009–2017.

Nassif, X., Beretti, J.-L., Lowy, J., Stenberg, P., O'Gaora, P., Pfeifer, J., Normark, S. and So, M. (1994) Roles of pilin and PilC in adhesion of *Neisseria meningitidis* to human epithelial and endothelial cells. *Proceedings of the National Academy of Sciences of the United States of America* 91, 3769–3773.

Nunn, D.N. and Lory, S. (1991) Product of the *Pseudomonas aeruginosa* gene *pilD* is a prepilin leader peptidase. *Proceedings of the National Academy of Sciences of the United States of America* 88, 3281–3285.

Orans, J., Johnson, M.D., Coggan, K.A., Sperlazza, J.R., Heiniger, R.W., Wolfgang, M.C. and Redinbo, M.R. (2010) Crystal structure analysis reveals *Pseudomonas* PilY1 as an essential calcium-dependent regulator of bacterial surface motility. *Proceedings of the National*

Academy of Sciences of the United States of America 107, 1065–1070.

Parge, H.E., Forest, K.T., Hickey, M.J., Christensen, D.A., Getzoff, E.D. and Tainer, J.A. (1995) Structure of the fibre-forming protein pilin at 2.6 Å resolution. *Nature* 378, 32–38.

Pelicic, V. (2008) Type IV pili: *e pluribus unum*? *Molecular Microbiology* 68, 827–837.

Proft, T. and Baker, E.N. (2009) Pili in Gram-negative and Gram-positive bacteria – structure, assembly and their role in disease. *Cellular and Molecular Life Sciences* 66, 613–635.

Pujol, C., Eugène, E., Marceau, M. and Nassif, X. (1999) The meningococcal PilT protein is required for induction of intimate attachment to epithelial cells following pilus-mediated adhesion. *Proceedings of the National Academy of Sciences of the United States of America* 96, 4017–4022.

Rudel, T., Scheuerpflug, I. and Meyer, T.F. (1995) *Neisseria* PilC protein identified as a type-4 pilus-tip located adhesin. *Nature* 373, 357–359.

Sampaleanu, L.M., Bonanno, J.B., Ayers, M., Koo, J., Tammam, S., Burley, S.K., Almo, S.C., Burrows, L.L. and Howell, P.L. (2009) Periplasmic domains of *Pseudomonas aeruginosa* PilN and PilO form a stable heterodimeric complex. *Journal of Molecular Biology* 394, 143–159.

Satyshur, K.A., Worzalla, G.A., Meyer, L.S., Heiniger, E.K., Aukema, K.G., Misic, A.M. and Forest, K.T. (2007) Crystal structures of the pilus retraction motor PilT suggest large domain movements and subunit cooperation drive motility. *Structure* 15, 363–376.

Strom, M.S. and Lory, S. (1991) Amino acid substitutions in pilin of *Pseudomonas aeruginosa*. Effect on leader peptide cleavage, amino-terminal methylation, and pilus assembly. *Journal of Biological Chemistry* 266, 1656–1664.

Strom, M.S., Nunn, D.N. and Lory, S. (1993) A single bifunctional enzyme, PilD, catalyzes cleavage and N-methylation of proteins belonging to the type IV pilin family. *Proceedings of the National Academy of Sciences of the United States of America* 90, 2404–2408.

Szabo, Z., Stahl, A.O., Albers, S.V., Kissinger, J.C., Driessen, A.J. and Pohlschroder, M. (2007) Identification of diverse archaeal proteins with class III signal peptides cleaved by distinct archaeal prepilin peptidases. *Journal of Bacteriology* 189, 772–778.

Szeto, T.H., Dessen, A. and Pelicic, V. (2011) Structure/function analysis of *Neisseria meningitidis* PilW, a conserved protein playing multiple roles in type IV pilus biology. *Infection and Immunity* 79, 3028–3035.

Tammam, S., Sampaleanu, L.M., Koo, J., Sundaram, P., Ayers, M., Chong, P.A., Forman-Kay, J.D., Burrows, L.L. and Howell, P.L. (2011) Characterization of the PilN, PilO and PilP type IVa pilus subcomplex. *Molecular Microbiology* 82, 1496–1514.

Tomich, M., Planet, P.J. and Figurski, D.H. (2007) The *tad* locus: postcards from the widespread colonization island. *Nature Reviews Microbiology* 5, 363–375.

Trindade, M.B., Job, V., Contreras-Martel, C., Pelicic, V. and Dessen, A. (2008) Structure of a widely conserved type IV pilus biogenesis factor that affects the stability of secretin multimers. *Journal of Molecular Biology* 378, 1031–1039.

Winther-Larsen, H.C., Wolfgang, M., Dunham, S., van Putten, J.P., Dorward, D., Lovold, C., Aas, F.E. and Koomey, M. (2005) A conserved set of pilin-like molecules controls type IV pilus dynamics and organelle-associated functions in *Neisseria gonorrhoeae. Molecular Microbiology* 56, 903–917.

Wolfgang, M., Lauer, P., Park, H.S., Brossay, L., Hébert, J. and Koomey, M. (1998a) PilT mutations lead to simultaneous defects in competence for natural transformation and twitching motility in piliated *Neisseria gonorrhoeae. Molecular Microbiology* 29, 321–330.

Wolfgang, M., Park, H.S., Hayes, S.F., van Putten, J.P. and Koomey, M. (1998b) Suppression of an absolute defect in type IV pilus biogenesis by loss-of-function mutations in *pilT*, a twitching motility gene in *Neisseria gonorrhoeae. Proceedings of the National Academy of Sciences of the United States of America* 95, 14973–14978.

Wolfgang, M., van Putten, J.P., Hayes, S.F., Dorward, D. and Koomey, M. (2000) Components and dynamics of fiber formation define a ubiquitous biogenesis pathway for bacterial pili. *EMBO Journal* 19, 6408–6418.

Yamagata, A., Milgotina, E., Scanlon, K., Craig, L., Tainer, J.A. and Donnenberg, M.S. (2012) Structure of an essential type IV pilus biogenesis protein provides insights into pilus and type II secretion systems. *Journal of Molecular Biology* 419, 110–124.

6 The *Pseudomonas aeruginosa* Type IV Pilus Assembly System in Three Dimensions

Stephanie Tammam[1], P. Lynne Howell[1] and Lori L. Burrows[2]

[1]*Program in Molecular Structure and Function, The Hospital for Sick Children, Toronto, Canada and* [2]*Department of Biochemistry and Biomedical Sciences, McMaster University, Hamilton, Canada*

6.1 Introduction

Based on the different functions they perform, one might not expect at first that the type IV pilus (T4P) and type II secretion (T2S) systems are related. T4P are long, retractable polymers involved in adhesion, motility, biofilm formation and DNA uptake/transfer (Craig and Li, 2008), while the T2S system exports folded proteins or protein complexes from the periplasm using a short pseudopilus (McLaughlin *et al.*, 2012). Emerging structures of several components of the two systems has given credence to early predictions – based on sequence homology – that the T4P and T2S machineries are descended from a common ancestor (Hobbs and Mattick, 1993). Further support comes from studies showing that T2S pseudopili can be manipulated to form long adhesive pilus-like fibres (Sauvonnet *et al.*, 2000; Vignon *et al.*, 2003), and that some T4P systems can secrete proteins (Hager *et al.*, 2006; Han *et al.*, 2007). Although it is now clear that many T4P and T2S components – even those with poor sequence identity – are structurally related and may operate using similar mechanisms (Ayers *et al.*, 2010), intriguing differences in the number and

arrangement of components hint at adaptations that may optimize them for their specific roles.

Understanding of structure–function relationships in the T2S/T4P superfamily is further complicated by divergence of T4P into distinct classes. T4P are generally divided into type IVa (T4a) and type IVb (T4b) pili, with the T4bP containing a monophyletic subclass called the tight adherence (Tad) pili (Burrows, 2012). T4aP are the best-studied group, and are associated with twitching motility, where cycles of pilus extension, adherence and retraction pull the cells along surfaces (Mattick, 2002; Burrows, 2012). This motility requires that the assembly system be able to withstand retraction forces of >100 pN (Merz *et al.*, 2000). While some assembly system components are conserved among all T4P and T2S systems, others are characteristic of their (sub)class. In general, both T4P and T2S systems have an inner membrane 'platform' protein that is thought to transduce conformational changes from one or more hexameric ATPases to pilin subunits in the inner membrane, resulting in pilin assembly (and disassembly, in the case of T4P); together, the platform protein and ATPase(s) form the motor subcomplex (Wolfgang *et al.*, 2000). T4P and T2S systems have a gated outer membrane portal called the secretin that allows the pilus or secreted proteins to cross the outer membrane (Russel, 1998). The 'alignment' subcomplex – composed of 3–4 less-conserved inner membrane proteins that connect the motor and secretin subcomplexes – may control access to, and gating of, the secretin (Ayers *et al.*, 2010; Burrows, 2012). The final subcomplex is the (pseudo)pilus containing the major and minor subunits (Winther-Larsen *et al.*, 2005; Giltner *et al.*, 2010; Cisneros *et al.*, 2012). In this chapter, we discuss recent structure–function insights into T4P and T2S components, highlighting those that might relate to the motility function of T4aP. With structural information in hand, we speculate on how and where the assembly systems are inserted into the cell envelope, and how assembly dynamics may be controlled.

6.2 Assembly of the PilMNOPQ Proteins across the Cell Envelope

In the T4aP system, a continuous protein-interaction network between products of the *pilMNOPQ* operon – encoding the alignment subcomplex and secretin – has been identified (Tammam *et al.*, 2013). The network begins in the cytoplasm with PilM, continues through the inner membrane and periplasm via PilNOP and ends in the outer membrane with PilQ. There are structural data available for soluble domains of four of the five proteins (PilM, PilO, PilP, PilQ), plus a homology model for PilN (Balasingham *et al.*, 2007; Sampaleanu *et al.*, 2009; Karuppiah and Derrick, 2011; Tammam *et al.*, 2011; Berry *et al.*, 2012), allowing us to propose a model for this transenvelope complex (Fig. 6.1).

PilM is a cytoplasmic, actin-like protein that is structurally related to the MreB and FtsA that coordinate cell shape and division

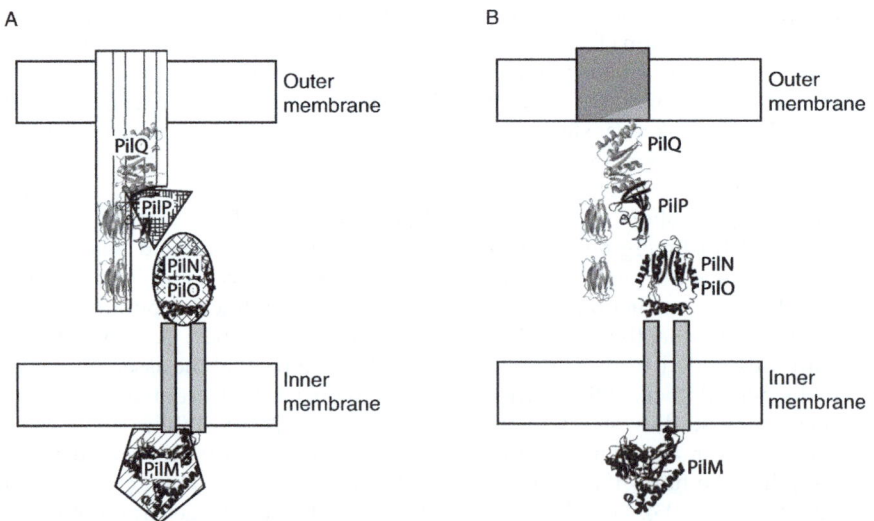

Fig. 6.1. Assembly of the transenvelope complex. (A) Schematic representation of the PilMNOPQ proteins arranged across the cell envelope. (B) Schematic showing available protein structures, and known protein interaction data to produce a model of the transenvelope complex. All structures and shapes are drawn to scale. The orientation of PilM relative to PilN is based on the known structure of PilM with the PilN N-terminal peptide (Karuppiah and Derrick, 2011), the PilNO heterodimer is from Sampaleanu *et al.* (2009) and the orientation of PilP with PilQ is based on the x-ray crystal structure of GspC-GspD heterodimer (Korotkov *et al.*, 2011b).

(Karuppiah and Derrick, 2011). Despite having very limited sequence identity to PilM, the BfpC and GspL components of the T4bP and T2S system, respectively, have similar cytoplasmic actin-like domains (Abendroth *et al.*, 2004; Ayers *et al.*, 2009; Gray *et al.*, 2011; Karuppiah and Derrick, 2011; Yamagata *et al.*, 2012). Unlike PilM, GspL and BfpC are connected to periplasmic domains by a single transmembrane segment (Karuppiah and Derrick, 2011; Yamagata *et al.*, 2012). Oddly, Tad pilus assembly systems lack an actin-like component, but instead have proteins that resemble the MinCD proteins involved in cell division (Burrows, 2012; Perez-Cheeks *et al.*, 2012; Xu *et al.*, 2012). Actin-like proteins typically have four subdomains labelled 1A, 1C, 2A and 2B (following FtsA nomenclature), with the nucleotide-binding site located between subdomains 1A and 2B (Karuppiah and Derrick, 2011). Although PilM binds ATP (Karuppiah and Derrick, 2011), the role of this activity is currently unclear. A hydrophobic cleft in PilM stably binds the short, highly conserved cytoplasmic N-terminus of PilN, creating the functional equivalent of GspL (Ayers *et al.*, 2009; Karuppiah and Derrick, 2011; Tammam *et al.*, 2013). Mutation of invariant residues in this segment of PilN impairs pilus function *in vivo* (Tammam *et al.*, 2013). Interestingly, although BfpC belongs to a T4bP assembly system, its actin-like domain is structurally more similar to that of GspL than PilM (Yamagata *et al.*, 2012). Also, BfpC has only a short periplasmic segment following the transmembrane helix, suggesting the function of PilN and the periplasmic domain of GspL is fulfilled by some other component in the T4bP system.

PilN and PilO each have a short cytoplasmic N-terminus, a single transmembrane segment and a periplasmic coiled-coil domain followed by a ferredoxin-like 'core' domain (Sampaleanu *et al.*, 2009). The N-terminus of PilN orthologues is highly conserved due to its interaction with PilM, while the N-terminus of PilO is more variable. The structure of part of the coiled-coil domain and the ferredoxin-like core domain of PilO was solved by X-ray crystallography

(Sampaleanu *et al.*, 2009). Based on similarity in their predicted secondary structures, and biophysical and biological data showing that PilO forms stable homodimers as well as heterodimers with PilN, the homodimeric PilO structure was used as a template to model PilN – which proved insoluble when expressed alone – and the heterodimeric PilNO complex (Sampaleanu *et al.*, 2009). The predicted PilNO interface involves the periplasmic core and coiled-coil domains (Sampaleanu *et al.*, 2009), and possibly the transmembrane segments. The equivalent interactions in the T2S system are likely those between the periplasmic portions of GspL and GspM, which are predicted to have similar secondary structures to PilN and PilO (Sampaleanu *et al.*, 2009). The PilNO interactions identified in *P. aeruginosa* were recently verified in the *Neisseria meningitidis* T4aP system (Georgiadou *et al.*, 2012).

The fourth member of the alignment subcomplex, PilP, is a periplasmic lipoprotein tethered to the inner membrane (Tammam *et al.*, 2011). PilP contains two distinct domains, a disordered N-terminal region and a C-terminal β-sandwich domain. The structure of the C-terminal domain of PilP was solved by both NMR spectroscopy and X-ray crystallography (Golovanov *et al.*, 2006; Tammam *et al.*, 2011; Gu *et al.*, 2012). PilP's 7-stranded β-sandwich fold is conserved in the GspC family of proteins in the T2S system (Korotkov *et al.*, 2006; Gu *et al.*, 2012). In some species (e.g. *Thermus* and *Deinococcus*), the *pilP* locus is replaced by *pilW*, encoding a protein with a single predicted transmembrane domain – rather than a lipidation site – and C-terminal β-domain that shows limited sequence identity with PilP (Rumszauer *et al.*, 2006). PilP forms a stable complex with PilNO via interactions between its disordered N-terminal domain and the periplasmic coiled-coil and core domains of PilNO (Tammam *et al.*, 2011), while its C-terminal domain interacts with PilQ (Berry *et al.*, 2012; Tammam *et al.*, 2013).

PilQ is a member of the secretin family (Genin and Boucher, 1994), and electron microscopy studies of PilQ suggest that it forms a homo-dodecamer in the outer membrane (Collins *et al.*, 2001; Burkhardt *et al.*, 2011; Berry *et al.*, 2012). Following the N-terminal signal peptidase I leader sequence, each *P. aeruginosa* monomer has a ~420 amino acid periplasmic region and a ~270 residue membrane-embedded 'secretin' domain. The periplasmic region is divided into four subdomains called SS1 ('species-specific'), SS2, N0 and N1 (Tammam *et al.*, 2013). The SS1 and SS2 subdomains are the least conserved between species but are predicted to have similar β-strand-rich architecture (Berry *et al.*, 2012; Tammam *et al.*, 2013). The N0 and N1 domains are conserved in a number of ring-forming proteins in both the T2S system (Korotkov *et al.*, 2009b) and Type 3 secretion (T3S) system (Spreter *et al.*, 2009). New NMR structures of *N. meningitidis* PilQ domains confirmed that N0 and N1 are structurally conserved, and highlighted the novelty of the SS1 and SS2 domains (Berry *et al.*, 2012). The N0 domain has two α-helices sandwiched between two β-sheets, one with 2 β-strands and the other with 3 β-strands (Korotkov *et al.*, 2009b). The linker connecting the N0 and N1 domains is disordered or helical, depending on the structure (Korotkov *et al.*, 2009b, 2011b; Spreter *et al.*, 2009; Tarry *et al.*, 2011). The N1 domain is a 2-layer sandwich formed by a 3-stranded β-sheet and a pair of α-helices (Korotkov *et al.*, 2009b). Recent studies showed that the C-terminal β-domain of PilP interacts directly with the N0 domain of PilQ (Berry *et al.*, 2012; Tammam *et al.*, 2013), linking the inner and outer membrane components of the trans-envelope complex. This interaction is conserved in the T2S system, where the 2-stranded β-sheet in the N0 domain of the secretin interacts with β-strands 6 and 7 of GspC (Korotkov *et al.*, 2011b; Wang *et al.*, 2012). The conservation of PilMN, PilNOP and PilPQ-type interactions in both T2S and T4P systems are supported by examples of gene fusions in some systems. For example, the PilM and PilN equivalents are part of a single GspL-like polypeptide in some T4aP-expressing species such as *Moraxella catarrhalis*, and PilO- and PilP-like domains are fused in *Pseudomonas putida* and related bacteria (Ayers *et al.*, 2010).

Interactions between members of the alignment subcomplex appear to be stable and long-lived, with the possible exception of the PilPQ interaction, which dissociates during size exclusion chromatography (Tammam *et al.*, 2013). Less stable associations may reflect a need to accommodate the large conformational changes proposed to occur during opening and closing of the secretin (below). We propose that *in vivo*, the PilMNOPQ complex remains as a stable, transenvelope assembly at the pole even when pili are not produced. Microscopy analyses using immunofluorescence or fluorescent protein fusions showed that the assembly components are localized to both poles of rod-shaped bacteria (Chiang *et al.*, 2005; Bulyha *et al.*, 2009), providing a mechanism for initiation of pilus biogenesis at either pole upon receipt of an appropriate signal. Data from a number of studies suggest that the stoichiometry of the PilMN, PilNOP and PilPQ complexes is 1:1 (Sampaleanu *et al.*, 2009; Karuppiah and Derrick, 2011; Tammam *et al.*, 2011; Berry *et al.*, 2012) with a single PilMNOP for every monomer of PilQ. However, a recent study of XcpQ from the T2S system of *P. aeruginosa* suggested that its basic unit is a dimer (Van der Meeren *et al.*, 2012), implying that a 1:1:1:1:2 stoichiometry of PilMNOPQ may also be possible.

6.3 Models for Assembly of the PilMNOPQ Complex at the Cell Pole

The assembly complex needs to cross the entire cell envelope – including the peptidoglycan (PG) layer – and localize to the poles in rod-shaped bacteria. PilMNOP are targeted to the inner membrane independently of other T4P proteins (Koo *et al.*, 2008; Ayers *et al.*, 2009; Tammam *et al.*, 2011). PilM is anchored to the inner membrane via its interaction with the N-terminus of PilN (Berry *et al.*, 2012; Tammam *et al.*, 2013), while PilN and PilO have typical Sec signal sequences preceded by positively charged residues that orient the proteins with their C-termini in the periplasm (Georgiadou *et al.*, 2012). PilP is a lipoprotein with an inner membrane targeting sequence, and is anchored in the outer leaflet of the inner membrane following lipidation by the signal peptidase II complex (Tammam *et al.*, 2011).

PilQ monomers first cross the inner membrane via the Sec translocon, then transit to the outer membrane with the help of an outer membrane lipoprotein called PilF in *P. aeruginosa*, PilW in *N. meningitidis* and Tgl in *M. xanthus* (Koo *et al.*, 2008). PilF is composed of 6 tetratricopeptide repeats (TPRs), a common protein–protein interaction motif (Koo *et al.*, 2008). In the absence of PilF, PilQ is not targeted to the outer membrane, does not multimerize, and T4P are not expressed on the cell surface (Koo *et al.*, 2008). Although the exact nature of the PilF–PilQ interaction is currently unknown, detailed mutagenesis studies suggest that it involves a hydrophobic groove on TPR1 of PilF (Koo *et al.*, 2013). It is not clear whether PilF interacts with PilQ during or after initial translocation of monomers by the Sec system, or if the interaction occurs prior to or during PilQ folding. One model proposes that after PilF encounters and interacts with a PilQ monomer, the Lol system recognizes and transports the PilFQ complex to the outer membrane, where PilQ is folded, multimerized and inserted in the outer membrane (Koo *et al.*, 2012). PilQ insertion appears to be independent of the Bam system (Hoang *et al.*, 2011), but outer membrane targeting of secretins from other systems ranges from fully independent of both Bam and Lol systems, to dependent on both (Koo *et al.*, 2012). Whether there are other roles for PilF in the T4P complex as has been proposed for *N. meningitidis* PilW (Szeto *et al.*, 2011), or if PilF remains associated with PilQ after secretin assembly in the outer membrane, is not yet clear.

6.3.1 Outside-in model for assembly

In some secretion systems, the outer membrane component is proposed to template the outside-in assembly of periplasmic and inner membrane components (Diepold *et al.*, 2010; McLaughlin *et al.*, 2012). In the T4P system,

PilQ might act as the 'nucleation protein' (McLaughlin *et al.*, 2012) following its trafficking to and assembly in the outer membrane (Fig. 6.2-I). The inner membrane PilMNOP complex may then associate with the assembled secretin by docking on the N0 domain of PilQ, completing the transenvelope complex (Fig. 6.2-II). Depending on the stoichiometry, a circular cage of 6 or 12 periplasm-spanning arms surrounding PilQ's SS domains would result.

The complex needs to negotiate the covalently closed PG layer. Folded proteins of up to 50 kDa can diffuse through PG, which has a porosity of ~2 nm (Demchick and Koch, 1996; Yao *et al.*, 1999; Pink *et al.*, 2000). Although it is possible that a complex of LolA, PilF and unfolded PilQ monomer may be able to transit the PG layer, the scale of available structures suggest that the periplasmic portion of the assembled secretin extends across the periplasm, nearly to the outer leaflet of the inner membrane. This arrangement suggests that secretin assembly might be coordinated with cell division, allowing for the formation of septal PG around the complex. The envisaged organization of the secretin's SS domains surrounded by PilMNOP arms connected to the secretin's N0 domains would thus provide a peptidoglycan-free lumen to accommodate the pilus. Interaction of the actin-like cytoplasmic component of the alignment subcomplex with the platform protein has been reported for the T2S system, hinting at one way in which the motor subcomplex and the secretin – which otherwise do not appear to directly contact one another – are connected. This arrangement would also anchor the assembly system in the cell envelope (below).

6.3.2 Inside-out model for assembly

One could also envision the opposite situation for polar localization, where the targeting of PilM to the pole might control positioning of the PilMNOPQ subcomplex. The strong structural resemblance of PilM to FtsA (Karuppiah and Derrick, 2011) might lead to transient interactions with FtsZ (Pichoff and

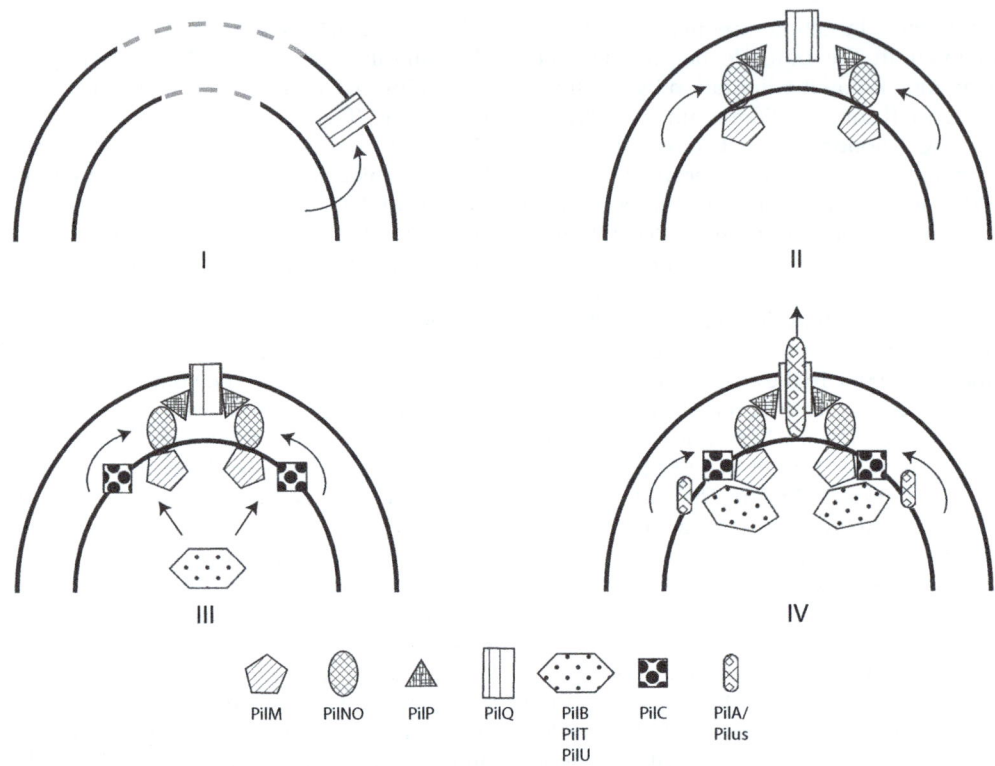

Fig. 6.2. Proposed model for the assembly of the pilus assembly and alignment complex based on the outside in model. (I) PilQ is assembled in the outer membrane (requires PilF – not shown), possibly at the time of cell division, prior to formation of the new pole (diagrammed by the dashes at the pole). (II) The PilMNOP complex is directed to the pole by the interaction of PilP to PilQ, and possibly PilM and the bacterial cytoskeleton in the cytoplasm (not shown). (III) PilC and the ATPases use the PilMNOPQ complex as a nucleation site. (IV) PilA encounters the fully assembled T4P assembly and alignment complex, and the fully assembled complex then polymerizes PilA monomers into a fully functional pilus.

Lutkenhaus, 2005), targeting the complex to the site of new pole formation for integration into the PG layer. This scenario is consistent with the lack of dedicated lytic trans-glycosylases associated with T4P and T2S assembly (Scheurwater and Burrows, 2011). The high local concentration of PilMNOP at the pole may subsequently drive polar localization of PilQ subunits via interactions with the C-terminus of PilP, positioning the secretin at the site of septal PG biosynthesis. The intriguing observation that Tad pilus systems lack an FtsA-like protein but have MinCD-like components suggests that their polar localization may proceed via an independent route (Burrows, 2012; Lutkenhaus, 2012), as the Min proteins

control localization of the FtsZ ring in dividing cells (Lutkenhaus, 2007).

6.3.3 Architectural considerations related to pilus assembly dynamics

Because of the dynamic nature of T4P assembly, sufficient space between the periplasmic arms of the assembly complex for the lateral diffusion of membrane-bound monomers into and out of the lumen of the assembly system is necessary. The system has to allow for disassembly (and presumably assembly) rates that are estimated to be of the order of 10^3 monomers/second (Merz *et al.*, 2000; Skerker and Berg, 2001). Conceptually, the T2S system should have a similar organization, as it assembles a pseudopilus and has structural and functional homologues of PilMNOPQ.

6.4 The Cytoplasmic Protein Interaction Network – PilM, PilC, PilBTU

The energy for pilus extension and retraction is provided by a set of three cytoplasmic ATPases, PilB, PilT and PilU. PilB is the extension ATPase (Nunn *et al.*, 1990) and while both PilT and PilU have been implicated in retraction (Whitechurch and Mattick, 1994; Wolfgang *et al.*, 1998a, 1998b), PilT appears to be the main retraction ATPase in most species. *N. meningitidis* has a fourth ATPase called PilT2 (Kurre *et al.*, 2012). The ATPases typically form homohexamers, although recent bacterial two-hybrid interaction data suggested *N. meningitidis* PilT2 can interact with both PilT and PilU (Georgiadou *et al.*, 2012). Structures of PilT from *T. thermophilus* and *P. aeruginosa* in apo- and nucleotide-bound states have been solved (Satyshur *et al.*, 2007; Misic *et al.*, 2010), as has the structure of the N-terminal domain of a PilB homologue, EpsE, from the T2S system of *Vibrio cholerae* (Abendroth *et al.*, 2005). Monomers have distinct N- and C-terminal domains connected by a flexible linker, and the hexameric structures show distinctly different interactions between subunits, depending on the nature of the bound nucleotide. The capture of both symmetric and asymmetric (2-fold pseudosymmetric) hexamers – in which N-terminal rotation of some subunits of up to 65° around the linker was observed – implies that large conformational changes occur upon nucleotide binding and hydrolysis (Savvides, 2007). Current models of assembly suggest that coupling of the ATPases to the cytoplasmic domains of membrane proteins that interact with pilin subunits could transduce the necessary force to add or remove pilins from a fibre (Satyshur *et al.*, 2007; McLaughlin *et al.*, 2012).

The cytoplasmic ATPases may interact with cytoplasmic domains of PilC and/or with PilM–PilN (Fig. 6.2-III). The platform protein

PilC is a polytopic inner membrane component with three predicted transmembrane helices (TMH) and two cytoplasmic domains sharing ~35% sequence similarity. The crystal structures of the N-terminal cytoplasmic domains both from PilC and its T2S system homologue EpsF reveal a compact bundle of six α-helices (Abendroth *et al.*, 2009; Karuppiah *et al.*, 2010). Domain movements of ATPase hexamers bound to these domains might be communicated through the TMHs of PilC to PilA monomers or polymerized PilA subunits at the base of the pilus (Fig. 6.2-IV). Alternately, interactions of the TMHs of PilA and PilC may trigger conformational changes that promote ATPase activity in bound hexamers. Based on protein interaction studies, extension ATPases such as PilB likely bind to the N-terminal cytoplasmic domain of the platform protein (Milgotina *et al.*, 2011), while PilT and/or PilU could similarly interface with the C-terminal domain, mediating depolymerization. Another possibility is that PilB interacts with the N-terminal domain of PilC, but when a retraction signal is activated, PilB is displaced by PilT (and/or PilU). The ATPases may also interact with PilM or its equivalent, as has been shown for the T2S system (Py *et al.*, 2001; Gray *et al.*, 2011; Yamagata *et al.*, 2012), but to date there is no evidence for such interactions in T4P systems (Georgiadou *et al.*, 2012).

6.5 Gating of the Secretin

Although the secretin is assumed to be in a closed state in the absence of the pilus – ensuring the integrity of the cell envelope – the stimuli and mechanism of secretin opening are unknown. Also, whether secretin opening is coordinated with intracellular or extracellular signals, and how these messages are transmitted across the cell envelope, need to be considered.

6.5.1 Gating by the pilus

It is possible that the pilus is responsible for opening PilQ (Fig. 6.3). In this scenario, the tip of the pilus, most likely a complex of low

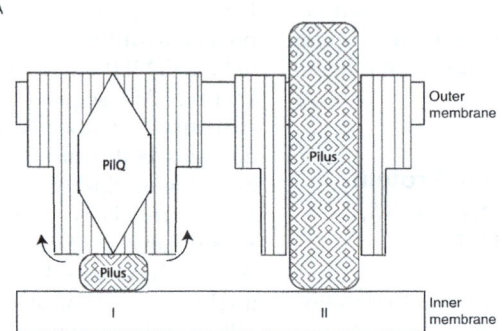

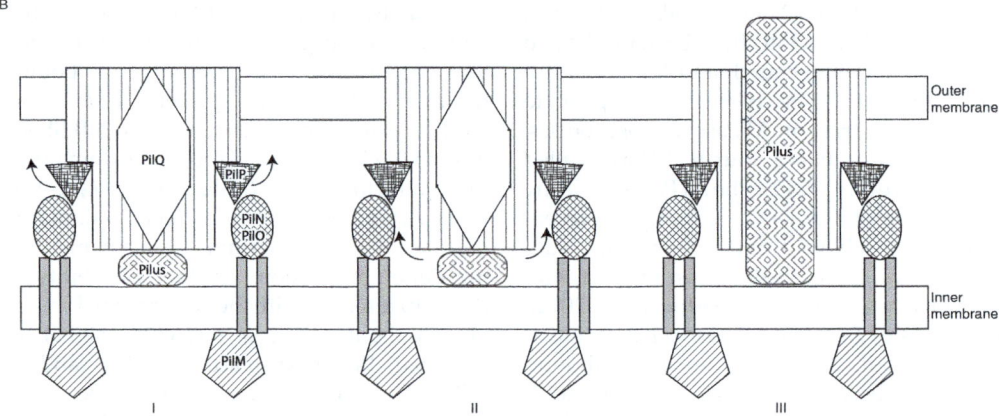

Fig. 6.3. Proposed mechanism for opening the PilQ pore. (A) Gating by the pilus. (I) The pilus tip encounters the SS1 domain of PilQ. (II) The PilQ pore rearranges upon this initial contact between the pilus tip, stimulating a structural change in the pore that opens it, facilitating complete polymerization of the pilus. In the absence of the pilus the pore returns to a closed position. (B) Gating via the PilMNOPQ complex. (I) A stimulus from the cytoplasm is somehow communicated to PilM, this communicates a signal to PilP through the PilMNOP complex. (II) This cascade causes PilP to cause a structural change in the PilQ pore, allowing the pilus to enter into the pore lumen. (III) The opening of the pore allows for the assembled pilus to exit through the outer membrane.

abundance minor pilins (Giltner *et al.*, 2010; Cisneros *et al.*, 2012), would contact the first periplasmic domain of PilQ, i.e. the SS1 domain. Such contact may stimulate opening, and subsequent pilus polymerization would plug the pore, maintaining cellular integrity. This model is supported by recent cryo-electron microscopy data (Berry *et al.*, 2012), where the secretin appeared closed at its periplasmic end, and the first species-specific domain was disordered as indicated by a lower-than-expected volume for the size of this domain. It is possible that the SS1 domain becomes ordered upon contact with the pilus

tip, opening the periplasmic gate. In this model, the secretin would remain in a closed state in the absence of a pilus, and upon complete retraction, would return to the closed state. This model benefits from its simplicity in that only one event (pilus assembly) needs to be tightly regulated. It is possible that once assembled, a pilus may never be fully disassembled upon retraction, leaving a short stub that would keep the secretin open, yet plugged. This scenario might increase the efficiency of pilus extension, potentially allowing for shorter turn-around times between extension events.

Optical tweezer experiments support the concept that pilus retraction is reversible without proceeding to complete depolymerization, as application of a stalling force on retracting pili can cause them to extend once again (Clausen *et al.*, 2009).

6.5.2 Secretin gating by the PilMNOP subcomplex

In an alternative scenario for secretin gating, a signal could be transmitted from the cytoplasm via PilM (potentially with input from ATPases and/or PilC) to the PilNOP complex, causing PilP to exert force on the N0 domain of PilQ (Fig. 6.3B) (Ayers *et al.*, 2010). In turn, the periplasmic gate of the PilQ multimer might transition into an open conformation while the growing pilus plugs the pore. In a practical sense, this requires that a 'pilus priming complex' would need to be assembled at all times, ready to accept monomers once the signal for secretin opening is received. A pseudohelical priming complex of minor subunits may form spontaneously, as has been demonstrated for the minor pseudopilins of the T2S system (Cisneros *et al.*, 2012).

The second model differs from the first in that the signal is propagated through PilMNOP to PilQ, but beyond that, the mechanism of secretin opening is likely similar in that the growing pilus would plug the open secretin. However, the mechanism for closing is likely more complicated. In this case the 'open' signal must switch to a 'close' signal, releasing the force of PilP on PilQ, and closing the secretin. The closing signal would need to be coordinated with the depolymerization event so that the lumen doesn't remain open in the absence of the pilus.

6.5.3 Dynamics of the pore: stochastic vs smooth

Does the secretin participate in extension/retraction events, or is it a passive conduit for the pilus? In a stochastic model, PilA addition – and hence pilus extension – is facilitated by repetitive structural changes in PilQ that pull on the periplasmic region of the pilus, making the addition of PilA molecules at the base of the pilus easier. In a similar fashion, upon retraction, PilQ closing signals could facilitate the removal of PilA monomers back into the inner membrane.

Alternatively, in a smooth model one could envision a situation where the secretin opens once, and subsequent pilus polymerization is driven by other factors while the secretin acts simply as a conduit. This scenario is more likely if secretin opening alone is regulated, and it returns naturally to a closed state in the absence of the pilus. The energy for pilus depolymerization would also need to come from the cytoplasmic motor components, as the secretin would not participate in the reaction.

A combination of the two scenarios may also exist, where the secretin acts as a passive conduit in one state but actively facilitates the other. Or, conformational changes in the secretin complex may optimize pilus assembly rates while counteracting the forces that the pilus encounters during twitching motility. Although the rate of extension is unknown, the estimated retraction rate of 10^3 monomers/second (Merz *et al.*, 2000; Skerker and Berg, 2001) is much faster than one would expect for an uncatalysed process.

6.6 Specific Adaptations in the T4aP vs T4bP and T2S Systems

Although it is clear that the T4aP, T4bP and T2S systems are related, it is worth considering how and why they are different. The T4bP system produces a long extracellular filament similar to the T4aP, but the assembly proteins and the pilus subunits have differences that may represent specific adaptations for this system, which is involved primarily in adhesion. The T2S system does not generate long extracellular fibres under normal circumstances, but rather a short periplasmic pseudopilus, raising the question of how the ancestral system evolved into one whose primary role is to secrete folded proteins into the extracellular milieu.

6.6.1 Comparison of T4aP and T4bP

Why have multiple subtypes of T4P evolved? The T4aP system produces a pilus capable of surface-associated retraction, motility, surface adhesion, epithelial and endothelial invasion, and biofilm maturation (Craig and Li, 2008; Burrows, 2012). Conversely, the T4bP has adherence and cell–cell aggregation roles, but is not associated with retraction or motility except for rare exceptions (Mazariego-Espinosa et al., 2010; Zahavi et al., 2011), and many T4bP systems have only a single extension ATPase (Roux et al., 2012). For those T4bP that can retract, it will be interesting to see if functional differences can be associated with their biophysical properties. The Tad subclass of T4bP lack a retraction ATPase, and it is not yet clear whether they are capable of depolymerization (Burrows, 2012; Roux et al., 2012).

The T2S system generates a short periplasmic pseudopilus that participates in the export of secreted virulence factors, possibly by controlling the opening/closing of the secretin (Forest, 2008; McLaughlin et al., 2012). The extent of structural similarity between T2S and T4aP systems has recently been reviewed (Ayers et al., 2010; McLaughlin et al., 2012; Korotkov et al., 2012), and the organization of many of the genes encoding key proteins of the transenvelope complex is conserved between the two systems. It is therefore interesting to consider the differences between the T2S and T4aP systems – why does one system generate long extracellular filaments while the other produces only short pseudopili?

One consideration is the stoichiometry of major versus minor pilin subunits. The T2S system can generate long extracellular pseudopili if the major pilin subunit is overexpressed (Sauvonnet et al., 2000; Vignon et al., 2003; Durand et al., 2005), suggesting that the concentration of major pilin subunits is a key driver of (pseudo)pilus length. Pilus length could therefore reflect regulation of subunit expression, or possibly differences in the affinities between subunits that might affect the rate of their assembly or disassembly.

A second consideration is the rate of polymerization versus depolymerization. Retraction may be tightly regulated, and invoked under specific conditions, allowing the pilus to polymerize until a specific retraction signal is received. In the T2S system, there is a single extension ATPase and a low concentration of pilin subunits that likely limits pseudopilus length. The lack of a retraction ATPase suggests that an alternative way to reset the system probably exists. It is possible that T2S system subunits preferentially populate the monomeric membrane-embedded state over the polymerized state, or that they are inherently unstable and retraction occurs via unfolding or degradation (Durand et al., 2005). Most T4P subunits have a C-terminal disulfide bond stabilizing the 'D-region', a critical structural element, while the same region of major pseudopilins is stabilized by non-covalent coordination of Ca^{++} ions (Korotkov et al., 2009a; Giltner et al., 2012). Treatment of assembled pili with reducing agents causes their rapid disintegration, showing that stability of subunits' C-terminal architecture is important for fibre integrity (Li et al., 2012).

A third consideration is the secretin, and its influence on fibre dynamics. T2S system secretin monomers lack the SS domains and instead have 2–3 repeats of the N1 domain, C-terminal to the single N0 domain found in T2S and T4P systems (Korotkov et al., 2011a). The additional N1 domains in the T2S system likely serve as multimerization motifs, and orient the N0 domain in such a way that it can interact with the GspC component of the T2S system alignment complex. In addition to potentially participating in gating of the pore (above), the SS domains in the T4aP secretin could help stabilize the polymeric state of the pilin subunits.

Finally, during twitching motility, the T4aP apparatus can experience forces in excess of 100 pN on a single fibre (Merz et al., 2000), and these forces are additive when pili form laterally associated bundles (Biais et al., 2008). Significant strain can thereby be generated on the assembly system, and requires that it be firmly anchored in the cell envelope. It may seem contradictory that a fibre capable of dynamic assembly–disassembly requires anchoring; however, it is possible that the assembly complex responds to externally applied forces, and that 'pulling' on the pilus changes the nature of the assembly such that cellular integrity is maintained. Propagation of forces through the assembly system upon tethering of a pilus may even act as a signalling mechanism, allowing the bacteria to interrogate the physical properties of nearby surfaces (Burrows, 2012).

FimV is a large inner membrane protein with an N-terminal periplasmic domain containing a PG-binding LysM motif, a single transmembrane domain, and a large acidic C-terminal cytoplasmic domain predicted to contain a number of protein–protein interaction motifs (Semmler *et al.*, 2000; Wehbi *et al.*, 2011). FimV mutants have defects in twitching motility, Type II secretion and PilQ secretin stability (Michel *et al.*, 2011; Wehbi *et al.*, 2011). In-frame deletion of the LysM domain led to formation of fewer secretins and a concomitant impairment in motility (Wehbi *et al.*, 2011), suggesting that FimV is involved in organization of the secretin in the cell wall. FimV may bind to and prevent the localized cross-linking of PG during secretin assembly and/or help to brace the assembly system against retraction forces through its PG interactions.

The actin-like protein PilM could also play a part in connecting the pilus to the bacterial cytoskeleton. PilM shares structural similarity to MreB (Karuppiah and Derrick, 2011), which was recently shown to be important for the localization of both T4P and PilT during growth on solid media (Cowles and Gitai, 2010). Interactions of PilM with cytoskeletal elements might provide additional stability to the T4P complex upon retraction.

6.7 Conclusions

Significant progress has been made in the last decade toward understanding the structures and functions of the components comprising the T4P and T2S assembly systems. The structures of the individual components and complexes that have been determined, coupled with extensive biochemical and microbiological data, provide much-needed clues about how these nanomachines are constructed and carry out their functions. However, much still remains to be determined. For example, atomic resolution structures of the remaining components of the T4aP system, especially PilQ, coupled with cryo-electron microscopy and other low-resolution hybrid approaches, such as small-angle X-ray scattering, of the intact machinery or sub-complexes is required. Such studies will help to clarify the stoichiometry of assembly system components, which will then aid our understanding of how the system forms. It will also be important to obtain detailed structural information on the less-well characterized T4bP and Tad pilus systems for comparison, as those data will help in development of hypotheses about the way in which these systems have evolved to serve their specialized roles. As experimental methods to isolate and purify these large complexes from the cell envelope improve, a molecular level understanding of these interesting nanomachines will emerge.

Acknowledgements

Work in the Howell and Burrows laboratories on T4P assembly systems is supported by Operating Grant MOP 93585 from the Canadian Institutes of Health Research (CIHR) to L.L.B. and P.L.H. S.T. was funded, in part, by a graduate scholarship from Cystic Fibrosis Canada. P.L.H. holds a Canada Research Chair in Structural Biology.

References

Abendroth, J., Bagdasarian, M., Sandkvist, M. and Hol, W.G.J. (2004) The structure of the cytoplasmic domain of EpsL, an inner membrane component of the type II secretion system of *Vibrio cholerae*: an unusual member of the actin-like ATPase superfamily. *Journal of Molecular Biology* 344, 619–633.

Abendroth, J., Murphy, P., Sandkvist, M., Bagdasarian, M. and Hol, W.G.J. (2005) The X-ray structure of the type II secretion system complex formed by the N-terminal domain of EpsE and the cytoplasmic domain of EpsL of

Vibrio cholerae. Journal of Molecular Biology 348, 845–855.

Abendroth, J., Mitchell, D.D., Korotkov, K.V., Johnson, T.L., Kreger, A., Sandkvist, M. and Hol, W.G. (2009) The three-dimensional structure of the cytoplasmic domains of EpsF from the type 2 secretion system of *Vibrio cholerae. Journal of Structural Biology* 166, 303–315.

Ayers, M., Sampaleanu, L., Tammam, S.D., Koo, J., Harvey, H., Howell, P.L. and Burrows, L.L. (2009) PilM/N/O/P proteins form an inner membrane complex that affects the stability of the *Pseudomonas aeruginosa* type IV pilus secretin. *Journal of Molecular Biology* 394, 128–142.

Ayers, M., Howell, P.L. and Burrows, L.L. (2010) Architectures of the type II secretion and type IV pilus machineries. *Future Microbiology* 5, 1203–1218.

Balasingham S.V., Collins R.F., Assalkhou R., Homberset H., Frye S.A., Derrick J.P., Tønjum T. (2007) Interactions between the lipoprotein PilP and the secretin PilQ in *Neisseria meningitidis. Journal of Bacteriology* 189, 5716–5727.

Berry, J.-L., Phelan, M.M., Collins, R.F., Adomavicus, T., Tonjum, T., Frye, S.A., Bird, L., Owens, R.J., Ford, R.C., Lian, L.-Y. and Derrick, J.P. (2012) structure and assembly of a trans-periplasmic channel for type IV pili in *Neisseria meningitidis. PLoS Pathogens* 8, e1002923.

Biais, N., Ladoux, B., Higashi, D., So, M. and Sheetz, M. (2008) Cooperative retraction of bundled type IV pili enables nanonewton force generation. *PLoS Biology* 6, e87

Bulyha, I., Schmidt, C., Lenz, P., Jakovljevic, V., Hne, A., Maier, B., Hoppert, M. and Sgaard-Andersen, L. (2009) Regulation of the type IV pili molecular machine by dynamic localization of two motor proteins. *Molecular Microbiology* 74, 691–706.

Burkhardt, J., Vonck, J. and Averhoff, B. (2011) Structure and function of PilQ, a secretin of the DNA transporter from the thermophilic bacterium *Thermus thermophilus* HB27. *Journal of Biological Chemistry* 286, 9977–9984.

Burrows, L.L. (2012) *Pseudomonas aeruginosa* twitching motility: type IV pili in action. *Annual Review of Microbiology* 66, 493–520.

Chiang, P., Habash, M. and Burrows, L.L. (2005) Disparate subcellular localization patterns of *Pseudomonas aeruginosa* type IV pilus ATPases involved in twitching motility. *Journal of Bacteriology* 187, 829–839.

Cisneros, D.A., Bond, P.J., Pugsley, A.P., Campos, M. and Francetic, O. (2012) Minor pseudopilin self-assembly primes type II secretion

pseudopilus elongation. *EMBO Journal* 31, 1041–1053.

Clausen, M., Koomey, M. and Maier, B. (2009) Dynamics of type IV pili is controlled by switching between multiple states. *Biophysical Journal* 96, 1169–1177.

Collins, R.F., Davidsen, L., Derrick, J.P., Ford, R.C. and Tonjum, T. (2001) Analysis of the PilQ secretin from *Neisseria meningitidis* by transmission electron microscopy reveals a dodecameric quaternary structure. *Journal of Bacteriology* 183, 3825–3832.

Cowles, K.N. and Gitai, Z. (2010) Surface association and the MreB cytoskeleton regulate pilus production, localization and function in *Pseudomonas aeruginosa. Molecular Microbiology* 76, 1411–1426.

Craig, L. and Li, J. (2008) Type IV pili: paradoxes in form and function. *Current Opinion in Structural Biology* 18, 267–277.

Demchick, P. and Koch, A.L. (1996) The permeability of the wall fabric of *Escherichia coli* and *Bacillus subtilis. Journal of Bacteriology* 178, 768–773.

Diepold, A., Amstutz, M., Abel, S., Sorg, I., Jenal, U. and Cornelis, G.R. (2010) Deciphering the assembly of the *Yersinia* type III secretion injectisome. *EMBO Journal* 29, 1928–1940.

Durand, E., Michel, G., Voulhoux, R., Kurner, J., Bernadac, A. and Filloux, A. (2005) XcpX controls biogenesis of the *Pseudomonas aeruginosa* XcpT-containing pseudopilus. *Journal of Biological Chemistry* 280, 31378–31389.

Forest, K.T. (2008) The type II secretion arrowhead: the structure of GspI–GspJ–GspK. *Nature Structural and Molecular Biology* 15, 428–430.

Genin, S. and Boucher, C.A. (1994) A superfamily of proteins involved in different scretion pathways in Gram-negative bacteria: modular structure and specificity of the N-terminal domain. *Molecular and General Genetics* 243, 112–118.

Georgiadou, M., Castagnini, M., Karimova, G., Ladant, D. and Pelicic, V. (2012) Large-scale study of the interactions between proteins involved in type IV pilus biology in *Neisseria meningitidis*: characterization of a subcomplex involved in pilus assembly. *Molecular Microbiology* 84, 857–873.

Giltner, C.L., Habash, M. and Burrows, L.L. (2010) *Pseudomonas aeruginosa* minor pilins are incorporated into type IV pili. *Journal of Molecular Biology* 398, 444–461.

Giltner, C.L., Nguyen, Y. and Burrows, L.L. (2012) Type IV pilin proteins: versatile molecular modules. *Microbiology and Molecular Biology Reviews: MMBR* 76, 740–772.

Golovanov, A.P., Balasingham, S., Tzitziolnis, C., Goult, B.T., Lian, L.-Y., Homberset, H., Tonjum, T. and Derrick, J.P. (2006) The solutions structure of a domain from the *Neisseria meningitidis* lipoprotein PilP reveals a new b-sandwich fold. *Journal of Molecular Biology* 364, 186–195.

Gray, M.D., Bagdasarian, M., Hol, W.G.J. and Sandkvist, M. (2011) *In vivo* cross-linking of EpsG to EpsL suggests a role for EpsL as an ATPase-pseudopilin coupling protein in the Type II secretion system of *Vibrio cholerae*. *Molecular Microbiology* 79, 786–798.

Gu, S., Kelly, G., Wang, X., Frenkiel, T., Shevchik, V.E. and Pickersgill, R.W. (2012) Solution structure of the HR domain of the type II secretion system. *Journal of Biological Chemistry* 287, 9072–9080.

Hager, A.J., Bolton, D.L., Pelletier, M.R., Brittnacher, M.J., Gallagher, L.A., Kaul, R., Skerrett, S.J., Miller, S.I. and Guina, T. (2006) Type IV pili-mediated secretion modulates *Francisella* virulence. *Molecular Microbiology* 62, 227–237.

Han, X., Kennan, R.M., Parker, D., Davies, J.K. and Rood, J.I. (2007) Type IV fimbrial biogenesis is required for protease secretion and natural transformation in *Dichelobacter nodosus*. *Journal of Bacteriology* 189, 5022–5033.

Hoang, H.H., Nickerson, N.N., Lee, V.T., Kazimirova, A., Chami, M., Pugsley, A.P. and Lory, S. (2011) Outer membrane targeting of *Pseudomonas aeruginosa* proteins shows variable dependence on the components of Bam and Lol machineries. *mBio* 2, e00246-11.

Hobbs, M. and Mattick, J.S. (1993) Common components in the assembly of type 4 fimbriae, DNA transfer systems, filamentous phage and protein-secretion apparatus: a general system for the formation of surface-associated protein complexes. *Molecular Microbiology* 10, 233–243.

Karuppiah, V. and Derrick, J.P. (2011) Structure of the PilM-PilN inner membrane type IV pilus biogenesis complex from *Thermus thermophilus*. *Journal of Biological Chemistry* 286, 24434–24442.

Karuppiah, V., Hassan, D., Saleem, M. and Derrick, J.P. (2010) Structure and oligomerization of the PilC type IV pilus biogenesis protein from *Thermus thermophilus*. *Proteins* 78, 2049–2057.

Koo, J., Tammam, S.D., Ku, S.-Y., Sampaleanu, L., Burrows, L.L. and Howell, P.L. (2008) PilF is an outer membrane lipoprotein required for multimerization and localization of the *Pseudomonas aeruginosa* type IV pilus secretin. *Journal of Bacteriology* 190, 6961–6969.

Koo, J., Burrows, L.L. and Howell, P.L. (2012)

Decoding the roles of pilotins and accessory proteins in secretin escort services. *FEMS Microbiology Letters* 328, 1–12.

Koo, J., Tang, T., Harvey, H., Tammam, S., Sampaleanu, L., Burrows, L.L. and Howell, P.L. (2013) Functional mapping of PilF and PilQ in the *Pseudomonas aeruginosa* type IV pilus system. *Biochemistry* 52, 2914–2923.

Korotkov, K.V., Krumm, B., Bagdasarian, M. and Hol, W.G.J. (2006) Structural and functional studies of EpsC, a crucial component of the type 2 secretion system from *Vibrio cholerae*. *Journal of Molecular Biology* 363, 311–321.

Korotkov, K.V., Gray, M.D., Kreger, A.C., Turley, S., Sandkvist, M. and Hol, W.G.J. (2009a) Calcium is essential for the major pseudopilin in the type 2 secretion system. *Journal of Biological Chemistry* 284, 25466–25470.

Korotkov, K.V., Pardon, E., Steyaert, J. and Hol, W.G.J. (2009b) Crystal structure of the N-terminal domain of the secretin GspD from ETEC determined with the assistance of a nanobody. *Structure* 17, 255–265.

Korotkov, K.V., Gonen, T. and Hol, W.G.J. (2011a) Secretins: dynamic channels for protein transport across membranes *Trends in Biochemical Sciences* 36, 433–443.

Korotkov, K.V., Johnson, T.L., Jobling, M.G., Pruneda, J., Pardon, E., Héroux, A., Turley, S., Steyaert, J., Holmes, R.K., Sandkvist, M. and Hol, W.G.J. (2011b) Structural and functional studies on the interaction of GspC and GspD in the type II secretion system. *PLoS Pathogens* 7, e1002228.

Korotkov, K.V., Sandkvist, M. and Hol, W.G. (2012) The type II secretion system: biogenesis, molecular architecture and mechanism. *Nature Reviews. Microbiology* 10, 336–351.

Kurre, R., Hone, A., Clausen, M., Meel, C. and Maier, B. (2012) PilT2 enhances the speed of gonococcal type IV pilus retraction and of twitching motility. *Molecular Microbiology* 86, 857–865.

Li, J., Egelman, E.H. and Craig, L. (2012) Structure of the *Vibrio cholerae* type IVb pilus and stability comparison with the *Neisseria gonorrhoeae* type IVa pilus *Journal of Molecular Biology* 418, 47–64.

Lutkenhaus, J. (2007) Assembly dynamics of the bacterial MinCDE system and spatial regulation of the Z ring. *Annual Review of Biochemistry* 76, 539–562.

Lutkenhaus, J. (2012) The ParA/MinD family puts things in their place. *Trends in Microbiology* 20, 411–418.

Mattick, J.S. (2002) Type IV pili and twitching motility. *Annual Review of Microbiology* 56, 289–314.

Mazariego-Espinosa, K., Cruz, A., Ledesma, M.A., Ochoa, S.A. and Xicohtencatl-Cortes, J. (2010) Longus, a type IV pilus of enterotoxigenic *Escherichia coli*, is involved in adherence to intestinal epithelial cells. *Journal of Bacteriology* 192, 2791–2800.

McLaughlin, L.S., Haft, R.J. and Forest, K.T. (2012) Structural insights into the Type II secretion nanomachine. *Current Opinion in Structural Biology* 22, 208–216.

Merz, A.J., So, M. and Sheetz, M.P. (2000) Pilus retraction powers bacterial twitching motility. *Nature* 407, 98–102.

Michel, G.P.F., Aguzzi, A., Ball, G., Soscia, C., Bleves, S. and Voulhoux, R. (2011) Role of FimV in type II secretion system-dependent protein secretion of *Pseudomonas aeruginosa* on solid medium. *Microbiology (Reading, England)* 157, 1945–1954.

Milgotina, E.I., Lieberman, J.A. and Donnenberg, M.S. (2011) The inner membrane subassembly of the enteropathogenic *Escherichia coli* bundle-forming pilus machine. *Molecular Microbiology* 81, 1125–1127.

Misic, A.M., Satyshur, K.A. and Forest, K.T. (2010) *P. aeruginosa* PilT structures with and without nucleotide reveal a dynamic type IV pilus retraction motor. *Journal of Molecular Biology* 400, 1011–1021.

Nunn, D., Bergman, S. and Lory, S. (1990) Products of three accessory genes, *pilB*, *pilC*, and *pilD*, are required for biogenesis of *Pseudomonas aeruginosa* pili. *Journal of Bacteriology* 172, 2911–2919.

Perez-Cheeks, B.A., Planet, P.J., Sarkar, I.N., Clock, S.A., Xu, Q. and Figurski, D.H. (2012) The product of *tadZ*, a new member of the parA/minD superfamily, localizes to a pole in *Aggregatibacter actinomycetemcomitans*. *Molecular Microbiology* 83, 694–711.

Pichoff, S. and Lutkenhaus, J. (2005) Tethering the Z ring to the membrane through a conserved membrane targeting sequence in FtsA. *Molecular Microbiology* 55, 1722–1734.

Pink, D., Moeller, J., Quinn, B., Jericho, M. and Beveridge, T. (2000) On the architecture of the gram-negative bacterial murein sacculus. *Journal of Bacteriology* 182, 5925–5930.

Py, B., Loiseau, L. and Barras, F. (2001) An inner membrane platform in the type II secretion machinery of Gram-negative bacteria. *EMBO Reports* 2, 244–248.

Roux, N., Spagnolo, J. and De Bentzmann, S. (2012) Neglected but amazingly diverse type IVb pili. *Research in Microbiology* 163, 659–673.

Rumszauer, J., Schwarzenlander, C. and Averhoff, B. (2006) Identification, subcellular localization and functional interactions of PilMNOWQ and PilA4 involved in transformation competency and pilus biogenesis in the thermophilic bacterium *Thermus thermophilus* HB27. *FEBS Journal* 273, 3261–3272.

Russel, M. (1998) Macromolecular assembly and secretion across the bacterial cell envelope: type II protein secretion systems. *Journal of Molecular Biology* 279, 485–499.

Sampaleanu, L.M., Bonanno, J.B., Ayers, M., Koo, J., Tammam, S., Burley, S.K., Almo, S.C., Burrows, L.L. and Howell, P.L. (2009) Periplasmic domains of *Pseudomonas aeruginosa* PilN and PilO form a stable heterodimeric complex. *Journal of Molecular Biology* 394, 143–159.

Satyshur, K.A., Worzalla, G.A., Meyer, L.S., Heiniger, E.K., Aukema, K.G., Misic, A.M. and Forest, K.T. (2007) Crystal structures of the pilus retraction motor PilT suggest large domain movements and subunit cooperation drive motility. *Structure* 15, 363–376.

Sauvonnet, N., Vignon, G., Pugsley, A.P. and Gounon, P. (2000) Pilus formation and protein secretion by the same machinery in *Escherichia coli*. *EMBO Journal* 19, 2221–2228.

Savvides, S.N. (2007) Secretion superfamily ATPases swing big. *Structure* 15, 255–257.

Scheurwater, E.M. and Burrows, L.L. (2011) Maintaining network security: how macromolecular structures cross the peptidoglycan layer. *FEMS Microbiology Letters* 318, 1–9.

Semmler, A.B., Whitchurch, C.B., Leech, A.J. and Mattick, J.S. (2000) Identification of a novel gene, fimV, involved in twitching motility in *Pseudomonas aeruginosa*. *Microbiology (Reading, England)* 146 (Pt 6), 1321–1332.

Skerker, J.M. and Berg, H.C. (2001) Direct observation of extension and retraction of type IV pili. *PNAS USA* 98, 6901–6904.

Spreter, T., Yip, C.K., Sanowar, S., Andre, I., Kimbrough, T.G., Vuckovic, M., Pfuetzner, R.A., Deng, W., Yu, A.C., Finlay, B.B., Baker, D., Miller, S.I. and Strynadka, N.C. (2009) A conserved structural motif mediates formation of the periplasmic rings in the type III secretion system. *Nature Structural Molecular Biology* 16, 468–476.

Szeto, T.H., Dessen, A. and Pelicic, V. (2011) Structure/function analysis of *Neisseria meningitidis* PilW, a conserved protein that plays multiple roles in type IV pilus biology. *Infection and Immunity* 79, 3028–3035.

Tammam, S., Sampaleanu, L.M., Koo, J., Sundaram, P., Ayers, M., Chong, P.A., Forman-Kay, J.D., Burrows, L.L. and Howell, P.L. (2011) Characterization of the PilN, PilO and PilP type

IVa pilus subcomplex. *Molecular Microbiology* 82, 1496–1514.

Tammam, S., Sampaleanu, L.M., Koo, J., Manoharan, K., Daubaras, M., Burrows, L.L. and Howell, P.L. (2013) PilMNOPQ from the *Pseudomonas aeruginosa* type IV pilus system form a transenvelope protein interaction network that interacts with PilA. *Journal of Biological Chemistry* 195, 2126–2135.

Tarry, M., Jääskeläinen, M., Paino, A., Tuominen, H., Ihalin, R. and Högbom, M. (2011) The extramembranous domains of the competence protein HofQ show DNA binding, flexibility and a shared fold with type I KH domains. *Journal of Molecular Biology* 409, 642–653.

Van Der Meeren, R., Wen, Y., Van Gelder, P., Tommassen, J., Devreese, B. and Savvides, S.N. (2012) New insights into the assembly of bacterial secretins: structural studies of the periplasmic domain of XcpQ from *Pseudomonas aeruginosa*. *Journal of Biological Chemistry* 288, 1214–1225.

Vignon, G., Kohler, R., Larquet, E., Giroux, S., Prevost, M.C., Roux, P. and Pugsley, A.P. (2003) Type IV-like pili formed by the type II secreton: specificity, composition, bundling, polar localization, and surface presentation of peptides. *Journal of Bacteriology* 185, 3416–3428.

Wang, X., Pineau, C., Gu, S., Guschinskaya, N., Pickersgill, R.W. and Shevchik, V.E. (2012) Cysteine scanning mutagenesis and disulfide mapping analysis of the arrangement of GspC and GspD protomers within the T2SS. *Journal of Biological Chemistry* 287, 19082–19093.

Wehbi, H., Portillo, E., Harvey, H., Shimkoff, A.E., Scheurwater, E.M., Howell, P.L. and Burrows, L.L. (2011) The peptidoglycan-binding protein FimV promotes assembly of the *Pseudomonas aeruginosa* type IV pilus secretin. *Journal of Bacteriology* 193, 540–550.

Whitechurch, C.B. and Mattick, J.S. (1994) Characterization of a gene, *pilU*, required for twitching motility but not phage sensitivity in *Pseudomonas aeruginosa*. *Molecular Microbiology* 13, 1079–1091.

Winther-Larsen, H.C., Wolfgang, M., Dunham, S.A., Van Putten, J.P.M., Dorward, D., Lovold, C., Aas, F.E. and Koomey, M. (2005) A conserved set of pilin-like molecules controls type IV pilus dynamics and organelle-associated functions in *Neisseria gonorrhoeae*. *Molecular Microbiology* 56, 903–917.

Wolfgang, M., Lauer, P., Park, H.-S., Brossay, L., Hebert, J. and Koomey, M. (1998a) PilT mutations lead to simultaneous defects in competence for natural transformation and twitching motility in piliated *Neisseria gonorrhoeae*. *Molecular Microbiology* 29, 321–330.

Wolfgang, M.C., Park, H.-S., Hayes, S.F., Van Putten, J.P.M. and Koomey, M. (1998b) Suppression of an absolute defect in type IV pilus biogenesis by loss-of-function mutations in pilT, a twitching motility gene in *Neisseria gonorrhoeae*. *Proceedings of the National Academy of Sciences of the United States of America* 95, 14973–14978.

Wolfgang, M., Putten, J.P.M.V., Hayes, S.F., Dorward, D. and Koomey, M. (2000) Components and dynamics of fiber formation define a ubiquitous biogenesis pathway for bacterial pili. *EMBO Journal* 19, 6408–6418.

Xu, Q., Christen, B., Chiu, H.J., Jaroszewski, L., Klock, H.E., Knuth, M.W., Miller, M.D., Elsliger, M.A., Deacon, A.M., Godzik, A., Lesley, S.A., Figurski, D.H., Shapiro, L. and Wilson, I.A. (2012) Structure of the pilus assembly protein TadZ from *Eubacterium rectale*: implications for polar localization. *Molecular Microbiology* 83, 712–727.

Yamagata, A., Milgotina, E., Scanlon, K., Craig, L., Tainer, J.A. and Donnenberg, M.S. (2012) Structure of an essential type IV pilus biogenesis protein provides insights into pilus and type II secretion systems. *Journal of Molecular Biology* 419, 110–124.

Yao, X., Jericho, M., Pink, D. and Beveridge, T. (1999) Thickness and elasticity of gram-negative murein sacculi measured by atomic force microscopy. *Journal of Bacteriology* 181, 6865–6875.

Zahavi, E.E., Lieberman, J.A., Donnenberg, M.S., Nitzan, M., Baruch, K., Rosenshine, I., Turner, J.R., Melamed-Book, N., Feinstein, N., Zlotkin-Rivkin, E. and Aroeti, B. (2011) Bundle-forming pilus retraction enhances enteropathogenic *Escherichia coli* infectivity. *Molecular Biology of the Cell* 22, 2436–2447.

7 *Corynebacterium diphtheriae* Pili: Assembly, Structure and Function

I-Hsiu Huang[1,2] and Hung Ton-That[1]

[1]*Department of Microbiology and Molecular Genetics, University of Texas Health Science Center at Houston, USA;* [2]*Institute of Basic Medical Sciences and Department of Microbiology and Immunology, National Cheng Kung University, Tainan, Taiwan*

7.1 Introduction

Pili or fimbriae were thought to be limited mainly to Gram-negative bacteria, yet the first description of fibril-like structures on the surface of Gram-positive bacteria was in *Corynebacterium renale* (Yanagawa *et al.*, 1968). Nearly a decade later, pili were reported in several species of the genus *Corynebacterium*, including the diphtheria causative agent *C. diphtheriae* (Yanagawa and Honda, 1976). Perhaps overshadowed by the enormous attention paid to diphtheria toxin, there was little work on the function of corynebacterial pili, except the observation that they were linked to haemagglutination (Ermolayev *et al.*, 1987). Surprisingly, mechanistic studies of pilus assembly in *C. diphtheriae*, and in Gram-positive bacteria on the whole, seemingly were put on hold until the era of genomics research. It happened at the same time that the sortase gene *srtA* was discovered in *Staphylococcus aureus* (Mazmanian *et al.*, 1999). *srtA* encodes a transpeptidase enzyme that catalyses surface display, via cell-wall anchoring, of a group of surface proteins that contain a cell-wall sorting signal (CWSS) comprised of a LPXTG motif, a hydrophobic domain and a positively charged tail (Navarre and Schneewind, 1999). It was also discovered that the genome of *C. diphtheriae* contains six copies of sortase (Cerdeno-Tarraga *et al.*, 2003), five of which are distributed into three gene clusters encoding surface proteins with the CWSS (Ton-That and Schneewind, 2003). Homology and BLAST searches revealed that several of these proteins are homologous to FimA and FimP (Ton-That and Schneewind, 2003), the previously known fimbrial proteins of the actinobacterium *Actinomyces naeslundii* (Yeung, 1999). These observations led to the conjecture that sortase catalyses pilus assembly in Gram-positive bacteria. As initially described in *C. diphtheriae* (Ton-That and Schneewind, 2003) and valued over the years, the mechanism of sortase-mediated pilus assembly has been extensively studied in many species of Gram-positive bacteria including *Enterococcus faecalis*, *Bacillus cereus* and several species of streptococci (Scott and Zahner, 2006; Telford *et al.*, 2006; Hendrickx *et al.*, 2011). We describe here the current knowledge of the pilus assembly mechanism in *C. diphtheriae* as well as the importance of *C. diphtheriae* pili as adhesive determinants.

7.2 Sortase Catalyses Cell-Wall Anchoring of Surface Proteins

Over 40 years ago, Swedish researchers Sjoquist and colleagues showed that protein A of *S. aureus* was removed from the bacterial surface by lysostaphin, a glycyl-glycine endopeptidase that cleaves the pentaglycine cross-bridge of the staphylococcal cell wall (Sjoquist *et al.*, 1972). This insightful finding demonstrated that the cell wall is a platform for surface proteins to be anchored. Indeed, the identification of the CWSS in many Gram-positive bacteria cemented the idea that there is a conserved pathway in Gram-positive bacteria that displays surface proteins on the bacterial peptidoglycan (Schneewind *et al.*, 1992). Subsequently, by biochemical and genetic analyses of *S. aureus* mutants that failed to display a reporter surface protein, Schneewind and colleagues discovered the first sortase gene *srtA* in *S. aureus* (Mazmanian *et al.*, 1999). The mechanism of trans-peptidation catalysed by the sortase enzyme was then revealed as the following (Ton-That *et al.*, 1999, 2000, 2002). The membrane-bound sortase enzyme recognizes a protein precursor with the LPXTG motif and cleaves between the threonine and glycine residues of this motif, forming an acyl-enzyme intermediate with the substrate. A nucleophilic attack by a lipid II molecule resolves this intermediate, leading to the substrate anchoring to the peptidoglycan (Ton-That *et al.*, 2004a). For a detailed description of the mode of action by sortase enzymes, readers are referred to a review by Clancy *et al.* (2010).

In addition to SrtA, *S. aureus* also harbours a second copy of sortase termed SrtB, which catalyses cell-wall anchoring of a NPQTN-containing surface protein involved in iron acquisition (Mazmanian *et al.*, 2002). It turns out that there are at least six classes of sortase enzymes in Gram-positive bacteria, whose classification has been mainly based on their primary sequences (Comfort and Clubb, 2004; Dramsi *et al.*, 2005). A more detailed analysis of different sortase classes can be found in a recent review by Spirig and *et al.* (2011). Forming the largest group are sortases of class C or pilin-specific sortases that are essential for pilus polymerization

first described in *C. diphtheriae* (Ton-That and Schneewind, 2003).

7.3 Multiples Pilus Gene Clusters in *C. diphtheriae*

As mentioned above, there are three gene clusters in the genome of *C. diphtheriae*, i.e. strain NCTC13129 (Cerdeno-Tarraga *et al.*, 2003), which expresses five sortases, namely SrtA–E (Ton-That and Schneewind, 2003). Located elsewhere in the chromosome is *srtF*, coding for the so-called housekeeping sortase (Swaminathan *et al.*, 2007). Each gene cluster encodes three surface proteins with the CWSS, conveniently called SpaA-I (spa: sortase-mediated pilus assembly) (Fig. 7.1A). A combination of biochemical, genetic and immuno-electron microscopic analyses revealed that each gene cluster produces a distinct pilus structure composed of a pilus shaft, a tip pilin and a pilus base. Strain NCTC13129 assembles the SpaA-, SpaD- and SpaH-type pili, each of which was designated by the pilus shaft (Ton-That and Schneewind, 2003; Gaspar and Ton-That, 2006; Swierczynski and Ton-That, 2006). Consistent with the notion that corynebacterial pili are covalently linked, treatment with hot sodium dodecyl sulfate (SDS) and formic acid did not disassemble the pilus polymers (Ton-That and Schneewind, 2003). Similar to the protein A study by Sjoquist and colleagues, treatment of mutanolysin, a cell-wall hydrolase, released pili into the extra-cellular milieu, indicating that corynebacterial pili are linked to the cell wall (Ton-That and Schneewind, 2003). Importantly, deletion of sortase gene(s) in each pilus gene cluster abrogated the assembly of each pilus type (Ton-That and Schneewind, 2003; Gaspar and Ton-That, 2006; Swierczynski and Ton-That, 2006). These studies have formed the basis for the mechanism of sortase-mediated pilus assembly.

The presence of multiple Spa gene clusters in strain NCTC13129 has raised questions about pilus gene acquisition by horizontal gene transfer mechanisms (Swierczynski and Ton-That, 2006) and pilus variation. A recent pangenomic study of 12 *C.*

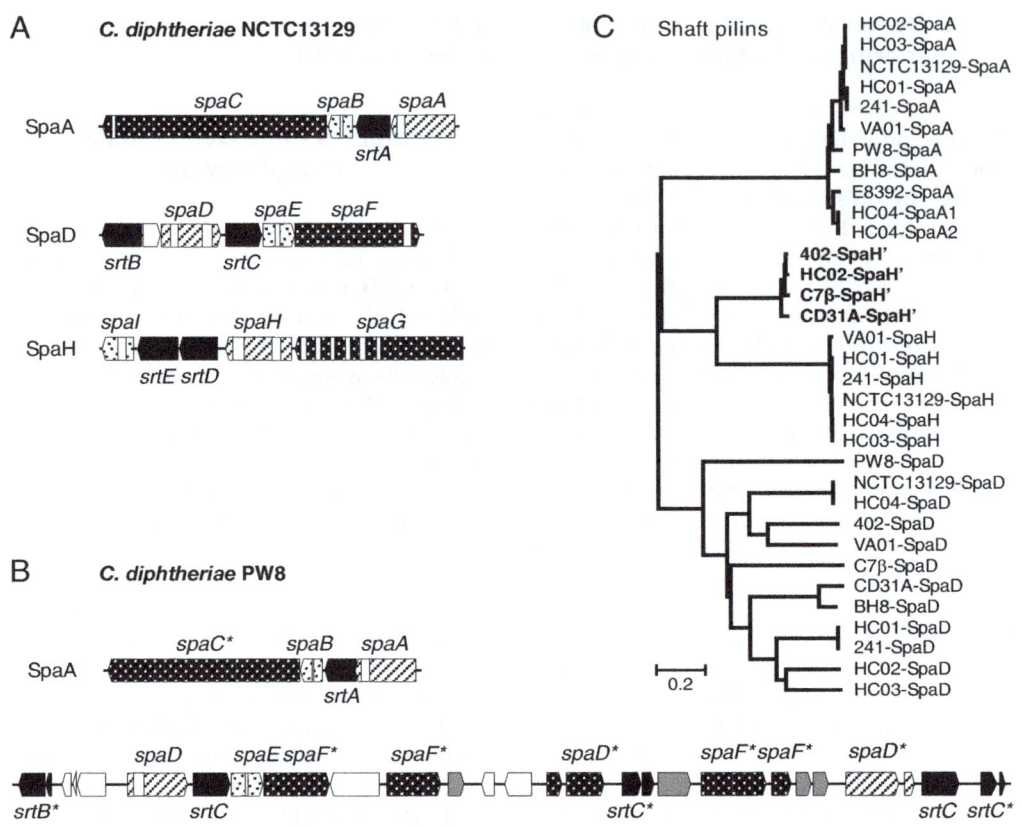

Fig. 7.1. Pilus gene clusters in *C. diphtheriae*. (A) The three pilus gene clusters encoding for SpaA-, SpaD-, SpaH-type pili of *C. diphtheriae* strain NCTC13129 and (B) the two gene clusters of the vaccine strain PW8. (C) A recent pangenomic analysis of 12 *C. diphtheriae* strains revealed a potential fourth pilus type, named SpaH′-type (bold), that forms a separate clade on the phylogenetic three and is more related to the SpaH-type pilus than the other two. Sortase genes are shown in black; Spa pilin genes are differently specified according to shaft, tip and base pilins; unknown genes are indicated in white; and mobile elements are marked as grey. Asterisks indicate disrupted genes. Cna domains within pilin genes are highlighted in white. (Adapted with permission from American Society for Microbiology, Trost *et al.*, 2012.)

diphtheriae clinical isolates – one-third of which are toxigenic strains including the vaccine production strain PW8 (Fig. 7.1B) – revealed several intriguing observations supportive of the above assertions (Trost *et al.*, 2012). First, there are at least two pilus gene clusters found in a given strain and these gene clusters are part of genomic islands that encodes for additional virulence factors. Second, the fourth pilus-type, namely SpaH′-type, potentially exists in strains 402, HC02, C7(β) and CD31A (Fig. 7.1C).

Phylogenetic analysis of major shaft and minor pilins revealed that these pilins form separate clades, which are closer to SpaH pilins than SpaA pilins on the phylogenetic tree. Consistently, their cognate pilin-specific sortases form separate clades that display the same trend. Last but not least, pili are not limited to toxigenic corynebacteria. This is expected since nontoxigenic strains of *C. diphtheriae* are subjected to lysogenic conversion to become toxigenic (Holmes and Barksdale, 1969).

7.4 The Archetype SpaA Pilus and Sortase-mediated Pilus Assembly

As the most prevalent and conserved pilus type in *C. diphtheriae* (Trost *et al.*, 2012), the SpaA pilus has been studied extensively, providing an experimental model for sortase-mediated pilus assembly. The SpaA pilus is expressed from the gene cluster *spaA-srtA-spaB-spaC* (Ton-That and Schneewind, 2003), which is presumably transcribed from a promoter upstream of the *spaA* gene. All genes in the cluster appear to be constitutively expressed (E. Rogers, Houston, 2012, personal communication). Encoded by *spaA*, the SpaA pilin precursor is predicted to have a molecular weight of 56 kDa, while SpaB is approximately 22 kDa in weight and SpaC being the largest at over 202 kDa. Immuno-electron microscopic studies demonstrated that SpaA forms the pilus shaft and SpaC is located at the pilus tip; in strains that SpaA was overexpressed, SpaB was found along the pilus shaft (Ton-That and Schneewind, 2003). Like other Gram-positive pilins, all three pilins harbour an N-terminal signal peptide and a C-terminal CWSS. Unlike SpaC and SpaB pilins, SpaA contains a pilin motif with a conserved sequence of WxxxVxVYPKN. Alanine-substitution of the conserved lysine residue within the SpaA pilin motif or the threonine residue of the SpaA LPXTG motif abrogated pilus assembly (Ton-That and Schneewind, 2003). Consistently, the pilin motif and the CWSS were shown to be necessary and sufficient to promote pilus polymers that are catalysed by a pilin-specific sortase (Ton-That *et al.*, 2004b). In this experiment, a hybrid protein was generated from staphylococcus enterotoxin B (SEB) that was fused between the N-terminal region of SpaA encompassing the signal peptide and the pilin motif and the C-terminal CWSS. When expressed in *C. diphtheriae*, the SEB hybrid protein was polymerized by sortase SrtA and dependent on the conserved lysine residue for sortase-catalysed polymerization (Ton-That *et al.*, 2004b).

The sortase-mediated pilus polymerization exhibits a remarkable resemblance of the transpeptidation reaction modelled around staphylococcal sortase SrtA, which crosslinks surface proteins to the bacterial cell wall (Ton-That *et al.*, 2004a). Basically, in the initial stage of pilus assembly in *C. diphtheriae*, Spa pilin precursors are exported across the cell membrane through the general Sec secretion machinery and embedded into the membrane via a hydrophobic domain and a positively charged tail. The membrane bound pilin-specific sortase SrtA recognizes and cleaves the LPXTG motif of the pilin precursors between the threonine and the glycine residues resulting in an acyl-enzyme intermediate, in which the cleaved threonine residue forms a thioester bond with the catalytic cysteine residue of SrtA. Pilus polymerization begins with a nucleophilic attack from the pilin motif lysine residue of a highly active SpaA–SrtA acyl enzyme intermediate to the thioester linkage of another one. This chain reaction continues to extend the pilus structure. It was thought that the same transpeptidation reaction generates the isopeptide linkage that joins SpaC at the pilus tip to SpaA (Fig. 7.2). Of note, although this bond has not been biochemically shown in *C. diphtheriae*, in *Bacillus cereus* the same linkage between the tip pilin and the pilus shaft was demonstrated (Budzik *et al.*, 2009). In support of the existence of the above intermediates, biochemical data has demonstrated in *C. diphtheriae* that SrtA forms a stable membrane-bound complex with pilin subunits that requires the transmembrane domain of SrtA (Guttilla *et al.*, 2009).

While the assembly of the pilus tip SpaC and pilus shaft SpaA was evident, the incorporation of SpaB into the pilus structure appeared to be promiscuous at first. It was not clear how pilus association of SpaB is mediated. Intriguingly, SpaA contains a conserved motif YxLxETxAPxGY (designated as the E-box) upstream of its LPXTG motif, and alanine substitution of the conserved Glu residue resulted in disrupted SpaA polymers and severely affected SpaB incorporation into the pilus (Ton-That *et al.*, 2004b). It was speculated that the E-box may be involved in structural stability of SpaA pilin, which potentially affects SpaB–SpaA

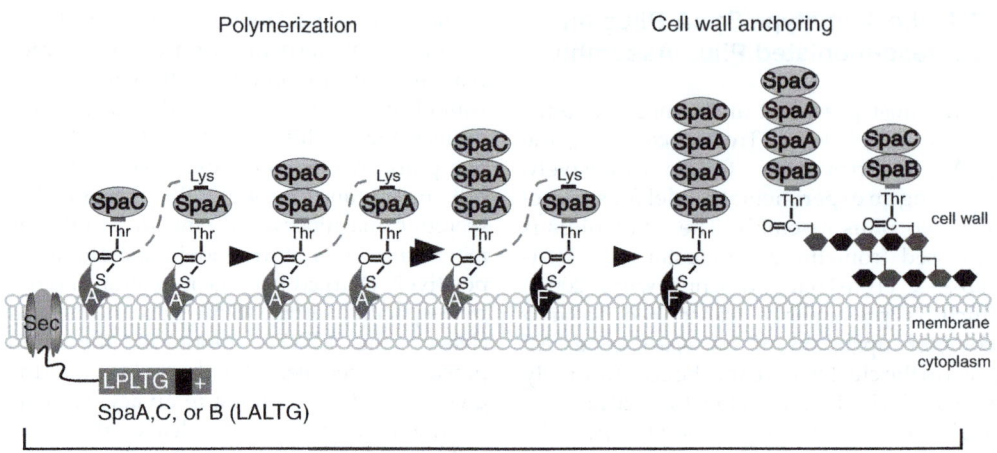

Fig. 7.2. Biphasic mode of pilus assembly in *Corynebacterium diphtheriae*. The model is based on studies of the SpaA-type pilus in *C. diphtheriae*. As proposed, after translation across the cytoplasmic membrane by the Sec apparatus, Spa pilin precursors are embedded into the membrane, where they form acyl-enzyme intermediates with sortase enzymes. Pilin-specific sortase catalyses lysine-mediated transpeptidation reactions that covalently cross-link pilin subunits into high molecular pilus polymers. Polymerization ends when the housekeeping sortase SrtF catalyses cell-wall anchoring of the pilus base SpaB via Lipid II precursors. The pilus assembly centre or pilusosome presumably constitutes the secretion machinery, sortases and their cognate substrates, and *trans*-acting factors (see text for detail).

interaction, given the strategic position of the E-box that is upstream of the SpaA LPXTG motif.

7.4.1 Biphasic mode of pilus assembly

After a series of studies on pilus assembly in *Enterococcus faecalis*, several species of streptococci and *B. cereus*, it was clear that the lysine-mediated cross-linking of pilins catalysed by pilin-specific sortase is a common feature in the assembly mechanism of Gram-positive pili (Ton-That and Schneewind, 2004; Scott and Zahner, 2006; Telford *et al.*, 2006). In these organisms, including *C. diphtheriae*, pilin-specific sortase is specific for polymerization of pilin substrates encoded within a pilus gene cluster. Like many of these microbes, in addition to pilin-specific sortases, *C. diphtheriae* expresses the housekeeping sortase SrtF, presumably catalysing cell-wall anchoring of general surface proteins. It was surprising to find that deletion of *srtF* resulted in abundant secretion of Spa pilus polymers

into the extracellular milieu, and an isogenic strain that expressed only SrF, while other pilin-specific sortases were absent, failed to produce pilus polymers (Swaminathan *et al.*, 2007). Two conclusions were made from these findings. First, the housekeeping sortase is involved in cell-wall anchoring of pilus polymers, not in pilus polymerization. Second, pilus polymerization is followed by cell-wall anchoring of pilus polymers. Importantly, this two-step mechanism of pilus assembly has been supported from studies of *B. cereus*, *Streptococcus agalactiae* and *S. pyogenes* pili (Budzik *et al.*, 2007; Nobbs *et al.*, 2008; Smith *et al.*, 2010).

Questions that arise from the above biphasic model are what terminates pilus polymerization and consequently how the growing pilus is anchored to the cell wall. The answers came from the promiscuous SpaB pilin, which harbours the LAFTG motif proposed to be favourable for the housekeeping sortase SrtF (Ton-That and Schneewind, 2003). Recall that without *srtF*, SpaA polymers were largely detached from the cell wall (Swaminathan *et al.*, 2007); in this mutant, accumulation of SpaB was found in the released SpaA polymers, consistent with the proposal above that SpaB is a preferred substrate of SrtF (Mandlik *et al.*, 2008a). Remarkably, deletion of *spaB* led to complete secretion of SpaA pilus polymers into the culture medium, and these polymers were increased in length (Mandlik *et al.*, 2008a), supporting the notion that SpaB is a molecular switch for pilus termination, hence controlling the pilus length. Indeed, this was the case, as over-expression of SpaB resulted in shorter pili than the wild-type pili. Furthermore, when the SpaB sequence was subjected to mutation, it was revealed that SpaB contains a C-terminal lysine residue (K139), whose substitution to alanine generated the same phenotype as the *spaB* deletion mutant (Mandlik *et al.*, 2008a). It was then proposed that the lysine residue of SpaB provides a nucleophilic attack for the transpeptidation reaction that incorporates SpaB at the pilus base, similar to the transpeptidation reaction mediated by the pilin motif lysine of SpaA that promotes pilus extension (Mandlik *et al.*, 2008b).

According to the above model (see Fig. 7.2), the reactive SpaB molecule is in the form of SpaB–SrtF acyl intermediate. Pilus polymerization catalysed by pilin-specific sortase SrtA is switched to cell-wall anchoring when the SpaB Lys139 residue within the reactive SpaB attacks the thioester bond between the last SpaA subunit of the growing pilus polymer and the SrtA enzyme. As a result, the pilus polymer is then transferred to SrtF, polymerization reaction is terminated and SrtF catalyses cell-wall anchoring of the pilus polymers via the lipid II precursor. As SpaB, a preferred substrate of the housekeeping sortase SrtF, it is possible that once a SpaB–SrtF complex is formed, SpaB can be anchored to the cell wall without pili associated. Indeed, SpaB monomers and SpaB–SpaC dimers have been observed on the cell wall of *C. diphtheriae* (Chang *et al.*, 2011) (see Fig. 7.2).

Not all Gram-positive pili contain a pilus base like SpaB of *C. diphtheriae*. Sortase-catalysed pili of *Actinomyces oris* and *B. cereus* are heterodimeric – a tip pilin and a pilus shaft (Budzik *et al.*, 2007; Mishra *et al.*, 2007). In these cases, the last subunit of the pilus shaft may function as the pilus base; in fact, by biochemical methods Budzik and colleagues showed in *B. cereus* that the major pilin protein BcpA is directly anchored to the cell wall, a process that is catalysed by the housekeeping sortase named SrtA (Budzik *et al.*, 2008). This raises an interesting conundrum: what prevents a shaft pilin from becoming a stop signal or a pilus base in a heterotrimeric pilus like the SpaA of *C. diphtheriae*? Looking closer at the *srtF* deletion mutant of *C. diphtheriae*, one would notice that few SpaA polymers were still anchored to the cell wall; however, these polymers did not contain any SpaB (Mandlik *et al.*, 2008a). It can be interpreted that the pilin-specific sortase SrtA was able to anchor the SpaA polymers to the cell wall, although this process was not efficient. A previous quantitative study with a mutant expressing only SrtA also confirmed this phenomenon (Swaminathan *et al.*, 2007). Thus, one can speculate that in a heterotrimeric pilus, incorporation of a pilin like SpaB of *C. diphtheriae* into the pilus base – by virtue of its substrate preference by the housekeeping sortase – terminates pilus polymerization, switching to cell-wall anchoring. When SpaB enters the pilus base is still an intriguing question, worthy of future investigation.

7.4.2 Pilusosome: the pilus assembly centre

As all sortases and pilin precursors contain a signal peptide, it is anticipated that the pilus machinery is closely associated with the secretion machinery, e.g. SecA. Guttilla and colleagues demonstrated by immuno-electron microscopy with corynebacterial thin sections that SecA is in close proximity with SpaA and SrtA; sortase SrtA and its cognate pilins were also found in close proximity (Guttilla *et al.*, 2009). Consistently, in *E. faecalis* the housekeeping sortase SrtA and pilin-specific sortase SrtC were found to co-localize with SecA at single foci (Kline *et al.*, 2009). Given the biphasic mode of pilus assembly, the existence of a pilus assembly centre or pilusosome for efficient pilus polymerization and cell-wall anchoring by tandem sortase enzymes has been speculated (Guttilla *et al.*, 2009).

It is surprising, however, that a SpaB mutant lacking its CWSS for membrane-retention is able to incorporate into the secreted pilus polymer (Mandlik *et al.*, 2008a). A simple explanation is the close proximity of sortase enzymes, their cognate substrates and the secretion machinery, as if they are part of an assembly line or pilus assembly and pilin secretion are coupled. It remains to be seen if a pilusosome can be isolated.

7.5 Three-dimensional Structural Insights into Pilus Assembly in *C. diphtheriae*

Structural biology has been instrumental to elucidate the assembly mechanisms of Gram-negative pili (Allen *et al.*, 2012), and it is equally important for Gram-positive counter-parts. A three-dimensional (3D) structure of Gram-positive pilins was first solved in *S. agalactiae* with the basal pilin GBS52 (Krishnan *et al.*, 2007). It was shortly followed by the crystal structure of the shaft pilin Spy0128 of Group A *Streptococcus* (GAS) pili (Kang *et al.*, 2007). Two common features found in the two structures are IgG-like fold and lysine-asparagine isopeptide bonds. Although not much sequence similarity or homology between them is observed, remarkably, the crystal structure of the *C. diphtheriae* major pilin SpaA solved in 2009 contains these conserved features (Kang *et al.*, 2009). SpaA is composed of three tandem Ig-like domains in which the middle (M) domain and the C-terminal (C) domain are arranged linearly connected by an extended strand (Fig. 7.3A). The N-terminal (N) domain sits on the M domain at an angle of ~20° to the long axis of the molecule.

Furthermore, the SpaA molecules appeared to pack in columns within the crystal where the N-domain of each molecule was in close contact with the C-domain of the next. The N- and C-domains of SpaA have an inverse IgG-fold, which was first described

for the CnaB domains of the *S. aureus* collagen binding protein Cna (Deivanayagam *et al.*, 2000). In general, these Ig-like folds are composed of a β-sandwich of seven strands held together by loops. The middle domain of SpaA has a CnaA fold, first observed in the N2 domain of *S. aureus* CnaA consisting of a β-barrel structure made up of nine β-strands. The pilin motif of SpaA is found to be located on the last strand of the N-domain, with the conserved Lys190 near the interface between the N- and M-domains pointing toward the C-terminus of the next SpaA molecule. Importantly, mass spectrometry analysis on native SpaA pili confirmed the intermolecular isopeptide bond formed between the pilin motif Lys190 and the LPXTG threonine residue of the next SpaA molecule as proposed (Fig. 7.2).

In addition to the Ig-like folds found in three domains of SpaA, another common feature discovered from the crystallographic study of SpaA is the presence of internal isopeptide bonds formed between Lys and Asn residues within the M- and the C-domains (Fig. 7.3B,C) (Kang *et al.*, 2009). As mentioned above, similar Lys–Asn bonds were first observed in the crystal structure of *S. pyogenes* Spy0128 and subsequently identified in many other pilin subunits as well as other MSCRAMMS, and these bonds have been shown to confer thermodynamic, proteolytic and mechanical stability to shaft pilins (Kang and Baker, 2011, 2012). Formation of these bonds in SpaA is proposed to be the result of a lower pKa of the Lys residue in the hydrophobic environment, which enables a nucleophilic attack from its unprotonated amino group on the carboxyamide group (Kang *et al.*, 2009). Interestingly, the carboxyl group of the conserved Glu446 residue within the SpaA E-box sequence serves as the proton shuttle for the isopeptide bond in the C-domain (Fig. 7.3C). Recall the defective phenotypes caused by E446A mutation (Ton-That *et al.*, 2004b). Conceivably, since the E-box resides upstream of the LPXTG motif, E446A mutation destabilizes the SpaA pilin, thus affecting SpaA polymerization and SpaB incorporation. Consistent with this, in *S. agalactiae* mutation of the conserved Glu

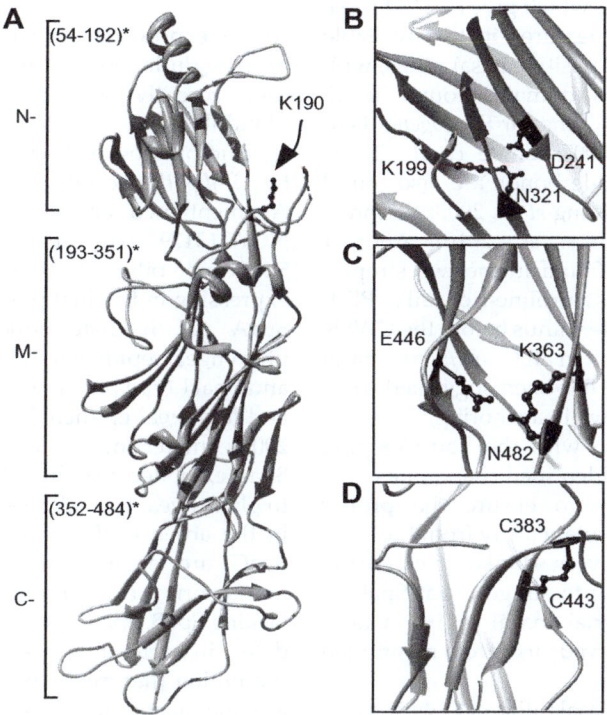

Fig. 7.3. Crystal structure of SpaA. (A) SpaA is made of three tandem Ig-like domains termed N-terminal (N-), middle (M-) and C-terminal (C-). The N-domain resembles the fold of *S. aureus* CnaA, while the M- and C- domains have the staphylococcal CnaB fold. Relative positions of each domain are labelled. The pilin motif lysine residue K190 is shown in balls and sticks. (B) In the M-domain, an isopeptide bond is formed between K199 and N321, with the catalytic residue D241. (C) The C-domain isopeptide bond is formed between K363 and N482, with the catalytic residue E446 (E-box). (D) Also found in the C-domain is the disulfide bond formed between C383 and C443. (Adapted with permission from PNAS, Kang *et al.*, 2009)

residue within the E-box of the shaft protein BP-2a severely affected the folding of BP-2a, regardless of isopeptide bond formation (Cozzi *et al.*, 2012).

It is intriguing to note that the C-domain of SpaA contains a disulfide bond formed between Cys383 and Cys443 (Fig. 7.3D), predictably contributing to pilin stabilization (Kang *et al.*, 2009). As no disulfide bonds have been observed in other Gram-positive pilin structures solved to date, except for FimA and FimP of actinobacterium *A. oris* (Mishra *et al.*, 2011; Persson *et al.*, 2012), disulfide bonds may be a unique feature of actinobacterial pilins. Additionally, all three SpaA, FimA and FimP structures hold a divalent cation. In the SpaA structure, a Ca^{2+}

binding site is located at the M-domain, formed by the loop linking strands H and I (Kang *et al.*, 2009), whereas five residues of a loop in the C-domain of FimP coordinate a Ca^{2+} binding site in *A. oris* (Persson *et al.*, 2012), and a Zn^{2+} binding site is found in the middle domain of FimA (Mishra *et al.*, 2011). Metal binding sites may be another stabilizing feature of these pilins.

7.5.1 Structural features of minor pilins in *C. diphtheriae*

While 3D-structures of *C. diphtheriae* basal pilins have not been solved, domain analysis of all corynebacterial basal pilins reveals the presence of at least one CnaB domain (Fig.

7.1). CnaB domains, which give rise to Ig-like fold, are common features in the available structures of Gram-positive basal pilins, with two tandem CnaB domains found in *S. agalactiae* GBS52 and one for *S. pyogenes* FctB (Krishnan *et al.*, 2007; Linke *et al.*, 2010). Obviously, isopeptide bonds are also found in these domains (Kang *et al.*, 2007; Krishnan *et al.*, 2007; Linke *et al.*, 2010). Another conserved feature found in the two strepto-coccal basal pilins is a proline-rich tail (a PPII-like helix) at the C-terminus before the CWSS, which is also present in all *C. diphtheriae* basal pilins. This feature has been suggested to be involved in cell-wall anchoring, perhaps through interaction with the housekeeping sortase or some unidentified chaperones, or it might likely serve to ensure the proper protrusion of the protein away from the thick cell-wall peptidoglycan (Linke *et al.*, 2010). Thus, it is tempting to suggest that SpaB or other corynebacterial basal pilins would contain all conserved features mentioned above.

Similar to the basal pilins of *C. diphtheriae*, the tip pilins, whose 3D-structures are not available, are predicted to contain one or more CnaB-like domains (Fig. 7.1). While SpaC and SpaF possess a CnaB domain near the C-terminus, the tip pilin SpaG astonishingly harbours six CnaB domains. Currently available are the structures of two tip pilins RrgA of *S. pneumoniae* and Spy0125 (Cpa) of *S. pyogenes*, each of which is made of four tandem domains (Izore *et al.*, 2010; Pointon *et al.*, 2010). The D1 and D4 domains of RrgA have IgG-like fold, whereas the D2 exhibits a CnaB-like domain and the D3 has a similar fold of the human A3 domain of von Willebrand factor (VWA) (Izore *et al.*, 2010). Of note, SpaC and SpaF are predicted to have a VWA domain. In the Spy0125 structure, the CnaB-like fold is also detected in the D3 and D2 domains (Pointon *et al.*, 2010). The D3 domain has been implicated in bacterial adhesion (Smith *et al.*, 2010).

7.6 Functions of *C. diphtheriae* Pili

Adhesion has been the only known attribute of *C. diphtheriae* pili thus far. More than two

decades ago, it was shown that *C. diphtheriae* strains expressing pili were associated with haemagglutination (Ermolayev *et al.*, 1987). More recently, using *ex-vivo* tissue cultures Mandlik and colleagues demonstrated specific binding or tissue tropism mediated by *C. diphtheriae* pili (Mandlik *et al.*, 2007). As mentioned above, *C. diphtheriae* strain NCTC13129 expresses SpaA-, SpaD- and SpaH-type pili. With isogenic mutant strains expressing individual pili, it was shown that SpaA pili mediate adherence to human pharyngeal epithelial cells, whereas SpaD- and SpaH-type pili are more specific for lung and laryngeal epithelial cells (Mandlik *et al.*, 2007). Interestingly, a mutant strain lacking SpaA exhibited a slight reduction in adhesion to pharyngeal epithelial cells; it turns out that in the absence of the shaft pilin, SpaB and SpaC are anchored to the cell wall in monomeric forms. In contrast, mutant strains lacking SpaB, SpaC or both displayed a severe defect in bacterial adhesion. Consistent with the notion that minor pilins SpaB and SpaB are the major adhesins of SpaA-type pili, antibodies against either SpaB or SpaC blocked corynebacterial adhesion. Sig-nificantly, latex beads conjugated with a SpaC fragment containing the VWA domain or a SpaB fragment encompassing a CnaB domain were shown to adhere to pharyngeal epithelial cells (Mandlik *et al.*, 2007). With the presence of the adhesive pilins SpaB and SpaC in the forms of pilus, monomers and heterodimers, it has been suggested that this dual adherence nature of SpaB and SpaC allows for initiation attachment by the long pili, followed by a more intimate adhesion that aids in the efficient delivery of toxin and other virulence factors (Rogers *et al.*, 2011).

7.7 Conclusions and Future Outlooks

Work over the past decade on *C. diphtheriae* pili has established the mechanism of sortase-mediated pilus assembly. We now know that the biphasic mode of assembly, as illustrated by studies of corynebacterial SpaA pili, is a common feature in Gram-positive pili; however, less known is how pilus poly-merization is terminated. Equally puzzling is the 'hand-off' mechanism of sortase enzymes, whereby a pilus polymer is transferred from a pilin-specific sortase to a non-polymerizing sortase for cell-wall anchoring. With the availability of a genetic tool box, biochemical methods and other microscopic technologies, *C. diphtheriae* continues to be an experimental model for pilus assembly. Also important is how pili contribute to corynebacterial pathogenesis, an area that is not well understood. Given the emergence of non-toxigenic strains of *C. diphtheriae* causing infections, it is crucial to continue elucidating the assembly mechanism of pili and to dissect their role in pathogen–host interactions.

Acknowledgements

We thank former and current lab members for their invaluable contributions to the pilus work, which was supported by grants AI061381 and DE017382 from the NIH to HTT.

References

Allen, W.J., Phan, G. and Waksman, G. (2012) Pilus biogenesis at the outer membrane of Gram-negative bacterial pathogens. *Current Opinion in Structural Biology* 22, 500–506.

Budzik, J.M., Marraffini, L.A. and Schneewind, O. (2007) Assembly of pili on the surface of *Bacillus cereus* vegetative cells. *Molecular Microbiology* 66, 495–510.

Budzik, J.M., Oh, S.Y. and Schneewind, O. (2008) Cell wall anchor structure of BcpA pili in *Bacillus anthracis*. *Journal of Biological Chemistry* 283, 36676–36686.

Budzik, J.M., Oh, S.Y. and Schneewind, O. (2009) Sortase D forms the covalent bond that links BcpB to the tip of *Bacillus cereus* pili. *Journal of Biological Chemistry* 284, 12989–12997.

Cerdeno-Tarraga, A.M., Efstratiou, A., Dover, L.G., Holden, M.T., Pallen, M., Bentley, S.D., Besra, G.S., Churcher, C., James, K.D., De Zoysa, A., Chillingworth, T., Cronin, A., Dowd, L., Feltwell, T., Hamlin, N., Holroyd, S., Jagels, K., Moule, S., Quail, M.A., Rabbinowitsch, E., Rutherford, K.M., Thomson, N.R., Unwin, L., Whitehead, S., Barrell, B.G. and Parkhill, J. (2003) The complete genome sequence and analysis of

Corynebacterium diphtheriae NCTC13129. *Nucleic Acids Research* 31, 6516–6523.

Chang, C., Mandlik, A., Das, A. and Ton-That, H. (2011) Cell surface display of minor pilin adhesins in the form of a simple heterodimeric assembly in *Corynebacterium diphtheriae*. *Molecular Microbiology* 79, 1236–1247.

Clancy, K.W., Melvin, J.A. and Mccafferty, D.G. (2010) Sortase transpeptidases: insights into mechanism, substrate specificity, and inhibition. *Biopolymers* 94, 385–396.

Comfort, D. and Clubb, R.T. (2004) A comparative genome analysis identifies distinct sorting pathways in gram-positive bacteria. *Infection and Immunity* 72, 2710–2722.

Cozzi, R., Nuccitelli, A., D'Onofrio, M., Necchi, F., Rosini, R., Zerbini, F., Biagini, M., Norais, N., Beier, C., Telford, J.L., Grandi, G., Assfalg, M., Zacharias, M., Maione, D. and Rinaudo, C.D. (2012) New insights into the role of the glutamic acid of the E-box motif in group B *Streptococcus* pilus 2a assembly. *FASEB Journal* 26, 2008–2018.

Deivanayagam, C.C., Rich, R.L., Carson, M., Owens, R.T., Danthuluri, S., Bice, T., Hook, M. and Narayana, S.V. (2000) Novel fold and assembly of the repetitive B region of the *Staphylococcus aureus* collagen-binding surface protein. *Structure* 8, 67–78.

Dramsi, S., Trieu-Cuot, P. and Bierne, H. (2005) Sorting sortases: a nomenclature proposal for the various sortases of Gram-positive bacteria. *Research in Microbiology* 156, 289–297.

Ermolayev, A.V., Fish, N.G., Birger, M.O., Sanzhakova, I.E. and Lobanova, A.N. (1987) A study on the adhesive properties of *Cor. diphtheriae* and *Cor. parvum* in the direct haemagglutination reaction. *Journal of Hygiene, Epidemiology, Microbiology and Immunology* 31, 313–319.

Gaspar, A.H. and Ton-That, H. (2006) Assembly of distinct pilus structures on the surface of *Corynebacterium diphtheriae*. *Journal of Bacteriology* 188, 1526–1533.

Guttilla, I.K., Gaspar, A.H., Swierczynski, A., Swaminathan, A., Dwivedi, P., Das, A. and Ton-That, H.(2009) Acyl enzyme intermediates in sortase-catalyzed pilus morphogenesis in gram-positive bacteria. *Journal of Bacteriology* 191, 5603–5612.

Hendrickx, A.P., Budzik, J.M., Oh, S.Y. and Schneewind, O. (2011) Architects at the bacterial surface – sortases and the assembly of pili with isopeptide bonds. *Nature Reviews Microbiology* 9, 166–176.

Holmes, R.K. and Barksdale, L. (1969) Genetic analysis of tox+ and tox– bacteriophages of

Corynebacterium diphtheriae. Journal of Virology 3, 586–598.

Izore, T., Contreras-Martel, C., El Mortaji, L., Manzano, C., Terrasse, R., Vernet, T., Di Guilmi, A.M. and Dessen, A. (2010) Structural basis of host cell recognition by the pilus adhesin from *Streptococcus pneumoniae. Structure* 18, 106–115.

Kang, H.J. and Baker, E.N. (2011) Intramolecular isopeptide bonds: protein crosslinks built for stress? *Trends in Biochemical Sciences* 36, 229–237.

Kang, H.J. and Baker, E.N. (2012) Structure and assembly of Gram-positive bacterial pili: unique covalent polymers. *Current Opinion in Structural Biology* 22, 200–207.

Kang, H.J., Coulibaly, F., Clow, F., Proft, T. and Baker, E.N. (2007) Stabilizing isopeptide bonds revealed in gram-positive bacterial pilus structure. *Science* 318, 1625–1628.

Kang, H.J., Paterson, N.G., Gaspar, A.H., Ton-That, H. and Baker, E.N. (2009) The *Corynebacterium diphtheriae* shaft pilin SpaA is built of tandem Ig-like modules with stabilizing isopeptide and disulfide bonds. *Proceedings of the National Academy of Sciences of the USA* 106, 16967–16971.

Kline, K.A., Kau, A.L., Chen, S.L., Lim, A., Pinkner, J.S., Rosch, J., Nallapareddy, S.R., Murray, B.E., Henriques-Normark, B., Beatty, W., Caparon, M.G. and Hultgren, S.J. (2009) Mechanism for sortase localization and the role of sortase localization in efficient pilus assembly in *Enterococcus faecalis. Journal of Bacteriology* 191, 3237–3247.

Krishnan, V., Gaspar, A.H., Ye, N., Mandlik, A., Ton-That, H. and Narayana, S.V. (2007) An IgG-like domain in the minor pilin GBS52 of *Streptococcus agalactiae* mediates lung epithelial cell adhesion. *Structure* 15, 893–903.

Linke, C., Young, P.G., Kang, H.J., Bunker, R.D., Middleditch, M.J., Caradoc-Davies, T.T., Proft, T. and Baker, E.N. (2010) Crystal structure of the minor pilin FctB reveals determinants of Group A streptococcal pilus anchoring. *Journal of Biological Chemistry* 285, 20381–20389.

Mandlik, A., Das, A. and Ton-That, H. (2007) *Corynebacterium diphtheriae* employs specific minor pilins to target human pharyngeal epithelial cells. *Molecular Microbiology* 64, 111–124.

Mandlik, A., Swierczynski, A., Das, A. and Ton-That, H. (2008a) The molecular switch that activates the cell wall anchoring step of pilus assembly in gram-positive bacteria. *Proceedings of the National Academy of Sciences of the USA* 105, 14147–14152.

Mandlik, A., Swierczynski, A., Das, A. and Ton-That, H. (2008b) Pili in Gram-positive bacteria: assembly, involvement in colonization and biofilm development. *Trends in Microbiology* 16, 33–40.

Mazmanian, S.K., Liu, G., Ton-That, H. and Schneewind, O. (1999) *Staphylococcus aureus* sortase, an enzyme that anchors surface proteins to the cell wall. *Science* 285, 760–763.

Mazmanian, S.K., Ton-That, H., Su, K. and Schneewind, O. (2002) An iron-regulated sortase anchors a class of surface protein during *Staphylococcus aureus* pathogenesis. *Proceedings of the National Academy of Sciences of the USA* 99, 2293–2298.

Mishra, A., Das, A., Cisar, J.O. and Ton-That, H. (2007) Sortase-catalyzed assembly of distinct heteromeric fimbriae in *Actinomyces naeslundii. Journal of Bacteriology* 189, 3156–3165.

Mishra, A., Devarajan, B., Reardon, M.E., Dwivedi, P., Krishnan, V., Cisar, J.O., Das, A., Narayana, S.V. and Ton-That, H. (2011) Two autonomous structural modules in the fimbrial shaft adhesin FimA mediate *Actinomyces* interactions with streptococci and host cells during oral biofilm development. *Molecular Microbiology* 81, 1205–1220.

Navarre, W.W. and Schneewind, O. (1999) Surface proteins of gram-positive bacteria and mechanisms of their targeting to the cell wall envelope. *Microbiology and Molecular Biology Reviews* 63, 174–229.

Nobbs, A.H., Rosini, R., Rinaudo, C.D., Maione, D., Grandi, G. and Telford, J.L. (2008) Sortase A utilizes an ancillary protein anchor for efficient cell wall anchoring of pili in *Streptococcus agalactiae. Infection and Immunity* 76, 3550–3560.

Persson, K., Esberg, A., Claesson, R. and Stromberg, N. (2012) The pilin protein FimP from *Actinomyces oris*: crystal structure and sequence analyses. *PLoS One* 7, e48364.

Pointon, J.A., Smith, W.D., Saalbach, G., Crow, A., Kehoe, M.A. and Banfield, M.J. (2010) A highly unusual thioester bond in a pilus adhesin is required for efficient host cell interaction. *Journal of Biological Chemistry* 285, 33858–33866.

Rogers, E.A., Das, A. and Ton-That, H. (2011) Adhesion by pathogenic corynebacteria. *Advances in Experimental Medicine and Biology* 715, 91–103.

Schneewind, O., Model, P. and Fischetti, V.A. (1992) Sorting of protein A to the staphylococcal cell wall. *Cell* 70, 267–281.

Scott, J.R. and Zahner, D. (2006) Pili with strong attachments: Gram-positive bacteria do it differently. *Molecular Microbiology* 62, 320–330.

Sjoquist, J., Meloun, B. and Hjelm, H. (1972) Protein A isolated from *Staphylococcus aureus* after digestion with lysostaphin. *European Journal of Biochemistry* 29, 572–578.

Smith, W.D., Pointon, J.A., Abbot, E., Kang, H.J., Baker, E.N., Hirst, B.H., Wilson, J.A., Banfield, M.J. and Kehoe, M.A. (2010) Roles of minor pilin subunits Spy0125 and Spy0130 in the serotype M1 *Streptococcus pyogenes* strain SF370. *Journal of Bacteriology* 192, 4651–4659.

Spirig, T., Weiner, E.M. and Clubb, R.T. (2011) Sortase enzymes in Gram-positive bacteria. *Molecular Microbiology* 82, 1044–1059.

Swaminathan, A., Mandlik, A., Swierczynski, A., Gaspar, A., Das, A. and Ton-That, H. (2007) Housekeeping sortase facilitates the cell wall anchoring of pilus polymers in *Corynebacterium diphtheriae*. *Molecular Microbiology* 66, 961–974.

Swierczynski, A. and Ton-That, H. (2006) Type III pilus of corynebacteria: pilus length is determined by the level of its major pilin subunit. *Journal of Bacteriology* 188, 6318–6325.

Telford, J.L., Barocchi, M.A., Margarit, I., Rappuoli, R. and Grandi, G. (2006) Pili in Gram-positive pathogens. *Nature Reviews Microbiology* 4, 509–519.

Ton-That, H. and Schneewind, O. (2003) Assembly of pili on the surface of *Corynebacterium diphtheriae*. *Molecular Microbiology* 50, 1429–1438.

Ton-That, H. and Schneewind, O. (2004) Assembly of pili in Gram-positive bacteria. *Trends in Microbiology* 12, 228–234.

Ton-That, H., Liu, G., Mazmanian, S.K., Faull, K.F. and Schneewind, O. (1999) Purification and characterization of sortase, the transpeptidase that cleaves surface proteins of *Staphylococcus aureus* at the LPXTG motif. *Proceedings of the National Academy of Sciences of the USA* 96, 12424–12429.

Ton-That, H., Mazmanian, S.K., Faull, K.F. and Schneewind, O. (2000) Anchoring of surface proteins to the cell wall of *Staphylococcus aureus*. Sortase catalyzed *in vitro* transpeptidation reaction using LPXTG peptide and NH(2)-Gly(3) substrates. *Journal of Biological Chemistry* 275, 9876–9881.

Ton-That, H., Mazmanian, S.K., Alksne, L. and Schneewind, O. (2002) Anchoring of surface proteins to the cell wall of *Staphylococcus aureus*. Cysteine 184 and histidine 120 of sortase form a thiolate-imidazolium ion pair for catalysis. *Journal of Biological Chemistry* 277, 7447–7452.

Ton-That, H., Marraffini, L.A. and Schneewind, O. (2004a) Protein sorting to the cell wall envelope of Gram-positive bacteria. *Biochimica et Biophysica Acta* 1694, 269–278.

Ton-That, H., Marraffini, L.A. and Schneewind, O. (2004b) Sortases and pilin elements involved in pilus assembly of *Corynebacterium diphtheriae*. *Molecular Microbiology* 53, 251–261.

Trost, E., Blom, J., De Castro Soares, S., Huang, I.H., Al-Dilaimi, A., Schroder, J., Jaenicke, S., Dorella, F.A., Rocha, F.S., Miyoshi, A., Azevedo, V., Schneider, M.P., Silva, A., Camello, T.C., Sabbadini, P.S., Santos, C.S., Santos, L.S., Hirata, R. Jr, Mattos-Guaraldi, A.L., Efstratiou, A., Schmitt, M.P., Ton-That, H. and Tauch, A. (2012) Pangenomic study of *Corynebacterium diphtheriae* that provides insights into the genomic diversity of pathogenic isolates from cases of classical diphtheria, endocarditis, and pneumonia. *Journal of Bacteriology* 194, 3199–3215.

Yanagawa, R. and Honda, E. (1976) Presence of pili in species of human and animal parasites and pathogens of the genus corynebacterium. *Infection and Immunity* 13, 1293–1295.

Yanagawa, R., Otsuki, K. and Tokui, T. (1968) Electron microscopy of fine structure of *Corynebacterium renale* with special reference to pili. *Japanese Journal of Veterinary Research* 16, 31–37.

Yeung, M.K. (1999) Molecular and genetic analyses of *Actinomyces* spp. *Critical Reviews in Oral Biology and Medicine* 10, 120–138.

8 Three-dimensional Structures of Pilin Subunits and their Role in Gram-positive Pilus Assembly and Stability

Neil G. Paterson[1] and Edward N. Baker[2]

[1]Diamond Light Source Ltd, UK; [2]School of Biological Sciences, University of Auckland, New Zealand

8.1 Introduction

The surfaces of Gram-positive bacteria are decorated with a wide range of virulence factors that modulate interaction with the host environment and promote adhesion, colonization, infection and invasion. Of these, the assemblies known as pili are arguably the most intriguing in terms of elegant structure and functionality.

Bacterial pili are proteinaceous filaments that extend from the cell surface where they mediate a range of cellular activities including host tissue recognition/adhesion, motility and genetic transfer (Proft and Baker, 2009). These functionally diverse fibres have been extensively studied and characterized in Gram-negative bacteria, where they are built from helical bundles of non-covalently associated protein subunits (Craig et al., 2004; Waksman and Hultgren, 2009). In contrast, little was known about the molecular nature or assembly of the pili expressed by Gram-positive bacteria until the past 10 years, following the discovery of sortases and the availability of bacterial genome sequences. The discovery of small gene clusters, encoding sortases and sortase substrates, enabled the production of recombinant proteins and specific antibodies. Immunoelectron micros-copy studies then led to the visualization of

pili on Gram-positive organisms and gene knockouts revealed the roles of their component proteins (e.g. Ton-That and Schneewind, 2003; Mora et al., 2005; Barocchi et al., 2006).

Subsequent studies have shown that Gram-positive bacterial pili are not tubular, as are the pili of Gram-negative bacteria, and are not involved in genetic transfer or motility. Rather, they comprise chains of covalently linked pilin subunits, resulting in long (3–5 μm) and very thin (~3 nm) fibres (Hilleringmann et al., 2009; Kang and Baker, 2012), whose primary role is adhesion, both in colonization and as an aid to biofilm formation (Konto-Ghiorghi et al., 2009). In this sense, these structures more closely con-form to the definition of fimbriae, although the term pili has been widely adopted.

Here we focus on insights that have come from structural studies of the individual pilin subunits, primarily crystallographic analyses, but complemented by electron microscopy (EM) studies of the fibres themselves. These studies have converged on a common model in which Gram-positive pili comprise a series of covalently-linked sub-units arranged in tandem like beads on a string. They also reveal totally unexpected stabilizing factors, in the form of intra-molecular isopeptide bond crosslinks that help to explain the robustness of these fascinating biological polymers.

8.2 Modular Nature of Pilus Structure and Assembly

Gram-positive bacterial pili usually comprise 2 or 3 components (Oh *et al.*, 2008); a major pilin that forms the polymeric pilus shaft, an adhesive minor pilin situated at the tip and often, but not always, a second minor pilin located at the base. In this terminology, minor and major refer to the relative abundance of each subunit in the overall structure rather than their respective sizes, since the adhesive domains tend to be significantly larger than the other components (Ton-That and Schneewind, 2003).

8.2.1 Pilus assembly by sequential addition of pilin subunits

The process of Gram-positive pilus assembly has been well described in several reviews (Scott and Zahner, 2006; Telford *et al.*, 2006; Hendrickx *et al.*, 2011). Assembly occurs on the cell surface, orchestrated by membrane-anchored cysteine transpeptidase enzymes called sortases (Marraffini *et al.*, 2006). As each pilin subunit emerges from the cell membrane, a pilus-specific sortase recognizes an LPXTG-like motif near the pilin C-terminus, cleaves the polypeptide between threonine and glycine, and ligates the newly-formed threonine carboxyl group to the e-amino group of a lysine residue on an incoming pilin subunit. We refer to this lysine herein as the linking lysine. Pilus assembly thus involves addition of pilin subunits from the bottom, like beads on a string, with successive subunits linked by covalent isopeptide (amide) bonds (Ton-That and Schneewind, 2004; Spirig *et al.*, 2011). The result is a unique covalent polymer.

Pilus formation evidently begins with the adhesin, which is destined to be located at the tip. This is ligated to a major pilin subunit, and the construct is then extended out from the cell surface as additional copies of major pilin are incorporated from the bottom. For 3-component pili, the addition of a basal minor pilin terminates pilus extension (Mandlik *et al.*, 2008), the assembled pilus then being covalently linked by a house-keeping sortase to the cell wall. For 2-component pili, as in *Bacillus cereus*, a major pilin is instead directly ligated to the cell wall by the housekeeping sortase (Budzik *et al.*, 2008a).

Several aspects of pilus assembly remain to be understood. In at least one organism, *Streptococcus pyogenes*, another protein SipA is required for assembly (Zahner and Scott, 2008), but what its role is and whether this implies some kind of assembly complex on the cell membrane is not known. It is conceivable that other organisms may require additional factors, but none have yet been identified. Also unknown are how the 'correct' order of subunits is orchestrated, what determines the length of the pilus and whether assembly is a regulated process or simply stochastic.

8.2.2 Pilin subunits have modular structures

The paradigm of a modular pilus is reflected in the structures of individual pilin subunits. All the Gram-positive pilins structurally characterized to date are built from domains that are topologically similar to immunoglobulin (Ig) domains and are usually arranged in tandem. These Ig-like domains can be classified into two structurally similar types: CnaA-like and CnaB-like (Fig. 8.1). These domains take their names from the A and B domains found in the collagen-binding adhesin Cna from *Staphylococcus aureus* (Patti *et al.*, 1993) and are widely used in other cell-surface proteins. The archetypal CnaA fold comprises 9 antiparallel β-strands with a jelly roll-like topology, forming a 4-stranded and a 5-stranded β-sheet, packed against each other in a β-sandwich (Symersky *et al.*, 1997; Zong *et al.*, 2005). The CnaB fold comprises 7 antiparallel β-strands, forming 3- and 4-stranded β-sheets, also packed as a β-sandwich (Deivanayagam *et al.*, 2000). Despite their common folds, however, the CnaA-type and CnaB-type domains used in

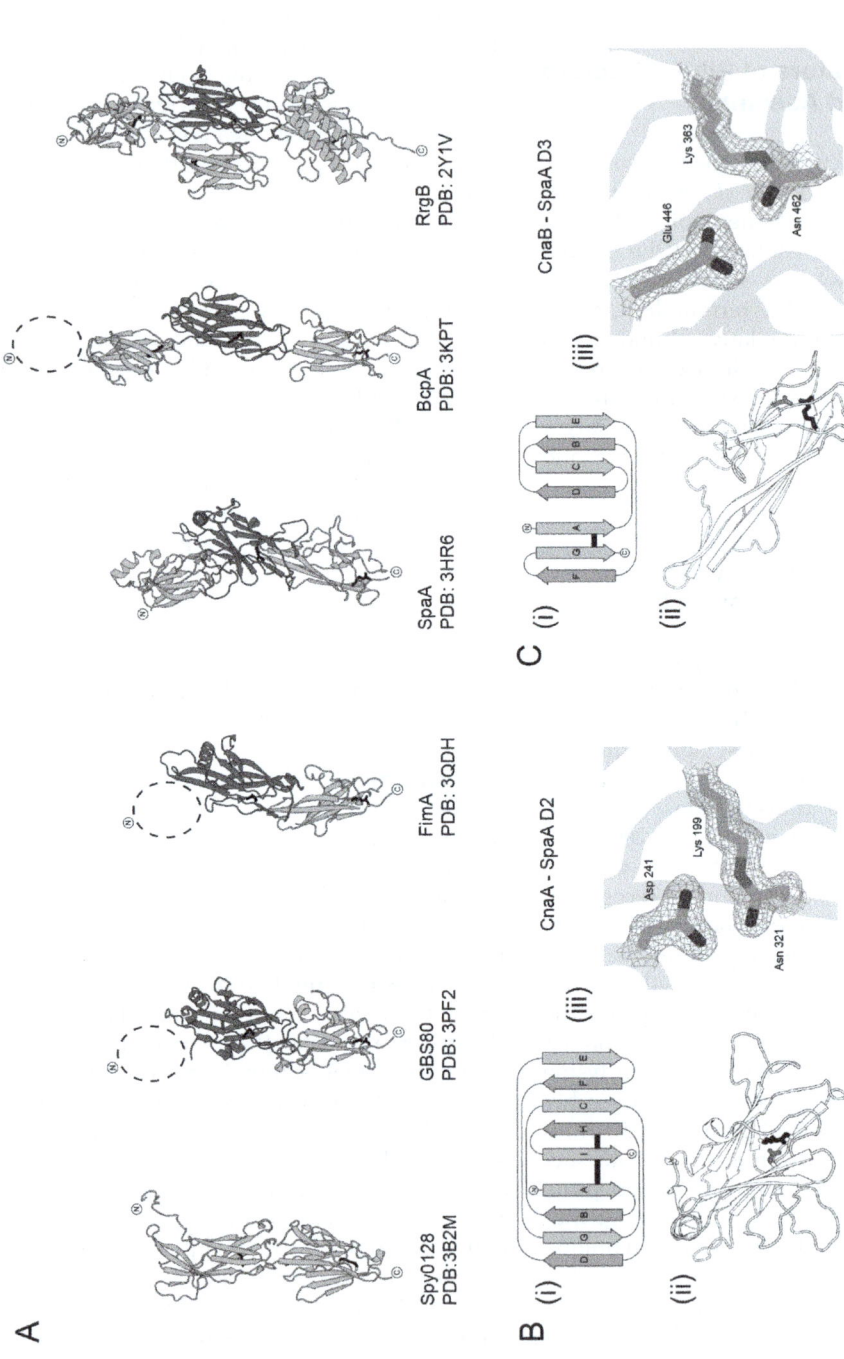

Fig. 8.1. Pilin domains and major pilin structures. (A) Cartoon diagrams and PDB codes for currently available major pilin structures. CnaA-like domains are shaded dark grey and CnaB-like domains are light grey. Isopeptide bond crosslinks are shown as black sticks. Dashed ovals indicate domains missing from the crystal structures, with N-termini indicated. (B) and (C) show typical CnaA-like and CnaB-like domains respectively, showing (i) the domain topology, with isopeptide crosslinks indicated by a black bar, (ii) representative examples from the *C diphtheriae* SpaA major pilin, with isopeptide bonds highlighted by black sticks and catalytic acidic residues in each case with associated electron sticks and catalytic acidic residues in grey and (iii) close-up views of the isopeptide bond and catalytic acidic residue in each case with associated electron density (2Fo–Fc contoured at 1.5 σ).

pilins share low sequence similarity and show many surface variations, probably reflecting host immune pressure. This makes sequence alignments and sequence-based structure predictions highly problematical and emphasizes the need for experimental structural data.

8.2.3 Intramolecular isopeptide bonding

An iconic feature of the Ig-like pilin domains is the consistent presence of intramolecular isopeptide bonds that form stabilizing internal covalent cross-links (Kang et al., 2007; Kang and Baker, 2011). Although the enzyme-catalysed formation of intermolecular iso-peptide bonds was well known, for example in ubiquitylation (Pickart, 2001) and in pilus assembly, the autocatalytic formation of intra-molecular isopeptide bonds was unknown until discovered in the Gram-positive pilins, except for one instance in the capsid of the bacteriophage HK97 (Wikoff et al., 2000). These bonds link a lysine side chain to an asparagine (or occasionally aspartic acid) side chain. Bond formation requires a nearby acidic residue, Asp or Glu, and evidently occurs as a single-turnover reaction that occurs when the three residues are brought together in the protein interior on folding. The bonds are clearly visible in crystal structures (Fig. 8.1) and can be confirmed by mass spectrometry (Kang et al., 2009a). Since their initial discovery in pilins they have been found in a number of other Gram-positive cell-surface adhesins (reviewed in Kang and Baker, 2011) and may be an evolutionary alternative to disulfide bonds.

The proposed reaction mechanism for isopeptide bond formation (Hagan et al., 2010; Hu et al., 2011; Kang and Baker, 2011) involves nucleophilic attack by the lysine ε-amino group on the γ-carbon of Asn (or Asp) with subsequent elimination of ammonia (or water). The associated acidic residue acts as a proton shuttle, and the reaction is promoted by modification of the pKas of the reacting residues by the surrounding hydrophobic environment; the lysine ε-amino group must be non-protonated for the reaction to occur.

In pilin domains, these intramolecular bonds are found at characteristic sites; in CnaA domains, they typically crosslink the first and second-last strands, forming a bridge between the two sheets of the β-sandwich (Fig. 8.1B), whereas in CnaB domains, they crosslink the first and last β-strands (Fig. 8.1C). These bonds are strategically placed to provide resistance to mechanical stress in the long, thin pilus poly-mers. They are also important for structural stability; loss of an isopeptide bond in the *B. cereus* major pilin by mutation of the catalytic Glu abolishes fibre formation (Budzik et al., 2008b). Likewise a Glu residue in the *Corynebacterium diphtheriae* major pilin SpaA, essential for minor pilin incorporation (Ton-That et al., 2004), is the catalytic residue for an internal isopeptide bond (Kang et al., 2009b).

8.3 Major Pilins Have Modular Folds

Crystal structures of major pilins give a graphic view of the modular nature of pilus structure (Fig. 8.1A). Currently, structures are available for three full-length major pilins: Spy0128 from *S. pyogenes* (Kang et al., 2007), SpaA from *C. diphtheriae* (Kang et al., 2009b) and RrgB from *S. pneumoniae* (Spraggon et al., 2010; Paterson and Baker, 2011; El Mortaji et al., 2012a). Crystal structures have also been determined for three truncated major pilins, lacking their N-terminal domains: GBS80 from *S. agalactiae* (Vengadesan et al., 2011), FimA from *Actinomyces oris* (Mishra et al., 2011) and BcpA from *B. cereus* (Budzik et al., 2009). These structures, all built of tandem CnaB and CnaA domains, range in size from two to four domains and are typically 100–120 Å in length and 30–40 Å in width. Domains share conserved core topology but with many surface variations.

The first structure solved, Spy0128 from *S. pyogenes*, is also the simplest, with two tandem CnaB domains (Kang et al., 2007). The two domains are arranged head-to-tail, giving a molecule ~100 Å long and 30–40 Å in width, similar to the estimated width of the pili themselves (Mora et al., 2005). Mass spectrometry of native pili was used to confirm the identity of the linking lysine residue involved in intermolecular isopeptide bond formation and two internal isopeptide bonds, one in each domain (Kang *et al.*, 2007, 2009). In addition to being the simplest major pilin, Spy0128 is also slightly anomalous in that it lacks the conserved YPKN pilin common to most of the other major pilins – see below.

SpaA, GBS80 and FimA show a modular arrangement similar to Spy0128, but each has three tandem domains, with a CnaA-type domain inserted between two CnaB domains. SpaA is the only one of these major pilins to be visualized in full, however, as the N-terminal domains of GBS80 and FimA were removed by proteolysis to facilitate crystallization (Mishra *et al.*, 2011; Vengadesan *et al.*, 2011). The missing domains are pre-dicted to be CnaB-like, with each containing an isopeptide bond, though successful proteolytic removal of these domains sug-gests these bonds were unformed in the recombinant proteins. The CnaA domains in these structures have internal Lys-Asn isopeptide bonds like those in the CnaB domains, but intriguingly use Asp as the catalytic acid, in contrast to the Glu of CnaB domains.

BcpA and RrgB are both four-domain major pilins but differ markedly in their architecture. BcpA follows the pattern common to most major pilins in having its four domains in tandem: two CnaB domains followed by a central CnaA domain and a C-terminal CnaB domain (Budzik *et al.*, 2009). RrgB, however, has an N-terminal CnaB domain (D1), a central CnaA domain (D2) and a C-terminal CnaB domain (D4), but has a large insertion between the two final strands of D2, which forms a complete CnaB domain (D3) that packs side-by-side with D2 (Spraggon *et al.*, 2010; Paterson and Baker, 2011; El Mortaji *et al.*, 2012a). The final domain D4 also differs in having two long helices as decorations, inserted between the second and third strands. Except for the N-terminal domains, discussed below, all domains contain internal isopeptide bond cross-links that follow the usual pattern.

8.3.1 Structural elements in polymerization of major pilins

Pilus assembly involves two key recognition events; the pilus-specific sortase must first recognize the C-terminal LPXTG-type motif on one pilin subunit and then the correct lysine on the next. The major pilin structures give important insights into these events, aided by an intriguing feature of the full-length structures, Spy0128, SpaA and RrgB. In each case, molecules in the crystal pack end-on-end, forming columns in which the C-terminal domain of one molecule abuts the N-terminal domain of the next (Kang *et al.*, 2007; El Mortaji *et al.*, 2012a). This pilus-like packing suggests an inherent tendency of major pilins to dock in this fashion. The LPXTG sorting motif generally follows just after the end of the C-terminal domain, perfectly placed to associate with the N-terminal domain of the next pilin molecule.

In most major pilins, the linking lysine belongs to a conserved YPKN pilin motif, first identified by mutagenesis (Ton-That and Schneewind, 2003). These residues are miss-ing from the truncated major pilins, but can be located in the full-length SpaA and RrgB structures, on the final β-strand of the N-terminal domain, just before the interface with the second domain. The key lysine projects into a surface groove partly formed by an extended loop located between the first two strands of the N-terminal domain. In the full-length RrgB structure solved by El Mortaji *et al.* (2012a) the end-to-end packing of the molecules allows the IPQTG sorting motif to lie in this groove, oriented towards the pilin motif lysine, in a good position for sortase-mediated ligation. The extended loop that fashions the groove is flexible in the crystal structures of RrgB and SpaA, which may facilitate docking of the sortase. This loop is found only on the N-terminal CnaB domains, not in other CnaB domains of these structures.

The one aberrant major pilin, Spy0128, lacks a YPKN pilin motif and lacks a similar groove. Instead, the linking lysine is located on an omega loop that interrupts the final β-strand in the N-terminal domain. This places it in a similar location, albeit with greater solvent exposure. This may be due to the use by *S. pyogenes* of a class B sortase to perform pilus assembly (Kang *et al.*, 2011), in contrast to the class C sortases utilized by other bacteria (Manzano *et al.*, 2008; Neiers *et al.*, 2009).

8.3.2 Correlation between internal isopeptide bond formation and pilus assembly

A common strategy to facilitate crystallization of major pilins has been the proteolytic removal of their N-terminal domains, as in the truncated RrgB, BcpA, GBS80 and FimA structures (Budzik *et al.*, 2009; Spraggon *et al.*, 2010; Vengadesan *et al.*, 2011; Mishra *et al.*, 2011). These domains were expected to form internal isopeptide bonds, based on sequence alignment and analysis of native pili (Budzik *et al.*, 2009). Their absence from the re-combinant major pilins led to the suggestion that the intermolecular linkage may need to be formed first (Budzik *et al.*, 2009).

Subsequent structures of RrgB have shed light on this phenomenon. A solution structure for the isolated D1 domain (Gentile *et al.*, 2011) and a full-length crystal structure (Paterson and Baker, 2011) both show D1 to be a flexible CnaB-like domain with no isopeptide bond. The expected isopeptide-forming residues are in close proximity but a change in hydrogen bonding between strands A and G, starting at the proline of the YPKN pilin motif and probably associated with steric repulsion by this residue, positions the asparagine residue too far from its predicted lysine partner for isopeptide bond formation. In contrast, a second full-length RrgB that includes its sorting motif (El Mortaji *et al.*, 2012a) shows an intact isopeptide bond in D1, with the sorting motif of a neighbouring molecule located in the groove leading to the intermolecular linking lysine. It appears that the pilus-like packing in the crystal and docking of the sorting motif in the groove stabilize the structure, overcome an energy barrier and restore hydrogen bonding between the first and last strands of D1, allowing isopeptide bond formation to occur.

An attractive inference is that flexibility in the D1 domain may facilitate productive docking of the sortase (Gentile *et al.*, 2011). Once this has occurred, however, and the intermolecular linkage has been formed, the domain can be stabilized by internal isopeptide bond formation. It should be noted, however, that some major pilins, such as SpaA (Kang *et al.*, 2009b), have no D1 domain isopeptide bond, and that the position of this bond when it does occur means that it cannot influence the overall load-bearing stability of the pilus.

8.3.3 Stabilization of the pilus shaft

The internal isopeptide bond cross-links provide a critical element in pilus stability. Their strategic positioning (Fig. 8.1) has the effect of reducing the shortest covalent-bonded path through each domain to a fraction of its previous value, and provides a covalently-bonded core that spans the length of the pilin subunit and extends through the inter-pilin linkage (Yeates and Clubb, 2007; Kang and Baker, 2011; see Fig. 8.2D). This core makes pilins essentially inextensible, as shown by atomic force microscopy (Alegre-Cebollada *et al.*, 2010), a feature that may enable the pilus to withstand substantial shear forces during adhesion. In addition to mechanical strength, isopeptide bond formation in pilin subunits also imparts a marked increase in resistance to proteolysis and thermal/chemical denaturation (Budzik *et al.*, 2009; Kang and Baker, 2009; Izore *et al.*, 2010; El Mortaji *et al.*, 2010).

The robustness of major pilins is further enhanced in some bacterial species by additional stabilizing features. SpaA and FimA both contain internal disulfide bonds that cross-link a similar ~55-residue section of D3. This cross-link is not part of the load-bearing core of either major pilin and appears to be present purely for additional proteolytic/thermal stability (Kang *et al.*, 2009b; Mishra *et al.*, 2011). Calcium binding sites are also present in SpaA (one site) and GBS80 (two sites). These are evidently high affinity sites, being fully occupied even without added calcium (Kang *et al.*, 2009b; Vengadesan *et al.*, 2011).

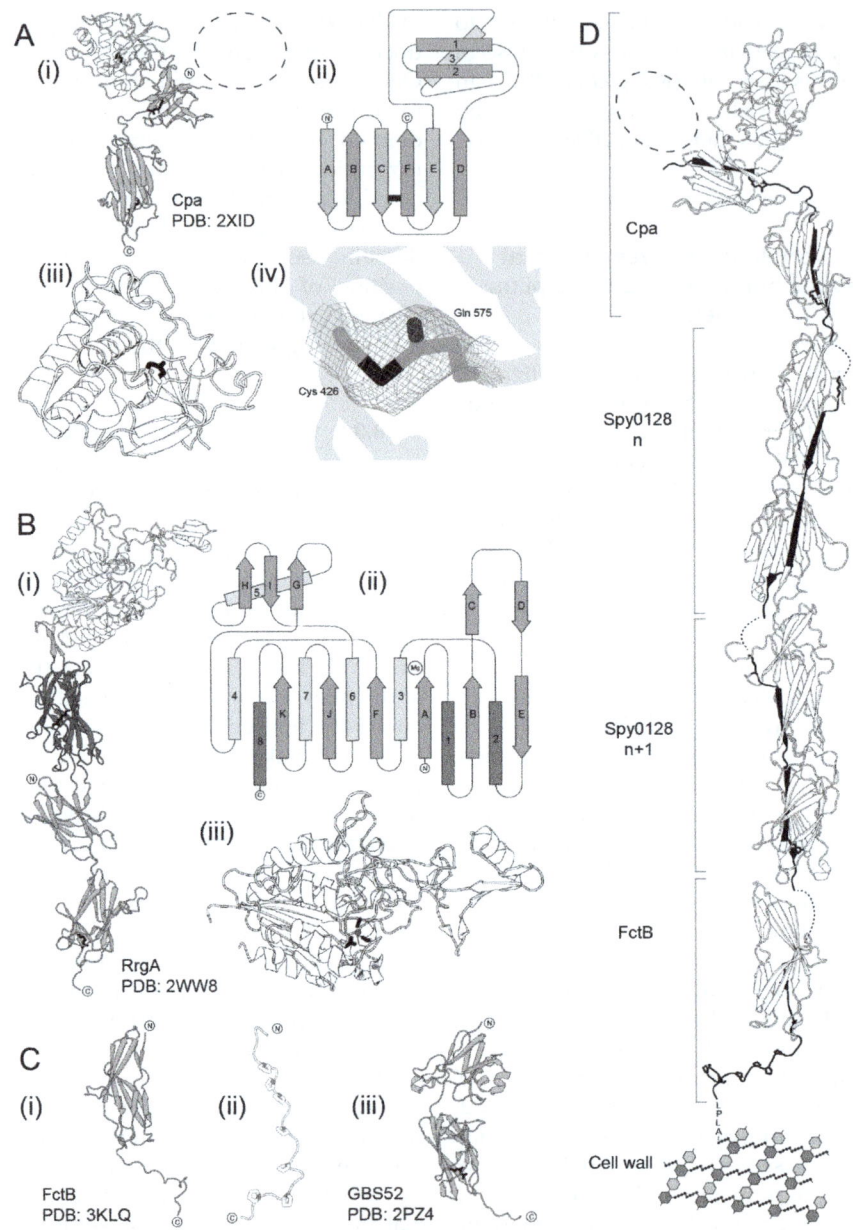

Fig. 8.2. Minor pilin structures and a model of an assembled pilus. (A) and (B) show the adhesins Cpa (*S. pyogenes*) and RrgA (*S. pneumoniae*) respectively, showing (i) cartoon diagrams of the available structures (with the missing D1 domain of Cpa as a dashed oval); (ii) the topology and (iii) a cartoon diagram of the adhesion domain in each case. Important features: the thioester bond in Cpa and the MIDAS site in RrgA are in black and (A) (iv) shows an expanded view of the Cpa D3 thioester bond and associated electron density (2Fo-Fc contoured at 1.5 σ). (C) Cartoon diagrams of the basal minor pilins (i) FctB (*S. pyogenes*) and (ii) GBS52 (*S. agalactiae*). A close up of the polyproline helical C-terminal tail region of FctB is shown in (iii). (D) Shows a model of the mature *S. pyogenes* pilus, comprising the adhesin Cpa, two copies of the major pilin Spy0128 and the basal pilin FctB attached to the bacterial cell wall. The load-bearing core of the pilus, formed from strands cross-linked by isopeptide bonds, is indicated in black and dashed lines show the location of the sortase-mediated intermolecular isopeptide linkages.

8.4 Minor Pilin Structure

8.4.1 Adhesins

The modular nature of pilin structure is continued in the adhesins that decorate the tip of Gram-positive bacterial pili and mediate binding to host cells. Current structural information on these moieties is limited to two crystal structures; the pneumococcal adhesin RrgA (Izore et al., 2010) and a partial structure of domains D2–D4 of Cpa from S. pyogenes (Pointon et al., 2010). These reveal strikingly different adhesin domains and approaches to host-cell interaction, while retaining other structural elements in common with the major pilins. An intriguing feature, considering their role in binding human cells, is that both adhesins appear to use mechanisms shared with host defence mechanisms.

RrgA is an extended protein of four tandem domains: three Ig-like domains with an integrin-like domain at the tip (Izore et al., 2010) (Fig. 8.2B). At ~195 Å in length, it is significantly longer than the major pilin RrgB (~115 Å). A striking feature of RrgA is its use of whole-domain insertions into otherwise-standard CnaA or CnaB domains. The overall domain arrangement runs D3–D2–D1–D4, with D3 at the tip of the pilus and D4 linking to the RrgB polymeric shaft. D1 has a CnaB fold into which D2 is inserted just before the final strand. D2 in turn has a CnaA fold into which the whole of D3 is inserted following its second β-strand. The two parts of D2 are cross-linked by an isopeptide bond. A crossover from D2 to D1 completes the final strand of the D1 CnaB fold and leads directly into the discrete CnaB-like D4 domain, which contains the second isopeptide bond and ends with the C-terminal sorting motif. This complicated series of insertions and crossovers continues the covalent core structure of the major pilins, with cross-links in D2 and D4 joined by a single strand running through D1. The presence of inserted domains also makes domain relationships impossible to predict from sequence alone.

Domain D3 of RrgA, at the tip of the pilus, is responsible for recognition and binding of respiratory epithelial cells (Nelson

et al., 2007; Hilleringmann et al., 2008). This domain (Fig. 8.2B) is structurally similar to von Willebrand factor domain A and integrin I domains (Lee et al., 1995; Qu and Leahy, 1995), with a parallel β-sheet core flanked by α-helices. At the C-terminal edge of the β-sheet is a bound Mg^{2+} ion in a site resembling the metal ion dependent adhesion site (MIDAS) of integrin domains, which functions in adhesion to host substrate (Michishita et al., 1993; Emsley et al., 2000; Springer et al., 2008). In RrgA, however, the core integrin-like fold is modified by two extended loops that cover one side of the MIDAS site and could be involved in recognition of the pneumococcal pilus substrates collagen, laminin and fibronectin (Hilleringmann et al., 2008).

The adhesin from S. pyogenes, Cpa (Spy0125), also has four domains, but differs radically from RrgA in its organization, cross-linking and, importantly, method of recognition and binding of host cells. Structural information is only available for domains D2–D4 (Fig. 8.2A), with the N-terminal domain D1, residues 1–285, absent from the construct used for structure solution (Pointon et al., 2010). Sequence alignment suggests that D1 and D3 have similar structures, but this has yet to be confirmed experimentally (Pointon et al., 2010). Cpa appears to have an overall Y-shaped structure, with arms D1 and D3 as arms branching out from either side of D2. D2 is a CnaB-like domain in which D3 is inserted between the two final β-strands. As in other CnaB domains, an isopeptide bond cross-links the first and last strands of D2, but interestingly this is a Lys–Asp crosslink as opposed to the usual Lys–Asn bond of other pilin domains. A similar Lys–Asp isopeptide bond has also been described in the S. pyogenes adhesin FbaB and others are predicted to exist in further cell-surface adhesin domains (Hagan et al., 2010). The final strand of D2 leads into the discrete CnaB-like domain D4, which has the usual Lys–Asn isopeptide cross-link between first and last strands, prior to the LPXTG sorting motif.

Domain D3 has been shown to mediate host-cell adhesion by the S. pyogenes pilus (Smith et al., 2010). It is formed from two distinct parts, a horseshoe-shaped pocket comprising two 3-stranded antiparallel β-sheets and a 3-helix bundle that is inserted between the first two strands of the second sheet. A striking feature of this domain is the presence of an internal Cys–Gln thioester bond (Fig. 8.2A (iv)) located in a surface cleft. The thioester bond, which has been verified by mass spectrometry, is labile and susceptible to nucleophilic attack; this regenerates the cysteine thiol and leaves the attacking nucleophile covalently linked to the gluta-mine side chain. Activated thioester groups perform a similar role in target recognition and binding in the complement system (Law and Dodds, 1997). This raises the intriguing suggestion that a mammalian system for attacking pathogens may have been hijacked by a pathogen to do the reverse.

8.4.2 Basal pilins

Many Gram-positive pili contain a third type of pilin subunit that appears to be located at the base, terminating pilus extension and facilitating transfer of the completed pilus to the cell wall (Nobbs *et al.*, 2008; Konto-Ghiorghi *et al.*, 2009; Smith *et al.*, 2010). Presumably there is some regulation of pilus length but the mechanics governing when this basal pilin is incorporated are unclear. Structural characterization of these subunits is limited to GBS52 from *S. agalactiae* (Krishnan *et al.*, 2007) and FctB from *S. pyogenes* (Linke *et al.*, 2010) and reveals a simple, modular arrangement reminiscent of the major pilins (Fig. 8.2C).

GBS52 is a 2-domain minor pilin com-prising two CnaB domains aligned head-to-tail (Krishnan *et al.*, 2007). The N-terminal domain, D1, has a deviant pilin motif, YPKI, which lacks the usual Asn residue and hence leaves D1 without an isopeptide bond cross-link. The C-terminal D2 domain, however, does have an isopeptide bond cross-link, joining the first and last strands as usual (Fig. 8.1). GBS52 also has a short proline-rich tail that extends from the end of D2 to the sorting motif. This suggests that GBS52 is a basal pilin (see below), although its position in the *S. agalactiae* pilus has not yet been confirmed.

Interestingly, GBS52 appears to show ad-hesive properties towards human lung epithelial cells, with this property located to domain D2 (Krishnan *et al.*, 2007). It is unclear how this interaction is mediated but it may be related to the fold similarities between GBS52 domains and other adhesins.

FctB, the basal pilin of *S. pyogenes* (Linke *et al.*, 2010; Smith *et al.*, 2010), is the simplest pilin characterized so far, with just a single CnaB-like domain. Structurally it mimics the N-terminal domain of its associated major pilin Spy0128; its fold is similar to the N-domain of Spy0128, and its linking lysine shares the same structural context, being presented from an omega loop inserted into the final strand of the domain (Linke *et al.*, 2010). FctB lacks an internal isopeptide bond but such a crosslink would be unnecessary from a load-bearing perspective, as a con-tinuous linear peptide chain joins the linking lysine to the sorting motif. Like GBS52, FctB has an extended proline-rich region at the C-terminus, immediately prior to the sortase motif, folded as a polyproline-II helix (Fig. 8.2C). Sequence analysis suggests this is a feature common to basal pilins, although its role is unclear. It has been suggested that the hydrophobic, proline-rich tail may be embedded in the peptidoglycan layer prior to covalent linkage by the sortase (Linke *et al.*, 2010).

8.5 The Assembled Pilus

Although immuno-EM studies enabled Gram-positive pili to be visualized as long, hair-like fibres extending from the cell wall, their limited resolution gave little information on the molecular nature of these structures. One very impressive cryo-EM study, of the pneumococcal pilus (Hilleringmann *et al.*, 2009), stands out, however. In these images, individual pilin subunits can be seen, enabling docking of the atomic structure of the major pilin into the EM density (Spraggon *et al.*, 2010; El Mortaji *et al.*, 2012a). This study also provided strong evidence that the pilus is not branched, the adhesin (RrgA) being localized to the tip and the other minor pilin RrgC to the base.

Branching cannot be entirely ruled out, however, since many lysine residues are present on the surfaces of pilin subunits. All that would be required is sufficient recognition of one of these lysines, perhaps adventitiously, by a sortase. Indeed, pneumococcal pili labelled with gold particles appeared to show clusters of RrgA (Hilleringmann *et al.*, 2008) and recent studies show sortase-mediated polymerization of RrgA localized to domain D4 (El Mortaji *et al.*, 2012b). D4 has numerous surface-exposed lysines, including several in structural contexts similar to the pilin motif lysine. If RrgA does polymerize to some extent through D4, this raises the intriguing possibility of a 'multi-headed' adhesin.

Aside from this question, convincing pseudo-atomic models of pilus structures can now be built, either by docking the individual pilin crystal structures into EM density, as for the pneumococcal pilus, or by using the docking of pilins in crystal structures as a guide. An example of such a model is shown in Fig. 8.2D for the *S. pyogenes* pilus, the only one for which the structures of all three pilin components are known.

The modular nature of the pilin components, and the juxtaposition of domains in the assembled pilus, correlate nicely with considerations of sortase specificity. In the *S. pyogenes* pilus, a single sortase, Spy0129, is used both to generate the shaft, by polymerization of the major pilin Spy0128, and to join this to Cpa at the tip and FctB at the bottom. The polymeric shaft is formed by joining the C-domain of each Spy0128 molecule to the N-domain of the next. Attachment of Cpa to the shaft requires its C-terminal D4 domain to be joined to the N-domain of Spy0128. This is facilitated because (a) Cpa D4 is remarkably similar in structure to the Spy0128 C-domain and (b) the N-domain lysine used to join Spy0128 to Cpa is the same one that is used for polymerization of the shaft (Quigley *et al.*, 2009; Smith *et al.*, 2010). The sorting motifs of Cpa and Spy0128 are also similar (VVPTG compared with EVPTG for M1 SF370 strain *S. pyogenes*). Thus the recognition elements for Cpa linkage are essentially the same as for shaft polymerization. Similar homology applies at the pilus base, where FctB has a very similar structure to the N-domain of Spy0128 and presents its linking lysine in the same way, explaining why Spy0129 can also join FctB to the C-domain of Spy0128. In contrast, FctB has a different C-terminal sorting motif LPSTG, in a very different structural context (following the proline-rich tail). This is consistent with the use of a different sortase, the housekeeping sortase, for ligation to the cell wall (Smith *et al.*, 2010).

Similar considerations are likely to apply for other pili, although structural data are less complete. For the pneumococcal pilus, the adhesin RrgA has a C-terminal domain D4 that is very similar to the C-terminal domain in RrgB, and both have the same interaction partner, the RrgB N-terminal domain D1. The situation is more complex here, however, as *S. pneumoniae* has three pilus-associated sortases with overlapping specificities (Manzano *et al.*, 2008; Neiers *et al.*, 2009). Moreover, neither the structure of the basal pilin RrgC, nor the identity of its linking lysine, are yet known.

8.6 Concluding Remarks

High-resolution structural studies of the individual pilin components have given remarkable new insights into the fine structure of Gram-positive bacterial pili. These have included the discovery of auto-catalytically-formed internal isopeptide bonds, which provide a completely unanticipated means of stabilizing these assemblies; the discovery of novel structures in the adhesin subunits that suggest how these may engage host cells; the recognition of the modular nature of pilin subunits, with repeated Ig-like domains; and testable models for the assembled pili.

Nevertheless, some very important questions remain. The adhesin structures hint at eukaryotic-like mechanisms, but we do not know what the molecular targets on host cells are, or exactly how they are engaged. This could be important knowledge for drug development. We do not yet understand the structural basis of sortase specificity; some creative chemistry may be needed to trap sortase–pilin complexes. And the events on the membrane, and whether other players are involved at the point of assembly, also remain to be elucidated.

Acknowledgements

We gratefully acknowledge research support from the Health Research Council of New Zealand and the New Zealand Marsden Fund.

References

Alegre-Cebollada, J., Badilla, C.L. and Fernandez, J.M. (2010) Isopeptide bonds block the mechanical extension of pili in pathogenic *Streptococcus pyogenes*. *Journal of Biological Chemistry* 285, 11235–11242.

Barocchi, M.A., Ries, J., Zogaj, X., Hemsley, C., Albiger, B., Kanth, A., Dahlberg, S., Fernebro, J., Moschioni, M., Masignani, V., Hultenby, K., Taddei, A.R., Beiter, K., Wartha, F., von Euler, A., Covacci, A., Holden, D.W., Normark, S., Rappuoli, R. and Henriques-Normark, B. (2006) A pneumococcal pilus influences virulence and host inflammatory responses. *Proceedings of the National Academy of Sciences of the United States of America* 103, 2857–2862.

Budzik, J.M., Oh, S.-Y. and Schneewind, O. (2008a) Cell wall anchor structure of BcpA pili in *Bacillus anthracis*. *Journal of Biological Chemistry* 283, 36676–36686.

Budzik, J.M., Marraffini, L.A., Souda, P., Whitelegge, J.P., Faull, K.F. and Schneewind, O. (2008b) Amide bonds assemble pili on the surface of bacilli. *Proceedings of the National Academy of Sciences of the United States of America* 105, 10215–10220.

Budzik, J.M., Poor, C.B., Faull, K.F., Whitelegge, J.P., He, C. and Schneewind, O. (2009) Intramolecular amide bonds stabilize pili on the surface of bacilli. *Proceedings of the National Academy of Sciences of the United States of America* 106, 19992–19997.

Craig, L., Pique, M.E. and Tainer, J.A. (2004) Type IV pilus structure and bacterial pathogenicity. *Nature Reviews. Microbiology* 2, 363–378.

Deivanayagam, C.C., Rich, R.L., Carson, M., Owens, R.T., Danthuluri, S., Bice, T., Hook, M. and Narayana, S.V. (2000) Novel fold and assembly of the repetitive B region of the *Staphylococcus aureus* collagen-binding surface protein. *Structure* 8, 67–78.

El Mortaji, L., Terrasse, R., Dessen, A., Vernet, T. and Di Guilmi, A.M. (2010) Stability and assembly of pilus subunits of *Streptococcus pneumoniae*. *Journal of Biological Chemistry* 285, 12405–12415.

El Mortaji, L., Contreras-Martel, C., Moschioni, M., Ferlenghi, I., Manzano, C., Vernet, T., Dessen, A. and Di Guilmi, A.M. (2012a) The full-length *Streptococcus pneumoniae* major pilin RrgB crystallizes in a fibre-like structure, which presents the D1 isopeptide bond and provides details on the mechanism of pilus polymerization. *Biochemical Journal* 441, 833–841.

El Mortaji, L., Fenel, D., Vernet, T. and Di Guilmi, A.M. (2012b) Association of RrgA and RrgC into the *Streptococcus pneumoniae* pilus by sortases C-2 and C-3. *Biochemistry* 51, 342–352.

Emsley, J., Knight, C.G., Farndale, R.W., Barnes, M.J. and Liddington, R.C. (2000) Structural basis of collagen recognition by integrin alpha2beta1. *Cell* 101, 47–56.

Gentile, M.A., Melchiorre, S., Emolo, C., Moschioni, M., Gianfaldoni, C., Pancotto, L., Ferlenghi, I., Scarselli, M., Pansegrau, W., Veggi, D., Merola, M., Cantini, F., Ruggiero, P., Banci, L. and Masignani, V. (2011) Structural and functional characterization of the *Streptococcus pneumoniae* RrgB pilus backbone D1 domain. *Journal of Biological Chemistry* 286, 14588–14597.

Hagan, R.M., Bjornsson, R., McMahon, S.A., Schomburg, B., Braithwaite, V., Buhl, M., Naismith, J.H. and Schwarz-Linek, U. (2010) NMR and theoretical analysis of a spontaneously formed Lys–Asp isopeptide bond. *Angewandte Chemie* 49, 8421–8425.

Hendrickx, A.P., Budzik, J.M., Oh, S.Y. and Schneewind, O. (2011) Architects at the bacterial surface – sortases and the assembly of pili with isopeptide bonds. *Nature Reviews Microbiology* 9, 166–176.

Hilleringmann, M., Giusti, F., Baudner, B.C., Masignani, V., Covacci, A., Rappuoli, R., Barocchi, M.A. and Ferlenghi, I. (2008) Pneumococcal pili are composed of protofilaments exposing adhesive clusters of Rrg A. *PLoS Pathogens* 4, e1000026.

Hilleringmann, M., Ringler, P., Muller, S.A., De Angelis, G., Rappuoli, R., Ferlenghi, I. and Engel, A. (2009) Molecular architecture of *Streptococcus pneumoniae* TIGR4 pili. *EMBO Journal* 28, 3921–3930.

Hu, X., Hu, H., Melvin, J.A., Clancy, K.W., McCafferty, D.G. and Yang, W. (2011) Autocatalytic intramolecular isopeptide bond formation in Gram-positive bacterial pili: a QM/

MM simulation. *Journal of the American Chemical Society* 133, 478–485.

Izore, T., Contreras-Martel, C., El Mortaji, L., Manzano, C., Terrasse, R., Vernet, T., Di Guilmi, A.M. and Dessen, A. (2010) Structural basis of host cell recognition by the pilus adhesin from *Streptococcus pneumoniae. Structure* 18, 106–115.

Kang, H.J. and Baker, E.N. (2009) Intramolecular isopeptide bonds give thermodynamic and proteolytic stability to the major pilin protein of *Streptococcus pyogenes. Journal of Biological Chemistry* 284, 20729–20737.

Kang, H.J. and Baker, E.N. (2011) Intramolecular isopeptide bonds: protein crosslinks built for stress? *Trends in Biochemical Sciences* 36, 229–237.

Kang, H.J. and Baker, E.N. (2012) Structure and assembly of Gram-positive bacterial pili: unique covalent polymers. *Current Opinion in Structural Biology* 22, 200–207.

Kang, H.J., Coulibaly, F., Clow, F., Proft, T. and Baker, E.N. (2007) Stabilizing isopeptide bonds revealed in gram-positive bacterial pilus structure. *Science* 318, 1625–1628.

Kang, H.J., Middleditch, M., Proft, T. and Baker, E.N. (2009a) Isopeptide bonds in bacterial pili and their characterization by X-ray crystallography and mass spectrometry. *Biopolymers* 91, 1126–1134.

Kang, H.J., Paterson, N.G., Gaspar, A.H., Ton-That, H. and Baker, E.N. (2009b) The *Corynebacterium diphtheriae* shaft pilin SpaA is built of tandem Ig-like modules with stabilizing isopeptide and disulfide bonds. *Proceedings of the National Academy of Sciences of the United States of America* 106, 16967–16971.

Kang, H.J., Coulibaly, F., Proft, T. and Baker, E.N. (2011) Crystal structure of Spy0129, a *Streptococcus pyogenes* class B sortase involved in pilus assembly. *PLoS One* 6, e15969.

Konto-Ghiorghi, Y., Mairey, E., Mallet, A., Dumenil, G., Caliot, E., Trieu-Cuot, P. and Dramsi, S. (2009) Dual role for pilus in adherence to epithelial cells and biofilm formation in *Streptococcus agalactiae. PLoS Pathogens* 5, e1000422.

Krishnan, V., Gaspar, A.H., Ye, N., Mandlik, A., Ton-That, H. and Narayana, S.V. (2007) An IgG-like domain in the minor pilin GBS52 of *Streptococcus agalactiae* mediates lung epithelial cell adhesion. *Structure* 15, 893–903.

Law, S.K. and Dodds, A.W. (1997) The internal thioester and the covalent binding properties of the complement proteins C3 and C4. *Protein Science* 6, 263–274.

Lee, J.O., Rieu, P., Arnaout, M.A. and Liddington, R. (1995) Crystal structure of the A domain from the alpha subunit of integrin CR3 (CD11b/CD18). *Cell* 80, 631–638.

Linke, C., Young, P.G., Kang, H.J., Bunker, R.D., Middleditch, M.J., Caradoc-Davies, T.T., Proft, T. and Baker, E.N. (2010) Crystal structure of the minor pilin FctB reveals determinants of Group A streptococcal pilus anchoring. *Journal of Biological Chemistry* 285, 20381–20389.

Mandlik, A., Das, A. and Ton-That, H. (2008) The molecular switch that activates the cell wall anchoring step of pilus assembly in gram-positive bacteria. *Proceedings of the National Academy of Sciences of the United States of America* 105, 14147–14152.

Manzano, C., Contreras-Martel, C., El Mortaji, L., Izore, T., Fenel, D., Vernet, T., Schoehn, G., Di Guilmi, A.M. and Dessen, A. (2008) Sortase-mediated pilus fiber biogenesis in *Streptococcus pneumoniae. Structure* 16, 1838–1848.

Marraffini, L.A., Dedent, A.C. and Schneewind, O. (2006) Sortases and the art of anchoring proteins to the envelopes of gram-positive bacteria. *Microbiology and Molecular Biology Reviews* 70, 192–221.

Michishita, M., Videm, V. and Arnaout, M.A. (1993) A novel divalent cation-binding site in the A domain of the beta 2 integrin CR3 (CD11b/CD18) is essential for ligand binding. *Cell* 72, 857–867.

Mishra, A., Devarajan, B., Reardon, M.E., Dwivedi, P., Krishnan, V., Cisar, J.O., Das, A., Narayana, S.V. and Ton-That, H. (2011) Two autonomous structural modules in the fimbrial shaft adhesin FimA mediate *Actinomyces* interactions with streptococci and host cells during oral biofilm development. *Molecular Microbiology* 81, 1205–1220.

Mora, M., Bensi, G., Capo, S., Falugi, F., Zingaretti, C., Manetti, A.G., Maggi, T., Taddei, A.R., Grandi, G. and Telford, J.L. (2005) Group A Streptococcus produce pilus-like structures containing protective antigens and Lancefield T antigens. *Proceedings of the National Academy of Sciences of the United States of America* 102, 15641–15646.

Neiers, F., Madhurantakam, C., Falker, S., Manzano, C., Dessen, A., Normark, S., Henriques-Normark, B. and Achour, A. (2009) Two crystal structures of pneumococcal pilus sortase C provide novel insights into catalysis and substrate specificity. *Journal of Molecular Biology* 393, 704–716.

Nelson, A.L., Ries, J., Bagnoli, F., Dahlberg, S., Falker, S., Rounioja, S., Tschop, J., Morfeldt, E., Ferlenghi, I., Hilleringmann, M., Holden, D.W., Rappuoli, R., Normark, S., Barocchi, M.A. and Henriques-Normark, B. (2007) RrgA is a pilus-associated adhesin in *Streptococcus pneumoniae*. *Molecular Microbiology* 66, 329–340.

Nobbs, A.H., Rosini, R., Rinaudo, C.D., Maione, D., Grandi, G. and Telford, J.L. (2008) Sortase A utilizes an ancillary protein anchor for efficient cell wall anchoring of pili in *Streptococcus agalactiae*. *Infection and Immunity* 76, 3550–3560.

Oh, S.-Y., Budzik, J.M. and Schneewind, O. (2008) Sortases make pili from three ingredients. *Proceedings of the National Academy of Sciences of the United States of America* 105, 13703–13704.

Paterson, N.G. and Baker, E.N. (2011) Structure of the full-length major pilin from *Streptococcus pneumoniae*: implications for isopeptide bond formation in Gram-positive bacterial pili. *PLoS One* 6, e22095.

Patti, J.M., Boles, J.O. and Hook, M. (1993) Identification and biochemical characterization of the ligand binding domain of the collagen adhesin from *Staphylococcus aureus*. *Biochemistry* 32, 11428–11435.

Pickart, C.M. (2001) Mechanisms underlying ubiquitination. *Annual Review of Biochemistry* 70, 503–533.

Pointon, J.A., Smith, W.D., Saalbach, G., Crow, A., Kehoe, M.A. and Banfield, M.J. (2010) A highly unusual thioester bond in a pilus adhesin is required for efficient host cell interaction. *Journal of Biological Chemistry* 285, 33858–33866.

Proft, T. and Baker, E. (2009) Pili in Gram-negative and Gram-positive bacteria – structure, assembly and their role in disease. *Cellular and Molecular Life Sciences* 66, 613–635.

Qu, A. and Leahy, D.J. (1995) Crystal structure of the I-domain from the CD11a/CD18 (LFA-1, alpha L beta 2) integrin. *Proceedings of the National Academy of Sciences of the United States of America* 92, 10277–10281.

Quigley, B.R., Zahner, D., Hatkoff, M., Thanassi, D.G. and Scott, J.R. (2009) Linkage of T3 and Cpa pilins in the Streptococcus M3 pilus. *Molecular Microbiology* 72, 1379–1394.

Scott, J.R. and Zahner, D. (2006) Pili with strong attachments: Gram-positive bacteria do it differently. *Molecular Microbiology* 62, 320–330.

Smith, W.D., Pointon, J.A., Abbot, E., Kang, H.J., Baker, E.N., Hirst, B.H., Wilson, J.A., Banfield, M.J. and Kehoe, M.A. (2010) Roles of minor pilin subunits Spy0125 and Spy0130 in the serotype M1 *Streptococcus pyogenes* strain SF370. *Journal of Bacteriology* 192, 4651–4659.

Spirig, T., Weiner, E.M. and Clubb, R.T. (2011) Sortase enzymes in Gram-positive bacteria. *Molecular Microbiology* 82, 1044–1059.

Spraggon, G., Koesema, E., Scarselli, M., Malito, E., Biagini, M., Norais, N., Emolo, C., Barocchi, M.A., Giusti, F., Hilleringmann, M., Rappuoli, R., Lesley, S., Covacci, A., Masignani, V. and Ferlenghi, I. (2010) Supramolecular organization of the repetitive backbone unit of the *Streptococcus pneumoniae* pilus. *PLoS One* 5, e10919.

Springer, T.A., Zhu, J. and Xiao, T. (2008) Structural basis for distinctive recognition of fibrinogen gammaC peptide by the platelet integrin alphaIIbbeta3. *Journal of Cell Biology* 182, 791–800.

Symersky, J., Patti, J.M., Carson, M., House-Pompeo, K., Teale, M., Moore, D., Jin, L., Schneider, A., DeLucas, L.J., Hook, M. and Narayana, S.V. (1997) Structure of the collagen-binding domain from a *Staphylococcus aureus* adhesin. *Nature Structural Biology* 4, 833–838.

Telford, J.L., Barocchi, M.A., Margarit, I., Rappuoli, R. and Grandi, G. (2006) Pili in gram-positive pathogens. *Nature Reviews Microbiology* 4, 509–519.

Ton-That, H. and Schneewind, O. (2003) Assembly of pili on the surface of *Corynebacterium diphtheriae*. *Molecular Microbiology* 50, 1429–1438.

Ton-That, H. and Schneewind, O. (2004) Assembly of pili in Gram-positive bacteria. *Trends in Microbiology* 12, 228–234.

Ton-That, H., Marraffini, L.A. and Schneewind, O. (2004) Sortases and pilin elements involved in pilus assembly of *Corynebacterium diphtheriae*. *Molecular Microbiology* 53, 251–261.

Vengadesan, K., Ma, X., Dwivedi, P., Ton-That, H. and Narayana, S.V. (2011) A model for group B *Streptococcus* pilus type 1: the structure of a 35-kDa C-terminal fragment of the major pilin GBS80. *Journal of Molecular Biology* 407, 731–743.

Waksman, G. and Hultgren, S.J. (2009) Structural biology of the chaperone–usher pathway of pilus biogenesis. *Nature Reviews Microbiology* 7, 765–774.

Wikoff, W.R., Liljas, L., Duda, R.L., Tsuruta, H., Hendrix, R.W. and Johnson, J.E. (2000) Topologically linked protein rings in the bacteriophage HK97 capsid. *Science* 289, 2129–2133.

Yeates, T.O. and Clubb, R.T. (2007) How some pili pull. *Science* 318, 1558–1559.

Zahner, D. and Scott, J.R. (2008) SipA is required for pilus formation in *Streptococcus pyogenes* serotype M3. *Journal of Bacteriology* 190, 527–535.

Zong, Y., Xu, Y., Liang, X., Keene, D.R., Hook, A., Gurusiddappa, S., Hook, M. and Narayana, S.V.L. (2005) A 'collagen hug' model for *Staphylococcus aureus* CNA binding to collagen. *EMBO Journal* 24, 4224–4236.

9 Sortase Structure and Specificity in Streptococci

Roberta Cozzi, Domenico Maione and Daniela Rinaudo

Novartis Vaccines and Diagnostics, Siena, Italy

9.1 Introduction

Sortases are a family of membrane-associated enzymes that either catalyse the covalent anchoring of surface proteins to the cell-wall envelope or polymerize protein sub-units (pilins) in covalently-linked pilus-like structures extending out the cell surface in Gram-positive bacteria (Marraffini *et al.*, 2006; Spirig *et al.*, 2011). These enzymes are cysteine transpeptidases, which recognize a conserved carboxylic cell-wall sorting signal (CWSS) consisting of a LPXTG-like sorting motif, where X is any amino acid, followed by a hydrophobic stretch of amino acids and a short positively charged tail (Mazmanian *et al.*, 2001; Ton-That *et al.*, 2004; Marraffini *et al.*, 2006).

The availability of complete bacterial genomes has led to the identification of sortase homologues in different bacterial species. The sequence of the prototypical well-characterized *Staphylococcus aureus* sortase A (SrtA) has been widely used to identify homologues in bacterial genomes, and a copious quantity of sortases have been identified in almost all Gram-positive bacteria. Multiple sortases are often found in the same genome and can be grouped into four or five classes based on their primary sequences, membrane topology, genomic localization and specificity for amino acid

sequence motifs (Comfort and Clubb, 2004; Dramsi *et al.*, 2005). All Gram-positive pathogens express a 'housekeeping' sortase, belonging to the class A sortase, which is responsible for the cell-wall anchoring of the majority of surface proteins and of poly-merized pilus-structures. The largest and most heterogeneous group of Gram-positive sortases is the C class (also defined as pilus-specific sortases). These specialized sortases are responsible for catalysing transpeptidation reactions linking pilin subunits in the pilus biogenesis mechanism (Ton-That and Schneewind, 2004). Once pilin subunits are synthetized in the bacterial cytoplasm and exported through the general Sec pathway, they are retained within the membrane via their C-terminal hydrophobic domain and their positively charged tail. Sortase enzymes, with their active Cysteinyl group, cleave the peptide bond between the threonine (T) and the glycine (G) residues of the LPXTG motif joining proteins to an amino group located on the next pilin subunits (Kang and Baker, 2012). The current biphasic model of pilus assembly derives from studies of the archetype SpaA-type pili in *Corynebacterium diphtheriae* (Ton-That and Schneewind, 2003) and outlines the basic steps of pilus poly-merization followed by the cell-wall anchoring step of the resulting polymer by the housekeeping sortase or, in some cases, by the pilus-specific C sortase itself (Ton-That and Schneewind, 2004).

This chapter will review data available thus far on pilus-specific sortases, focusing on their genomic organization, structure, substrate recognition specificity and regulation.

9.2 Genetic Organization of Pilus-specific Sortases

Sortases that build pili on the surface of Gram-positive bacteria have been described in many important human pathogens starting from C. diphtheriae (Ton-That and Schneewind, 2003), and afterwards in Streptococcus agalactiae (Lauer et al., 2005; Dramsi et al., 2006; Rosini et al., 2006), Streptococcus pyogenes (Mora et al., 2005), Streptococcus pneumoniae (Barocchi et al., 2006), Actinomyces naeslundii (Mishra et al., 2007), Bacillus cereus (Budzik et al., 2007) and anthracis (Budzik et al., 2008b), Enterococcus faecalis (Nallapareddy et al., 2006) and faecium (Hendrickx et al., 2008), Streptococcus suis (Fittipaldi et al., 2010) and recently also in non-pathogenic microbes including Lactococcus lactis (Oxaran et al., 2012).

Sortase (srt) genes are present in several copies in a genome and differently from the srtA gene, which is present in a monocistronic operon in most Gram-positive bacteria, occur in operons located in pathogenicity islands that also encode their substrates. The first insights into the genetic organization of pilus components and the mechanism of pili assembly in Gram-positive bacteria came from studies of the archetype SpaA-type pili in C. diphtheriae (Ton-That and Schneewind, 2003). This pathogen produces three distinct pilus structures, SpaA-, SpaD- and SpaH-type pili encoded by three pilus gene clusters, each expressing three pilus proteins named SpaA through SpaI (Spa for sortase-mediated pilus assembly), carrying a LPXTG-like motif and one or two class C sortases. Thus in C. diphtheriae a total of five sortases, named SrtA through SrtE are expressed (Ton-That and Schneewind, 2003, 2004; Gaspar and Ton-That, 2006; Swierczynski and Ton-That, 2006). A sixth sortase (SrtF) is present, but it is a

class D sortase homologue located at a different region of the chromosome, and is now referred to as the C. diphtheriae housekeeping sortase (Dramsi et al., 2005).

In S. agalactiae (also known as Group B Streptococcus or GBS) three structurally distinct types of pili have been identified, each encoded by a distinct genomic island. These genomic loci have been named as pilus island 1 (PI-1), pilus island 2a (PI-2a) and pilus island 2b (PI-2b) (Dramsi et al., 2006; Rosini et al., 2006). The overall organization of the three islands is similar, with each island containing genes encoding three LPXTG-like motif carrying proteins and two class C sortase enzymes (SrtC1 and SrtC2), catalysing pilus protein polymerization. Actually, PI-1 carries an additional gene predicted to code for a third sortase C enzyme (SrtC3) which seems not to be involved in pilus assembly (Buccato et al., 2006) and whose role needs to be elucidated.

Similarly, in some S. pneumoniae strains, the genes that encode the pilus 1 are contained in a 14-kb pathogenicity island (known as the rlrA islet) in which, in addition to three genes coding for the structural pilus subunits (rrgA, rrgB and rrgC), there are three srt genes (srtB, srtC and srtD) coding for SrtC-1, SrtC-2 and SrtC-3 sortase enzymes (Barocchi et al., 2006). In S. pneumoniae a second pilus island has been identified (PI-2), located in a chromosomal region of approximately 6.5 Kb in which two srt genes, srtG1 and srtG2, are present clustered with other three genes involved in some way in pilus assembly (Bagnoli et al., 2008).

Sortase genes are present in all of the 9 different pilus encoding islets identified so far in S. pyogenes (also known as Group A Streptococcus or GAS) (Mora et al., 2005; Falugi et al., 2008). These genomic regions are extremely variable in their gene composition (number and order of the encoded genes) ranging from 11 to 16 Kb in size and contain one, two or three srt genes per island (Falugi et al., 2008).

The four pilus clusters disclosed in S. suis contain either one or three class C sortase genes as well as several genes encoding cell-wall anchor family proteins and these pathogenic islets have been named srtBCD, *srtE*, *srtF* and *srtG* clusters according to class C sortase genes in the respective locus (Takamatsu *et al.*, 2009; Fittipaldi *et al.*, 2010; Okura *et al.*, 2011).

Pilus gene clusters containing one or two pilus-specific sortase gene(s) in association with genes encoding structural subunits have been identified in other Streptococcal (i.e. *S. sanguinis* (Zahner *et al.*, 2011); *S. gallolyticus* (Danne *et al.*, 2011)) and Enterococcal (i.e. *E. faecalis* (Nallapareddy *et al.*, 2006); *E. faecium* (Hendrickx *et al.*, 2008)) species. *B. cereus* forms pili on its cell surface, and the genes for pilus assembly reside in the *bcpA-srtD-bcpB* operon whose pilus-specific sortase (called Sortase D) are required for pilus formation (Budzik *et al.*, 2007). The housekeeping SrtA is involved in pilus anchoring to the cell wall (Budzik *et al.*, 2008b). *Actinomyces oris* can express two different types of pili: type-1 and type-2, encoded by two separate gene clusters. Each gene cluster contains three genes that encode a large putative adhesin, the pilus shaft protein and the pilus-specific sortase. The encoded pilin proteins are as follows: FimQ, FimP and SrtC-1 for type-1 and FimA, FimB and SrtC-2 for type-2. SrtC-1 and SrtC-2 share 42% sequence identity within the enzymatic domain (Yeung *et al.*, 1998; Chen *et al.*, 2007). Very recently, a putative pilus biogenesis cluster has been identified in the genome of *L. lactis* strain IL1403. It consists of a sortase C gene flanked by three LPxTG protein encoding genes (*yhgD*, *yhgE* and *yhhB*) (Oxaran *et al.*, 2012).

9.3 Structure and Regulation of Pilus-specific Sortases

The crystal structure of several pilus-related sortases from different Gram-positive bacteria have been reported so far, including three sortases from *S. pneumoniae* (SrtC-1, SrtC-2 and SrtC-3) (Manzano *et al.*, 2008; Neiers *et al.*, 2009), three sortases from *S. agalactiae* pilus type 2a and 1 (SrtC1-2a, SrtC1-1 and SrtC2-1) (Cozzi *et al.*, 2011, 2012; Khare *et al.*, 2011), AcSrtC-1 from *A. oris* (Persson, 2011), Spy129 from *Streptococcus pyogenes* (Kang *et al.*, 2011) and SrtC1 from *S. suis* (Lu *et al.*, 2011). The 3-D structures of pilus-specific sortases solved to date are represented in Fig. 9.1. All sortase trans-peptidases crystallized so far belong to class C, except for Spy129 from *S. pyogenes* whose three-dimensional structure has revealed a class B sortase. Class C sortases are predicted to carry two trans-membrane (TM) domains at the C- and N-terminal regions. These TM domains are essential for sortase membrane localization and biological function. The analysis of the available crystal structures shows that they can adopt a stable conformation even in the absence of the predicted TM domains (Manzano *et al.*, 2008; Neiers *et al.*, 2009; Cozzi *et al.*, 2011; Khare *et al.*, 2011). Structural analyses show that the overall fold is very similar among all sortase family members. They share a common catalytic domain, based on an eight-stranded β-barrel core and a highly conserved catalytic triad made of histidine, cysteine and arginine residues situated in the active-site at the end of a groove along one side of the β barrel and with key roles in catalysis. The β-sheet core is surrounded by an N-terminal region made of one, two or three α helices constituting the roof linked to the β-sheet core through a long loop called 'lid', covering the entrance of the substrate binding site in a close-inactive conformation (Gaspar and Ton-That, 2006; Chen *et al.*, 2007; Zahner *et al.*, 2011; Kang and Baker, 2012; Oxaran *et al.*, 2012). Whereas the central β-barrel is well conserved in different sortases, the connecting loops and helices vary widely. The structures of SrtC1 from *S. agalactiae* pilus 1 and *S. suis* were determined with the active-site in the 'open' conformation, while the other structures showed that the groove of the active site was occluded by the lid. The lid is a specific unique feature of pilus-specific sortases and contains a conserved DPX motif, where X can be any aromatic residue like tyrosine, phenylalanine or tryptophan. The sequence of the lid resembles the LPXTG-like sorting motif located at the C-terminal region of the structural pilins. The structures reveal that the electron density of the residues DPX of the lid is well defined, as they are strongly interacting with key catalytic residues (Fig. 9.2). The carboxylate group of the aspartate residue interacts with the side chain of the

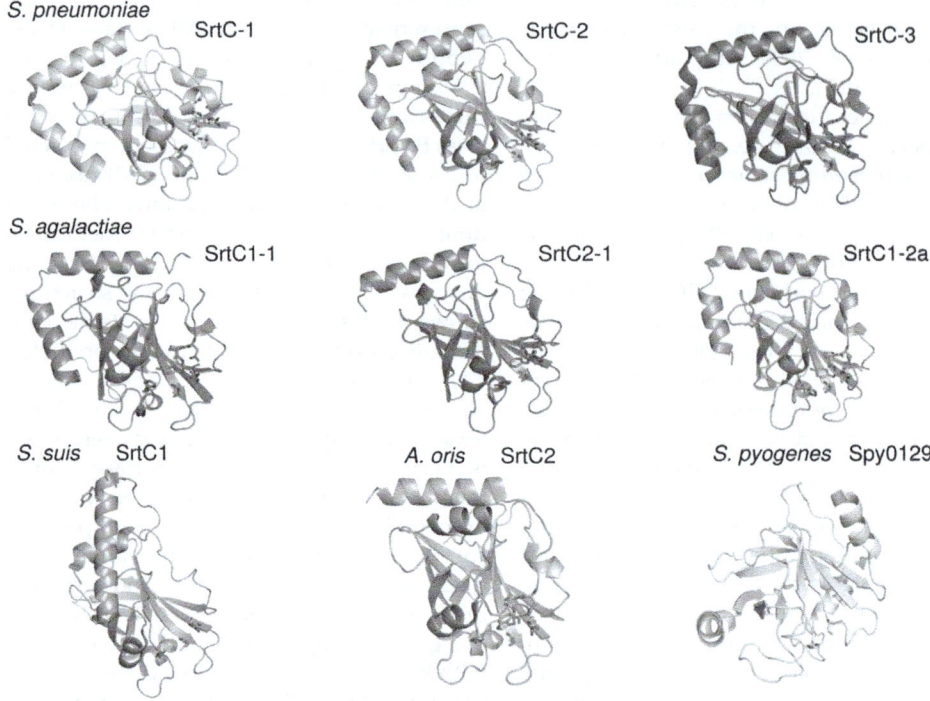

Fig. 9.1. Representative structures of pilus-specific sortases solved to date. Ribbon diagrams of *S. pneumoniae* sortases SrtC-1, SrtC-2 and SrtC-3 (PDB ID 2W1J, 3G66, 2W1K, respectively) (Manzano *et al.*, 2008; Neiers *et al.*, 2009); *S. agalactiae* SrtC1-1, SrtC2-1 and SrtC1-2a (PDB ID 4G1J, 4G1H, 3O0P, respectively) (Cozzi *et al.*, 2011, 2012); *S. suis* SrtC1 (PDB ID 3RE9) (Lu *et al.*, 2011), *A. oris* SrtC2 (PDB ID 2XWG) (Persson, 2011) and *S. pyogenes* Spy0129 (PDB ID 3PSQ) (Kang *et al.*, 2011).

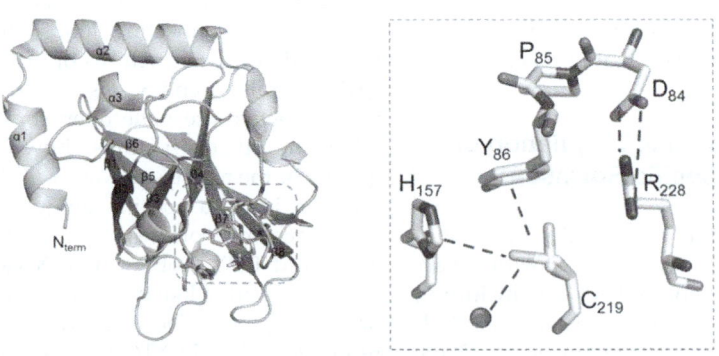

Fig. 9.2. Overall fold of *S. agalactiae* SrtC1-2a and active site organization. (A) Overall fold of SrtC1 of GBS Pilus Island 2a. SrtC1 is represented as cartoon. Residues forming the mobile lid and the active site are shown in sticks. (B) Active site of SrtC1-2a. Residues forming the mobile lid (Asp84, Tyr86) and the active site (His157, Cys219, Arg228) are shown as sticks. Two conformations of Cys219 are represented as seen in the crystal structure. Distances between atoms are labelled and shown as dashes and the water molecule is shown as a sphere.

catalytic arginine forming a salt bridge, while the hydrophobic ring of the lid aromatic residue (Tyr, Phe or Trp) forms an aromatic-sulfur interaction with the conserved catalytic cysteine suggesting that the lid is anchored to the active site through this interaction. Most of the pilus-related sortases described so far present a lid loop, except for Spy129 from *S. pyogenes* (Kang *et al.*, 2011) for which a canonical lid motif has not been identified. The poor electron density for most of the main and side chains of residues located within the lid (except for the DPY/W/F residues) suggests a high flexibility of this loop. Moreover, the entire N-terminal region, carrying the lid loop, displays B-factors relatively higher than the B-factor values for the overall structure (Manzano *et al.*, 2008; Neiers *et al.*, 2009; Cozzi *et al.*, 2011). Finally, in terms of sequence conservation, the helices and the lid in the N-terminal region are variable among different Gram-positive sortases, differently by the central β-barrel core harbouring the catalytic triad that is highly conserved (Cozzi *et al.*, 2011, 2012).

The lid function has just started to be elucidated. In SrtC1 from *S. agalactiae* pilus island 2a (SrtC1-2a) and in SrtC2 from *A. oris* the lid has been demonstrated to be dispensable for sortase activity *in vivo* (Cozzi *et al.*, 2011; Wu *et al.*, 2012). However, all structural analyses available indicated an important role of this flexible region in enzyme activation, specifically in the regulation of substrate accessibility to the active site. With respect to this aspect, recent evidence obtained in biochemical characterization studies of *S. agalactiae* pilus-related sortases will be further discussed.

9.3.1 *Streptococcus pneumoniae* pilus islet 1 sortases

The X-ray structures of SrtC-1 (called also SrtB), SrtC-2 (SrtC) and SrtC-3 (SrtD) from pneumococcal pilus 1 have been the first among pilus-related sortases to be determined (PDB ID 2W1J, 3G66, 2W1K, respectively) (Manzano *et al.*, 2008; Neiers *et al.*, 2009). The overall fold of all three enzymes is very

similar to other known sortases, corresponding to a β-barrel structure, composed of eight anti-parallel β-strands linked by multiple helices. The catalytic triad (constituted of His131, Cys193, Arg202 in SrtC1; His159, Cys221, Arg230 in SrtC2; His144, Cys206, Arg215 in SrtC3) within the substrate binding region is encapsulated by the lid, which maintain the active site in a closed conformation in the absence of substrate. The lid anchoring within the active site is through multiple interactions with key catalytic residues (Manzano *et al.*, 2008; Neiers *et al.*, 2009). Structural comparison of the three pilus-associated sortases revealed some slight differences in terms of flexibility, positioning and number of residues of the lid and B-factor values of the N-terminal helices. An additional helix in the C-terminal region is only present in SrtC-3. Some structural differences suggested a molecular explanation for the functional differences observed among these sortases, in terms of substrate specificity and incorporation of the ancillary pilins into pili. The observed differences in charge on the lid can represent an important aspect that explains why SrtC-1 and SrtC-2 act as pilus-polymerizing enzymes, in contrast to SrtC-3. Manzano *et al.* have also showed that site-specific mutations of the anchor residues in the lid region did not affect backbone protein recognition or the formation of the acyl-intermediate; however, the stability and the efficiency of the enzyme were negatively affected (Manzano *et al.*, 2009).

9.3.2 *Streptococcus agalactiae* (GBS) sortases

Three crystal structures of GBS class C sortases have been solved thus far. They are SrtC1 from pilus island 2a (SrtC1-2a, PDB ID 3O0P) and SrtC1 and SrtC2 from pilus island 1 (SrtC1-1, PDB ID 4G1J and SrtC2-1, PDB ID 4G1H) (Cozzi *et al.*, 2011, 2012; Khare *et al.*, 2011). All three structures share a similar fold with the other class C sortase members. Some differences have been identified in the number of N-terminal α-helices composing the roof. Three α-helices are present in SrtC1-2a, two in

SrtC1-1 and one in SrtC2-1. However, also the SrtC2-1 sortase is predicted to form two α-helices and the absence of electron density for the first N-terminal α-helix is likely due to an artefact of crystallization (Cozzi *et al.*, 2012). The three SrtC enzymes have been crystallized in presence of bivalent ions, however, the loop carrying the calcium-binding motif and identified in other sortase crystal structures is not present, and no functional calcium ions have been identified in GBS enzymes in agreement with NMR spectroscopy measurements (Cozzi *et al.*, 2011). Structural analyses of the three crystalized sortases show that the active site is positioned on one side of the β-barrel core in a hydrophobic environment and is made of a canonical catalytic triad (His157-Cys219-Arg228 in SrtC1-2a, His163-Cys225-Arg234 in SrtC1-1 and Hys156-Cys218-Arg227 in SrtC2-1). The catalytic residues are not accessible to pilin protein substrates, suggesting that the enzymes cannot be functionally active in this conformation due to the presence of the lid loop that sterically blocks the catalytic cysteine. Residues forming the mobile lid (Asp84-Pro85-Tyr86 in SrtC1-2a, Asp90-Tyr92 in SrtC1-1 and Asp84-Phe86 in SrtC2-1) directly interact with the catalytic residues of the active site, blocking the access of the substrate. Recombinant lid mutants of GBS sortases have been generated based on structural information and characterized in biochemical assays to measure the enzyme cleavage activity on synthetic peptides mimicking the LPXTG-like sorting motifs of different pilin subunits. Kinetic measurements of the activity of sortase lid mutants have shown that the mutant enzymes lacking the entire lid or a single mutation of the lid aromatic amino acid (Tyr86 in SrtC1-2a, Tyr92 in SrtC1-1 and Phe86 in SrtC2-1) performed even better than the wild-type enzymes in cleaving LPXTG-carrying peptides. These data strongly supported the hypothesis that the catalytic activity of C sortases can be induced through a displacement of the lid from the active site of the enzyme, probably as a result of the interaction with the substrate proteins and/or other unknown factors (Cozzi *et al.*, 2011, 2012). Very recently, the first direct experimental evidence that sortase enzymes are auto-inhibited by the presence of the lid has been reported (Cozzi *et al.*, 2013). Cozzi, Zerbini and co-workers demonstrated that an efficient pilus protein polymerization can be achieved *in vitro* by using a recombinant SrtC1-2a lid mutant enzyme carrying a single residue mutation in the lid region. It has been observed that, while the wild-type enzyme was totally inactive, the lid mutant SrtC1$_{Y86A}$ was able to efficiently assemble the backbone subunit in high-molecular weight polymers, clearly detectable by SDS-PAGE analysis (Fig. 9.3). Thus, a single residue can regulate the enzyme catalytic activity through the anchoring of the lid in the active site. The mutation of this crucial residue might break the interaction of the aromatic ring of Tyr86 with the catalytic cysteine, making the active site available for substrate binding. Moreover, biochemical analyses of recombinant wild-type and mutant SrtC1 have also revealed that the lid confers thermodynamic and proteolytic

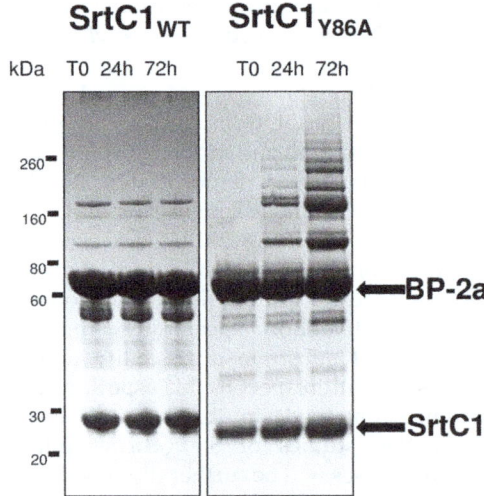

Fig. 9.3. GBS SrtC1$_{Y86A}$ lid mutant polymerizes the backbone protein subunit of PI-2a *in vitro*. Time course of reactions with 25 µM of the wild type or lid mutant (Tyr86Ala) sortase enzymes (SrtC1$_{WT}$ and SrtC1$_{Y86A}$) mixed with 100 µM of the recombinant backbone protein BP-2a. Aliquots of each reaction at different time points (0, 24 hours and 72 hours) were analysed by SDS-PAGE and Coomassie staining. A pattern of high-molecular-weight bands has been detected only in the reaction containing the SrtC1$_{Y86A}$ lid mutant.

stability to the sortase enzymes (Cozzi *et al.*, 2013).

The DPY(F\W) motif of the lid is located at the end of a highly flexible region of the structure, as indicated by the B-factor values of the N-terminal region, compared to the β-barrel core, suggesting that this entire portion can be displaced during enzyme activation (Manzano *et al.*, 2008; Cozzi *et al.*, 2011). The superimposition of *Staphylococcus aureus* SrtA structure with GBS sortase C enzymes crystal structure revealed that C sortases, differently by housekeeping sortase, are made of two functional domains (Cozzi *et al.*, 2012). The catalytic β-barrel structural core is conserved and overlapping in both SrtC and SrtA. Unlike SrtA, GBS sortase C enzymes contain an additional N-terminal region, composed of two main α-helices and a lid that blocks the access of substrates to the active site. Surprisingly, ligand-free SrtC structures match the active, peptide-bound, conformation of SrtA and not the apo structure of SrtA. These observations suggest that the conserved residues in the lid that interact with the active site of GBS sortase are a pseudo-substrate as they mimic the binding of the LPXTG motif in the catalytic site (Cozzi *et al.*, 2012).

Structural analysis and experiments performed *in vitro* with fluorogenic peptides and with N-terminal and lid deletion mutants of GBS sortase C enzymes showed that the entire N-terminus, and not just the lid, as shown for GBS PI-2a SrtC1, is disposable for catalytic activity indicating that the smallest sortase module retaining catalytic activity is the β-sheet core, present in the *Staphylococcus aureus* SrtA structure and common to all sortase family members (Cozzi *et al.*, 2012). It is possible that different function between SrtA and pilus-related sortases might be due to the presence in this specific class of sortases of a highly specialized N-terminal region and it has been proposed that the SrtC enzymes have two functional domains: (i) an N-terminal regulatory region that contains the inhibitory, pseudo-substrate lid, involved in enzyme regulation and (ii) the β-barrel core representing the catalytic-active region, containing the catalytic triad (Cozzi *et al.*, 2012).

It is thought that the activation of the enzymatic function of sortase C enzymes might occur by a conformational change, which includes the opening of the N-terminal region and subsequent rearrangement of the lid to expose the active site (Cozzi *et al.*, 2012). The lid function has just started to be elucidated and the mechanism by which the lid promotes either substrate accessibility or enzyme stability is not completely understood yet.

Multiple sequence alignment performed with all the available GBS sortase C enzyme sequences revealed that PI-2b sortases form a cluster far away from the others, sharing just a short central portion of the sequence. The pilus-associated sortases of pilus island 2b are shorter in terms of amino acid sequence, and even if the catalytic triad is conserved, they do not contain the conserved motif DPY(F/W). In particular, SrtC2-2b enzymes lack the predicted C-terminal transmembrane helix, but retain the one at the N-terminus, three cysteines are present instead of just the catalytic one, and the lid is unlikely to be present, since the main hydrophobic core is linked to the N-terminal transmembrane helix by a connecting sequence missing more than 30 residues compared with the other sortase C sequences (Cozzi *et al.*, 2011). Further structural investigations of these sortases will be necessary for a better understanding of PI-2b class C sortases in pilus biogenesis and how they are regulated in absence of the lid.

9.3.3 *Streptococcus pyogenes* sortase Spy0129

Sortase Spy0129 is involved in the pilus polymerization in *S. pyogenes* serotype M1. Despite its role in pilus assembly, the crystal structure of Spy0129 confirmed that it belongs to class B sortases as also indicated by the sequence similarity with class B sortases from *Staphylococcus aureus* and *B. anthracis*, which are involved in anchoring surface proteins containing the NPQTN motif to the cell wall (Zhang *et al.*, 2004; Zong *et al.*, 2004; Kang *et al.*, 2011). The overall structure of Spy0129 (PDB ID 3PSQ) conforms to the canonical sortase fold, but with some variations. The structure shows the conserved eight-stranded β-barrel core, whose surface is decorated with loops and helices, but it does not share the unique features of the other pilus-specific class C sortases, such as the flexible lid covering the active site and the C-terminal transmembrane anchor. Spy0129 has only an N-terminal hydrophobic anchor. The most distinctive features of the Spy0129 structure, however, are three long connecting loops β2/β3, β4/β5 and β6/β7, which shape the presumed substrate-binding region. The β6/β7 loop provides the most notable difference between Spy0129 and other sortases; its flexibility is supposed to play a significant role in binding the sortase recognition motif of the substrate protein. It has been proposed that the conformational movements in these long loops might correlate with the positions and orientations of the catalytic Cys and His residues, Cys221 and His126, and be important for enzyme function (Kang et al., 2011). Very interestingly, genetic experiments have demonstrated that Spy0129 is not sufficient alone to polymerize the backbone protein Spy0128, being the signal peptidase-like protein SipA, coded by the same genomic pilus operon, also required for Spy0128 polymers formation (Zahner and Scott, 2008). The need of a chaperone-like protein for pilus assembly in vivo could explain why a recombinant form of Spy0129 produced in vitro is not active in polymerization assay (Zahner and Scott, 2008).

9.4 Sortase Specificity Substrate Recognition

In Gram-positive bacteria, pili can be distinguished into two- or three-components polymers and the mechanism of pilus assembly/cell-wall anchoring can be mediated by one, two or three sortase enzymes. Pili are covalently-linked structures composed of a main structural protein, known as the backbone protein (BP) that assembled in multiple copies forms the pilus shaft and by one or two ancillary proteins (APs), currently understood to be located one at the pilus tip and the other at the base where it is involved

in pilus anchoring to the cell wall. However, evidence indicates a different localization of the ancillary proteins in the pilus structures (Ton-That and Schneewind, 2003; Mandlik et al., 2008) and the mechanism underlying the incorporation of the ancillary subunits remains quite controversial and is still not fully understood. A potential explanation of this disagreement could be sought in the promiscuity in the substrates that the pilus-specific sortases recognize and/or in the redundancy of their function. In fact, both the polymerization of the main structural subunit and the incorporation of the ancillary protein(s) into the pilus shaft are catalysed by the same sortases, through the recognition of specific motifs present in the structural pilins (Kang and Baker, 2012). A C-terminal LPXTG-like motif (where X represents any amino acid), typically conserved also in cell wall-anchored proteins, is present both in the backbone and ancillary pilin subunits and represents the main sortase recognition site. However, other sequence elements and/or residues, which are conserved among pilin subunits in different bacteria, are required for pilus assembly (Ton-That and Schneewind, 2004). Such motives include the pilin motif (consensus Wxxx-VxVYPK) that contain a specific lysine or asparagine residue, the side chain of which can participate in sortase-catalysed amide bond formation by reaction with the C-terminus of the next subunit molecule during pilus polymerization. Recent studies in different species have revealed that class C sortases are surprisingly promiscuous in their ability to recognize a variety of sorting signals and amino groups. For instance, in C. diphtheriae, the pilus-specific sortase, known as SrtA, that assembles the prototypical SpaA pilus can recognize two distinct sorting signals (LPLTG in the backbone protein SpaA and in the ancillary SpaC, and LAFTG in minor pilin SpaB) and specific lysine residues, that C sortase used as a nucleophile, present in different pilin proteins (either Lys190 within the pilin motif of SpaA or Lys139 in SpaB) (Mandlik et al., 2008). The pilus 'tip' protein SpaC doesn't carry a lysine residue/pilin motif, explaining why it is only found at the tip of the pilus. Moreover, when the housekeeping sortase SrtF, whose role is to mediate the cell-wall attachment of Spa-type pili, is absent, C sortase SrtA can also catalyse this step, although less efficiently than SrtF (Swaminathan et al., 2007).

Similarly, in B. cereus a single class C enzyme (called SrtD) is responsible both for the polymerization of the major pilin protein BcpA that forms the pilus shaft and for the incorporation in a two-component pilus of a single minor pilin, BcpB, at the tip of polymerized structure (Budzik et al., 2009). This sortase is able to recognize and cleave two different sorting signals (LPVTG within BcpA and IPNTG in BcpB) to join the threonine residues in each signal to the side-chain of Lys162 located within a pilin motif in BcpA (Budzik et al., 2008a). However, this C sortase is unable to attach the pilus to the cell wall (Budzik et al., 2008b). If C. diphtheriae Spa-type and B. cereus BcpA pili are examples of covalently-linked polymers assembled by a single sortase C enzyme, many other pili identified thus far are assembled by two class C sortases that appear to have partially redundant functions. This is the case of C. diphtheriae SpaD- and SpaH-type pili (Gaspar and Ton-That, 2006; Swierczynski and Ton-That, 2006) and of the three pilus types in S. agalactiae (Rosini et al., 2006). Each of these pili are made by three structural subunits, and both two sortases are able to recognize distinct sorting signals and nucleophiles within all three pilin proteins (Gaspar and Ton-That, 2006; Swierczynski and Ton-That, 2006). The basal pilin, SpaE into the SpaD-type, has also been found interspersed throughout the shaft pilus, again highlighting the substrate promiscuity of class C enzymes (Gaspar and Ton-That, 2006). In Streptococcus agalactiae two sortases (SrtC1 and SrtC2) mediate the polymerization of each of the three pilus types. Preliminary genetic studies performed in strains expressing pilus type 1 or 2a established the relative contribution of SrtC1 and SrtC2 sortases in pilus assembly. In these studies, each enzyme alone can efficiently polymerize the backbone protein in vivo and, although clearly exhibiting redundant functions, predominantly in-corporated into pili one of the two ancillary subunits, with significantly reduced ability to incorporate the other pilin (Rosini et al., 2006). However, more recent FRET (Fluorescence Resonance Energy Transfer)-based assays, used to monitor the transpeptidation activity in vitro of recombinant sortases, expressed in and purified from E. coli, on peptides containing the LPXTG-like sorting signals of different pilins have clearly demonstrated that this signal cannot be the sole determinant of C sortase specificity during pilin protein recognition. Indeed, both sortases from pilus type 1 were able to recognize and hydrolyse the various LPXTG-like motifs of the structural pilins of pilus-1 and pilus-2a (Cozzi et al., 2012), suggesting that GBS sortase C enzymes can recognize several and different sorting signals. The substrate promiscuity of class C sortases for pilin subunit incorporation into pili has been also confirmed by in vivo complementation studies of GBS strains in which the expression of a specific C sortase of one pilus type can polymerize pilus proteins belonging to other pili (Cozzi et al., 2012). These studies suggest that the promiscuous action shown by class C sortases on distinct substrates originates from their ability to cleave variable LPXTG-like sorting motifs. On the other hand, the preferential ancillary protein incorporation observed in in vivo genetic studies, not apparent in the cleavage reactions with peptides in vitro, suggests that the sortase substrate specificity may be due to recognition of more extensive structural determinants rather than a few specific residues. Thus, experiments performed in vitro involving only purified sortases in combination with LPXTG-like mimicking peptides are likely to be insufficient to define the molecular determinants of sortase specificity. In addition to the LPXTG-like motif there are obviously other factors that guide sortase function and their specific substrate recognition in the membrane environment in vivo. The assembly of the FCT-3 pilus in S. pyogenes serotype M3 represents an example of the involvement of accessory factors in pilus biogenesis. A single class B sortase polymerizes the backbone subunit FctA and incorporates the minor tip pilin Cpa. Nevertheless, the assembly of the FCT-3 pilus also requires the SipA2/LepA protein that shares sequence homology with signal peptidases and that functions as a chaperone (Zahner and Scott, 2008). The use of peptidase-like proteins as chaperones may be common to many pilus systems; pilus gene clusters in *A. naeslundii* contain genes encoding peptidase-like proteins (Mishra *et al.*, 2007), a SipA homologue is also required for the assembly of pilus type 2 in *S. pneumoniae* (Bagnoli *et al.*, 2008) and in *Streptococcus suis* the minor tip pilin (Sgp2) functions as a chaperone for the major backbone pilin (Sgp1) that forms the pilus shaft, since its presence together with the expression of the class C sortase (SrtG) is required to initiate the polymerization of Sgp1 through an unknown mechanism (Okura *et al.*, 2011).

The mechanism of assembly of *S. pneumoniae* pilus 1 appears to be more complicated. A particular feature of pneumococcal pilus islet 1 is the presence of three distinct sortases (SrtC-1, SrtC-2 and SrtC-3) required to covalently associate the three structural pilins: RrgB, forming the pilus backbone and the accessory pilins RrgA and RrgC that are distributed along the fibre, in single spots or in a cluster as observed for RrgA or, as reported in other studies, RrgA and RrgC at the distal and proximal ends of the RrgB shaft (Barocchi *et al.*, 2006; Falker *et al.*, 2008; Hilleringmann *et al.*, 2008, 2009; LeMieux *et al.*, 2008; Moschioni *et al.*, 2010). Although the redundant function of these sortases has been demonstrated in several studies, the relative contribution of each sortase enzyme in pilus formation remains quite controversial and fragmentary and no consensual overview has emerged in different groups. SrtC-1 is required for RrgB fibre pilus shaft formation by polymerization of RrgB, the major structural subunit. SrtC-1 is also involved in the association of the pilus with the cell wall by mediating the transpeptidation reaction between RrgB and the basal pilin RrgC (Manzano *et al.*, 2008; El Mortaji *et al.*, 2010). The substrate specificities of the two other sortases remain quite controversial. By biochemical and site-directed mutagenesis approaches using recombinant sortase proteins expressed in *E. coli*, El Mortaji *et al.* have demonstrated that both SrtC-2 and SrtC-3 are able to catalyse the formation of RrgB–RrgC and RrgB–RrgA complexes, although the tip pilin subunit RrgA appeared to be the preferred substrate for SrtC-2 (El Mortaji *et al.*, 2010). This sortase attaches RrgA to the pilin motif lysine 183 of RrgB, but is also able to catalyse the multimerization of RrgA. Similarly, SrtC-3 is also able to catalyse the incorporation of RrgA between RrgB subunits along the pilus shaft (El Mortaji *et al.*, 2012). With respect to biogenesis of *S. pneumoniae* pilus 1, the challenge will be now to solve the discrepancies reported in several studies trying to elucidate the molecular mechanism of the enzymatic activity of class C sortases in the native pili. The redundancy and the promiscuity in the substrates that pilus-specific sortases recognize might explain different and/or often discrepant experimental evidence reported in several studies, making it difficult to elucidate in detail the mechanism of pilus assembly in different Gram-positive bacteria.

9.5 Conclusions

Sortase enzymes have a critical role in Gram-positive bacteria pathogenesis due to their function to covalently link to the bacterial cell-wall surface proteins or pilus polymers. Pili structures have been discovered in the last years as important virulence factors as well as immunogenic vaccine candidates (Telford *et al.*, 2006). Because of the importance of their substrates for a successful bacterial infection, sortases represent an attractive antivirulence/therapeutic target.

Pilus-specific sortases are present in several copies in a genome and in different variants across species and strains, and are responsible for catalysing transpeptidation reactions linking pilin subunits. Gram-positive bacteria differ in their mechanism of pilus biogenesis with respect to the number of pilin subunits, number and class of sortases and type of substrates that these enzymes recognize (sorting signals and pilin motifs in the pilin subunits and cell-wall acceptors). Despite the variety of structures, substrate recognition motifs, nucleophile acceptors and physiological roles associated with this large enzyme family, sortases share a common enzymatic mechanism. However, these specific enzymes exhibit a redundant role and are surprisingly promiscuous in their ability to recognize a variety of sorting signals and amino groups.

In conclusion, although in the last decade new insights into the structure, mechanism of action, substrate specificity and regulation of sortases have been reported in different Gram-positive bacteria, several studies reported controversial and/or discrepant data and many issues remain unresolved, for example why some pili are associated with multiple sortases and what determines sortase substrate specificity in pilus assembly at the membrane environment during the establishment and persistence of infections by Gram-positive microorganisms.

References

Bagnoli, F., Moschioni, M., Donati, C., Dimitrovska, V., Ferlenghi, I., Facciotti, C., Muzzi, A., Giusti, F., Emolo, C., Sinisi, A., Hilleringmann, M., Pansegrau, W., Censini, S., Rappuoli, R., Covacci, A., Masignani, V. and Barocchi, M.A. (2008) A second pilus type in *Streptococcus pneumoniae* is prevalent in emerging serotypes and mediates adhesion to host cells. *Journal of Bacteriology* 190, 5480–5492.

Barocchi, M.A., Ries, J., Zogaj, X., Hemsley, C., Albiger, B., Kanth, A., Dahlberg, S., Fernebro, J., Moschioni, M., Masignani, V., Hultenby, K., Taddei, A.R., Beiter, K., Wartha, F., von Euler, A., Covacci, A., Holden, D.W., Normark, S., Rappuoli, R. and Henriques-Normark, B. (2006) A pneumococcal pilus influences virulence and host inflammatory responses. *Proceedings of the National Academy of Sciences of the United States of America* 103, 2857–2862.

Buccato, S., Maione, D., Rinaudo, C.D., Volpini, G., Taddei, A.R., Rosini, R., Telford, J.L., Grandi, G. and Margarit, I. (2006) Use of *Lactococcus lactis* expressing pili from group B *Streptococcus* as a broad-coverage vaccine against streptococcal disease. *Journal of Infectious Diseases* 194, 331–340.

Budzik, J.M., Marraffini, L.A. and Schneewind, O. (2007) Assembly of pili on the surface of *Bacillus cereus* vegetative cells. *Molecular Microbiology* 66, 495–510.

Budzik, J.M., Marraffini, L.A., Souda, P., Whitelegge, J.P., Faull, K.F. and Schneewind, O. (2008a) Amide bonds assemble pili on the surface of

bacilli. *Proceedings of the National Academy of Sciences of the United States of America* 105, 10215–10220.

Budzik, J.M., Oh, S.Y. and Schneewind, O. (2008b) Cell wall anchor structure of BcpA pili in *Bacillus anthracis*. *Journal of Biological Chemistry* 283, 36676–36686.

Budzik, J.M., Oh, S.Y. and Schneewind, O. (2009) Sortase D forms the covalent bond that links BcpB to the tip of *Bacillus cereus* pili. *Journal of Biological Chemistry* 284, 12989–12997.

Chen, P., Cisar, J.O., Hess, S., Ho, J.T. and Leung, K.P. (2007) Amended description of the genes for synthesis of *Actinomyces naeslundii* T14V type 1 fimbriae and associated adhesin. *Infection and Immunity* 75, 4181–4185.

Comfort, D. and Clubb, R.T. (2004) A comparative genome analysis identifies distinct sorting pathways in Gram-positive bacteria. *Infection and Immunity* 72, 2710–2722.

Cozzi, R., Malito, E., Nuccitelli, A., D'Onofrio, M., Martinelli, M., Ferlenghi, I., Grandi, G., Telford, J.L., Maione, D. and Rinaudo, C.D. (2011) Structure analysis and site-directed mutagenesis of defined key residues and motives for pilus-related sortase C1 in group B *Streptococcus*. *FASEB Journal: Official Publication of the Federation of American Societies for Experimental Biology* 25, 1874–1886.

Cozzi, R., Prigozhin, D., Rosini, R., Abate, F., Bottomley, M.J., Grandi, G., Telford, J.L., Rinaudo, C.D., Maione, D. and Alber, T. (2012) Structural basis for group B *Streptococcus* pilus 1 sortases C regulation and specificity. *PloS One* 7, e49048.

Cozzi, R., Zerbini, F., Assfalg, M., D'Onofrio, M., Biagini, M., Martinelli, M., Nuccitelli, A., Norais, N., Telford, J.L., Maione, D. and Rinaudo, C.D. (2013) Group B *Streptococcus* pilus sortase regulation: a single mutation in the lid region induces pilin protein polymerization *in vitro*. *FASEB Journal: Official Publication of the Federation of American Societies for Experimental Biology* Apr 30.

Danne, C., Entenza, J.M., Mallet, A., Briandet, R., Debarbouille, M., Nato, F., Glaser, P., Jouvion, G., Moreillon, P., Trieu-Cuot, P. and Dramsi, S. (2011) Molecular characterization of a *Streptococcus gallolyticus* genomic island encoding a pilus involved in endocarditis. *Journal of Infectious Diseases* 204, 1960–1970.

Dramsi, S., Trieu-Cuot, P. and Bierne, H. (2005) Sorting sortases: a nomenclature proposal for the various sortases of Gram-positive bacteria. *Research in Microbiology* 156, 289–297.

Dramsi, S., Caliot, E., Bonne, I., Guadagnini, S., Prevost, M.C., Kojadinovic, M., Lalioui, L.,

Poyart, C. and Trieu-Cuot, P. (2006) Assembly and role of pili in group B streptococci. *Molecular Microbiology* 60, 1401–1413.

El Mortaji, L., Terrasse, R., Dessen, A., Vernet, T. and Di Guilmi, A.M. (2010) Stability and assembly of pilus subunits of *Streptococcus pneumoniae. Journal of Biological Chemistry* 285, 12405–12415.

El Mortaji, L., Fenel, D., Vernet, T. and Di Guilmi, A.M. (2012) Association of RrgA and RrgC into the *Streptococcus pneumoniae* pilus by sortases C-2 and C-3. *Biochemistry* 51, 342–352.

Falker, S., Nelson, A.L., Morfeldt, E., Jonas, K., Hultenby, K., Ries, J., Melefors, O., Normark, S. and Henriques-Normark, B. (2008) Sortase-mediated assembly and surface topology of adhesive pneumococcal pili. *Molecular Microbiology* 70, 595–607.

Falugi, F., Zingaretti, C., Pinto, V., Mariani, M., Amodeo, L., Manetti, A.G., Capo, S., Musser, J.M., Orefici, G., Margarit, I., Telford, J.L., Grandi, G. and Mora, M. (2008) Sequence variation in group A *Streptococcus* pili and association of pilus backbone types with lancefield T serotypes. *Journal of Infectious Diseases* 198, 1834–1841.

Fittipaldi, N., Takamatsu, D., de la Cruz Dominguez-Punaro, M., Lecours, M.P., Montpetit, D., Osaki, M., Sekizaki, T. and Gottschalk, M. (2010) Mutations in the gene encoding the ancillary pilin subunit of the *Streptococcus suis* srtF cluster result in pili formed by the major subunit only. *PloS One* 5, e8426.

Gaspar, A.H. and Ton-That, H. (2006) Assembly of distinct pilus structures on the surface of *Corynebacterium diphtheriae. Journal of Bacteriology* 188, 1526–1533.

Hendrickx, A.P.A., Bonten, M.J.M., van Luit-Asbroek, M., Schapendonk, C.M.E., Kragten, A.H.M. and Willems, R.J.L. (2008) Expression of two distinct types of pili by a hospital-acquired *Enterococcus faecium* isolate. *Microbiology* 154, 3212–3223.

Hilleringmann, M., Giusti, F., Baudner, B.C., Masignani, V., Covacci, A., Rappuoli, R., Barocchi, M.A. and Ferlenghi, I. (2008) Pneumococcal pili are composed of protofilaments exposing adhesive clusters of Rrg A. *PLoS Pathogens* 4, e1000026.

Hilleringmann, M., Ringler, P., Muller, S.A., De Angelis, G., Rappuoli, R., Ferlenghi, I. and Engel, A. (2009) Molecular architecture of *Streptococcus pneumoniae* TIGR4 pili. *Embo Journal* 28, 3921–3930.

Kang, H.J. and Baker, E.N. (2012) Structure and assembly of Gram-positive bacterial pili: unique covalent polymers. *Current Opinion in Structural Biology* 22, 200–207.

Kang, H.J., Coulibaly, F., Proft, T. and Baker, E.N. (2011) Crystal structure of Spy0129, a *Streptococcus pyogenes* class B sortase involved in pilus assembly. *PloS one* 6, e15969.

Khare, B., Fu, Z.Q., Huang, I.H., Ton-That, H. and Narayana, S.V. (2011) The crystal structure analysis of group B *Streptococcus* sortase C1: a model for the 'lid' movement upon substrate binding. *Journal of Molecular Biology* 414, 563–577.

Lauer, P., Rinaudo, C.D., Soriani, M., Margarit, I., Maione, D., Rosini, R., Taddei, A.R., Mora, M., Rappuoli, R., Grandi, G. and Telford, J.L. (2005) Genome analysis reveals pili in Group B *Streptococcus. Science* 309, 105.

LeMieux, J., Woody, S. and Camilli, A. (2008) Roles of the sortases of *Streptococcus pneumoniae* in assembly of the RlrA pilus. *Journal of Bacteriology* 190, 6002–6013.

Lu, G., Qi, J., Gao, F., Yan, J., Tang, J. and Gao, G.F. (2011) A novel 'open-form' structure of sortase C from *Streptococcus suis. Proteins* 79, 2764–2769.

Mandlik, A., Das, A. and Ton-That, H. (2008) The molecular switch that activates the cell wall anchoring step of pilus assembly in gram-positive bacteria. *Proceedings of the National Academy of Sciences of the United States of America* 105, 14147–14152.

Manzano, C., Contreras-Martel, C., El Mortaji, L., Izore, T., Fenel, D., Vernet, T., Schoehn, G., Di Guilmi, A.M. and Dessen, A. (2008) Sortase-mediated pilus fiber biogenesis in *Streptococcus pneumoniae. Structure* 16, 1838–1848.

Manzano, C., Izore, T., Job, V., Di Guilmi, A.M. and Dessen, A. (2009) Sortase activity is controlled by a flexible lid in the pilus biogenesis mechanism of gram-positive pathogens. *Biochemistry* 48, 10549–10557.

Marraffini, L.A., Dedent, A.C. and Schneewind, O. (2006) Sortases and the art of anchoring proteins to the envelopes of gram-positive bacteria. *Microbiology and Molecular Biology Reviews* 70, 192–221.

Mazmanian, S.K., Ton-That, H. and Schneewind, O. (2001) Sortase-catalysed anchoring of surface proteins to the cell wall of *Staphylococcus aureus. Molecular Microbiology* 40, 1049–1057.

Mishra, A., Das, A., Cisar, J.O. and Ton-That, H. (2007) Sortase-catalyzed assembly of distinct heteromeric fimbriae in *Actinomyces naeslundii. Journal of Bacteriology* 189, 3156–3165.

Mora, M., Bensi, G., Capo, S., Falugi, F., Zingaretti, C., Manetti, A.G., Maggi, T., Taddei, A.R., Grandi, G. and Telford, J.L. (2005) Group A

Streptococcus produce pilus-like structures containing protective antigens and Lancefield T antigens. *Proceedings of the National Academy of Sciences of the United States of America* 102, 15641–15646.

Moschioni, M., Emolo, C., Biagini, M., Maccari, S., Pansegrau, W., Donati, C., Hilleringmann, M., Ferlenghi, I., Ruggiero, P., Sinisi, A., Pizza, M., Norais, N., Barocchi, M.A. and Masignani, V. (2010) The two variants of the *Streptococcus pneumoniae* pilus 1 RrgA adhesin retain the same function and elicit cross-protection *in vivo*. *Infection and Immunity* 78, 5033–5042.

Nallapareddy, S.R., Singh, K.V., Sillanpaa, J., Garsin, D.A., Hook, M., Erlandsen, S.L. and Murray, B.E. (2006) Endocarditis and biofilm-associated pili of *Enterococcus faecalis*. *Journal of Clinical Investigation* 116, 2799–2807.

Neiers, F., Madhurantakam, C., Falker, S., Manzano, C., Dessen, A., Normark, S., Henriques-Normark, B. and Achour, A. (2009) Two crystal structures of pneumococcal pilus sortase C provide novel insights into catalysis and substrate specificity. *Journal of Molecular Biology* 393, 704–716.

Okura, M., Osaki, M., Fittipaldi, N., Gottschalk, M., Sekizaki, T. and Takamatsu, D. (2011) The minor pilin subunit Sgp2 is necessary for assembly of the pilus encoded by the srtG cluster of *Streptococcus suis*. *Journal of Bacteriology* 193, 822–831.

Oxaran, V., Ledue-Clier, F., Dieye, Y., Herry, J.M., Pechoux, C., Meylheuc, T., Briandet, R., Juillard, V. and Piard, J.C. (2012) Pilus biogenesis in *Lactococcus lactis*: molecular characterization and role in aggregation and biofilm formation. *PloS One* 7, e50989.

Persson, K. (2011) Structure of the sortase AcSrtC-1 from *Actinomyces oris*. *Acta Crystallographica Section D, Biological Crystallography* 67, 212–217.

Rosini, R., Rinaudo, C.D., Soriani, M., Lauer, P., Mora, M., Maione, D., Taddei, A., Santi, I., Ghezzo, C., Brettoni, C., Buccato, S., Margarit, I., Grandi, G. and Telford, J.L. (2006) Identification of novel genomic islands coding for antigenic pilus-like structures in *Streptococcus agalactiae*. *Molecular Microbiology* 61, 126–141.

Spirig, T., Weiner, E.M. and Clubb, R.T. (2011) Sortase enzymes in Gram-positive bacteria. *Molecular Microbiology* 82, 1044–1059.

Swaminathan, A., Mandlik, A., Swierczynski, A., Gaspar, A., Das, A. and Ton-That, H. (2007) Housekeeping sortase facilitates the cell wall anchoring of pilus polymers in *Corynebacterium diphtheriae*. *Molecular Microbiology* 66, 961–974.

Swierczynski, A. and Ton-That, H. (2006) Type III pilus of corynebacteria: pilus length is determined by the level of its major pilin subunit. *Journal of Bacteriology* 188, 6318–6325.

Takamatsu, D., Nishino, H., Ishiji, T., Ishii, J., Osaki, M., Fittipaldi, N., Gottschalk, M., Tharavichitkul, P., Takai, S. and Sekizaki, T. (2009) Genetic organization and preferential distribution of putative pilus gene clusters in *Streptococcus suis*. *Veterinary Microbiology* 138, 132–139.

Telford, J.L., Barocchi, M.A., Margarit, I., Rappuoli, R. and Grandi, G. (2006) Pili in gram-positive pathogens. *Nature Reviews Microbiology* 4, 509–519.

Ton-That, H. and Schneewind, O. (2003) Assembly of pili on the surface of *Corynebacterium diphtheriae*. *Molecular Microbiology* 50, 1429–1438.

Ton-That, H. and Schneewind, O. (2004) Assembly of pili in Gram-positive bacteria. *Trends in Microbiology* 12, 228–234.

Ton-That, H., Marraffini, L.A. and Schneewind, O. (2004) Protein sorting to the cell wall envelope of Gram-positive bacteria. *Biochimica et Biophysica Acta* 1694, 269–278.

Wu, C., Mishra, A., Reardon, M.E., Huang, I.H., Counts, S.C., Das, A. and Ton-That, H. (2012) Structural determinants of *Actinomyces sortase* SrtC2 required for membrane localization and assembly of type 2 fimbriae for interbacterial coaggregation and oral biofilm formation. *Journal of Bacteriology* 194, 2531–2539.

Yeung, M.K., Donkersloot, J.A., Cisar, J.O. and Ragsdale, P.A. (1998) Identification of a gene involved in assembly of *Actinomyces naeslundii* T14V type 2 fimbriae. *Infection and Immunity* 66, 1482–1491.

Zahner, D. and Scott, J.R. (2008) SipA is required for pilus formation in *Streptococcus pyogenes* serotype M3. *Journal of Bacteriology* 190, 527–535.

Zahner, D., Gandhi, A.R., Yi, H. and Stephens, D.S. (2011) Mitis group streptococci express variable pilus islet 2 pili. *PloS One* 6, e25124.

Zhang, R., Wu, R., Joachimiak, G., Mazmanian, S.K., Missiakas, D.M., Gornicki, P., Schneewind, O. and Joachimiak, A. (2004) Structures of sortase B from *Staphylococcus aureus* and *Bacillus anthracis* reveal catalytic amino acid triad in the active site. *Structure* 12, 1147–1156.

Zong, Y., Mazmanian, S.K., Schneewind, O. and Narayana, S.V. (2004) The structure of sortase B, a cysteine transpeptidase that tethers surface protein to the *Staphylococcus aureus* cell wall. *Structure* 12, 105–112.

10 Pili of *Streptococcus pyogenes*

June R. Scott

Microbiology and Immunology, Emory University School of Medicine, Atlanta, USA

10.1 Introduction

Streptococcus pyogenes, also called Group A streptococcus (GAS), is one of the most common human pathogens. GAS can initiate infection in the oropharynx or on the skin. In addition to causing usually self-limiting infections, including pharyngitis ('strep throat') and pyoderma, GAS causes serious invasive diseases that destroy tissue and can be fatal. These include myositis and fasciitis as well as streptococcal toxic shock. The immune sequelae that sometimes follow GAS infections include rheumatic heart disease and glomerulonephritis and are further cause for concern.

There are many genetically different strains of GAS, which were first classified by Lancefield using serotyping of the surface M (Dochez *et al.*, 1919; Lancefield, 1928) and T (Lancefield, 1940) proteins. The M protein, so-called because its presence endows GAS colonies with a matt appearance, is the predominant surface protein of GAS. The M protein serves as an important virulence factor that affects the immune response to the bacterium and is involved in bacterial–bacterial association following attachment to the human host (Caparon *et al.*, 1991). The T protein was originally identified as a surface protein resistant to trypsin, and it has been used for over 75 years to classify GAS strains. Recently, it has been shown to constitute the shaft of peritrichous pili.

Although different GAS strains encode serologically different types of pili, and their structure and assembly differs, some general rules hold for all GAS strains. In all GAS strains, the proteins necessary for synthesis and assembly of pili are encoded in a single locus (Fig. 10.1). Pili are composed of many copies of the backbone protein (the T antigen), linked end-to-end by a sortase family transpeptidase, or pilin polymerase. All additional proteins present in pili are found in minor amounts and are therefore referred to as 'ancillary proteins' (APs) or 'minor pilins'. In many GAS strains, ancillary protein 1 (AP1) has been shown to be located at the pilus tip and acts as the adhesin. In some strains, ancillary protein 2 (AP2) serves to anchor the pilus to the cell wall, a process thought to be catalysed normally by the housekeeping sortase, encoded elsewhere in the genome. Because pili are involved in attachment of GAS to the human host (Mora *et al.*, 2005) and in biofilm formation (Manetti *et al.*, 2007, 2010; Köller *et al.*, 2010; Becherelli *et al.*, 2012; Kimura *et al.*, 2012), a greater understanding of their synthesis and assembly should greatly assist with rational vaccine and drug development.

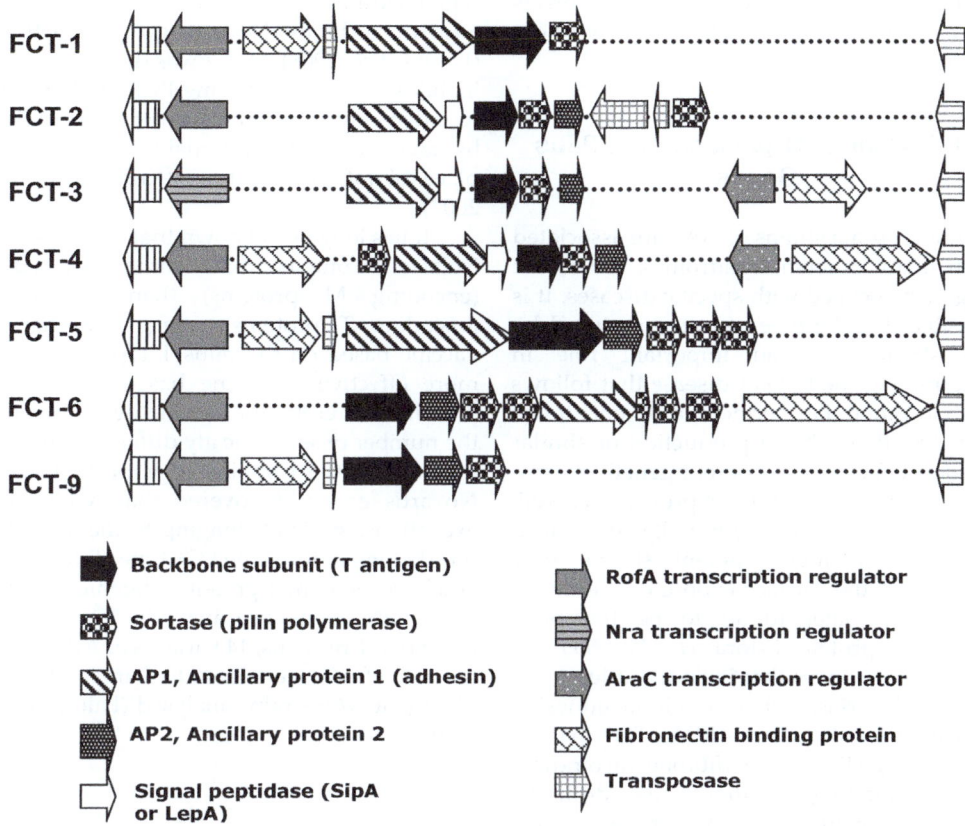

FCT-1

FCT-2

FCT-3

FCT-4

FCT-5

FCT-6

FCT-9

Backbone subunit (T antigen)

Sortase (pilin polymerase)

AP1, Ancillary protein 1 (adhesin)

AP2, Ancillary protein 2

Signal peptidase (SipA or LepA)

RofA transcription regulator

Nra transcription regulator

AraC transcription regulator

Fibronectin binding protein

Transposase

Fig. 10.1. Diagram of the FCT (pilus-encoding) genetic regions of different strains of GAS. These genes are sufficient for production of pili in a foreign Gram⁺ host, although not all are necessary for this process. In strains encoding the SipA protein it is essential for pilus polymerization, although its role has not been defined. (Adapted from a figure in Falugi *et al.* (2008), kindly supplied by Dr I. Margarit Y Ros.)

10.2 Identification of Pili in GAS

The recent identification of pili in group B streptococci (GBS) by their ability to produce mouse-protective antibodies (Lauer *et al.*, 2005; Maione *et al.*, 2005), and the genetic analyses of the genes encoding these and the pili of *Corynebacteria diphtheriae* (Ton-That and Schneewind, 2004), led the Novartis group to search for possible pilus-encoding loci in the available genome sequences of GAS (Mora *et al.*, 2005). They looked for genes encoding proteins predicted to contain C-terminal cell-wall sorting signal (CWSS) sequences preceded by a motif related to the conserved LPXTG (required for pilus polymerization in

other Gram⁺ pili) (Schneewind *et al.*, 1992) that lie in proximity to a gene predicted to encode a sortase-family transpeptidase. Such loci have now been identified in all GAS genomes (see below). The Novartis group demonstrated that this GAS locus encodes pili consisting of covalently associated subunits that are not dissociated by heating in SDS (Mora *et al.*, 2005). Surprisingly, they found that these pili are recognized by anti-T antibodies, and showed that the T antigen constitutes the multisubunit backbone of a pilus structure that can be visualized by immunogold EM. These authors suggested that pilus proteins are likely to be excellent vaccine candidates, as they are in GBS, since immunization with these proteins protects mice from mucosal challenge by virulent GAS.

10.3 Genetic Organization of Pilus Genes

Although some strains of GAS are associated with many different syndromes, others are largely associated with specific diseases. It is possible that the type of pilus produced by the strain plays an important role in determining the type of disease that follows GAS infection. Furthermore, different strains of GAS often achieve production of similar virulence factors by different genetic means, and this is true of pilus expression as well. Bessen and Kalia (2002) initially recognized that, in the four different strains they examined, the surface-exposed T antigen, which was later found to be the pilus backbone protein (Mora *et al.*, 2005), is encoded in a locus that includes at least one microbial surface cell recognition adhesion matrix molecule (MSCRAMM). They named this locus the FCT region (fibronectin binding, collagen binding, T antigen). As additional genome sequences of GAS strains became available, this original observation proved correct: all GAS strains examined encode a single FCT region located in the same position between the same two flanking genes (Fig. 10.1). In addition to the T antigen, the FCT region encodes: (1) one or more ancillary pilin proteins present in minor amounts in pili (which I will call AP1, AP2, etc. in order of appearance in the gene cluster, following the lead of Falugi (Falugi *et al.*, 2008)); (2) one or more transpeptidase(s), including a member of the sortase structural Class B (pilin polymerase); and (3) in the most common GAS types, a gene with weak homology to the bacterial signal peptidase. In addition, usually at least one transcriptional regulator is encoded in the FCT region and these are transcribed in the opposite direction from the pilus genes.

Analyses by the Bessen group of >100 strains representing a broad variety of variants of GAS led them to identify eight different arrangements of genes in the FCT region (Kratovac *et al.*, 2007), and a ninth was added later by the Novartis group (Fig 10.1) (Falugi *et al.*, 2008). FCT-7 was defined by the inability to detect *cpa* using PCR, and FCT-8, found in only three strains, differs from FCT-4 by having the transcriptional regulator *nra* in place of *rofA* (see below) (Kratovac *et al.*, 2007).

It has long been known that clinical GAS isolates encompass many more *emm* types (encoding M proteins) than *tee* types (encoding T proteins), suggesting that a vaccine based on the pilus T type might be more effective than one based on the M protein. Therefore, for vaccine development, the number of serologically different types of pili among GAS strains is important. The Novartis group discovered that, with two exceptions, strains belonging to the same M type belong to the same FCT type (Table 10.1) and have the same T protein antigenic variant. Using DNA sequence analysis, 15 different backbone T proteins, 14 variants of AP1 and 5 variants of AP2 were identified among the 57 divergent GAS strains analysed (Falugi *et al.*, 2008).

10.4 Role of FCT Genes in Pilus Assembly

The FCT regions encoding GAS pili are divergent in structure and function (Fig. 10.1). At this time, the best-studied are FCT-2 and FCT-3, which are highly homologous to each other, so these will be discussed first.

Table 10.1. Strains belonging to the same M type usually have pilus backbone proteins of the same T type and belong to the same FCT type. (From Falugi *et al.*, 2008.)

FCT type	M types
FCT-1	M6
FCT-2	M1
FCT-3	M3, 5, 18, 33, 49, 53, 77
FCT-4	M9, 11, 12, 22, 28, 50, 78, 89
FCT-5	M4
FCT-6	M2
FCT-9	M75

10.4.1 The major pilin protein: the T antigen

Proteins present in a pilus structure are called pilins. Pili are composed of multiple identical subunits that form the backbone or shaft of the structure, often referred to as 'the major pilus protein', and they may also be decorated with 'minor' or 'accessory' proteins. In GAS, for FCT-2 and FCT-3, it was shown that the backbone protein is the only pilin required for formation of the high molecular weight polymers present in extracts boiled in SDS (Barnett *et al.*, 2004; Mora *et al.*, 2005) and for the hairlike structures recognizable by immunogold EM (Mora *et al.*, 2005). The genes encoding the other pilins can be deleted; only the sortase and Sip protein (see below) are required for pilus polymerization in addition to the backbone pilin, which is the T antigen. The gene for the backbone pilin has often been referred to by its 'SPy' gene number, but this is confusing because such numbers are strain-specific and the same number is used for a different gene in each strain. Thus, it is more useful and appropriate to use the designation that recognizes that the backbone protein is the T antigen, e.g. '*tee1*' or '*tee3*'.

Like all other pilins of Gram⁺ bacteria, the T protein has an N-terminal signal sequence, removed on transport through the cell membrane, and a cell-wall sorting signal (CWSS), consisting of a hydrophobic region and a positively charged C-terminus. However, the T protein does not include the LPXTG motif at the start of the CWSS, which is conserved in pilins of most other Gram+ bacteria. Instead, in FCT-3, the motif in *tee3* is QVPTG and in the FCT-2 cluster *tee1* gene it is VPTGV. It was demonstrated genetically for *tee3* that the CWSS motif QVPTG is required for polymerization of the protein and when it is replaced with LPXTG the protein is not polymerized (Barnett *et al.*, 2004). Mass spectrometric (MS) analysis of T1 pili showed that the threonine in the VPTGV motif is linked to an ε amino group of a lysine in the next T1 subunit (Kang *et al.*, 2007), and the corresponding threonine in the QVPTG motif of T3 was independently shown genetically to be required for linkage of T3 to the next T3 subunit (Quigley *et al.*, 2009). In T3, this lysine

was identified genetically (Quigley *et al.*, 2009) and in T1, MS analysis showed that the homologous lysine forms the inter-molecular bond with the T in VPTGV (Kang *et al.*, 2007).

The elegant crystallographic study of Kang *et al.* (2007) suggested that the T1 pilins stack head to tail to form the pili. It also revealed the unprecedented presence of two intramolecular peptide bonds in the T1 protein, one in each of the two domains of the protein. The presence of these intramolecular isopeptide bonds was confirmed by MS analysis. These bonds lie between lysine and asparagine side chains and appear to be self-generated. Based on sequence comparisons, Kang *et al.* suggested that such intramolecular peptide bonds are a conserved feature of Gram⁺ pilins and this has proven correct. Their presence causes dimers and trimers of the T protein to migrate more slowly in SDS-PAGE and, using this as an assay, site-specific mutagenesis showed that formation of these bonds is not essential for pilin polymerization (Quigley *et al.*, 2009). However, they may be important for stability of the resulting structure, as suggested by Kang *et al.* (2007). Kang *et al.* proposed that these unusual intramolecular peptide bonds may be the functional equivalent in Gram⁺ proteins of disulfide bonds, which are formed in the periplasm in some Gram⁻ proteins but are rarely found in proteins in Gram⁺ bacteria. Subsequent work demonstrated that the intra-molecular isopeptide bonds are important in providing T proteins with resistance to trypsin, the attribute for which T proteins were named, and for heat denaturation (Kang and Baker, 2009). In addition, the intra-molecular bonds are required for the unusual resistance of the T1 protein to mechanical force (Alegre-Cebollada *et al.*, 2010). The importance of these physical attributes for colonization and pathogenesis has not yet been investigated.

10.4.2 Pilin polymerase

The sortase family transpeptidase responsible for covalently linking pilin subunits to each other is sometimes called the pilin poly-merase. In FCT-2 and -3, the gene encoding this protein is called *srtC1* or *srtC2*, respectively (Barnett *et al.*, 2004). This is the only sortase-encoding gene in FCT-3 and is required for polymerization of pili (Zahner and Scott, 2008; Quigley *et al.*, 2009). For FCT-2, although there is a second sortase-encoding gene with homology to pilin polymerases of structural class C (the class encompassing most pilin polymerases; Spirig *et al.*, 2011), this gene is not required for polymerization of pili and its role has not been further investigated (Abbot *et al.*, 2007).

Sortases contain an N-terminal transmembrane anchor and conserved motifs involved in recognition of the specific amino acids in the two substrates that they join covalently. It is important to remember that although each sortase has substrate preferences, excess enzyme may result in use of less-preferred proteins as substrates. In pili from most Gram⁺ bacteria, the pilin polymerase links the ε amino group of a lysine in the pilin motif of one pilus backbone protein to the threonine in the LPXTG motif found at the 5′ end of the CWSS of the next subunit. These motifs are necessary but not sufficient for recognition by the pilin polymerase. However, none of the GAS pilins has either a pilin motif or the LPXTG sequence. Therefore, it may not be surprising that the pilin polymerase of GAS differs structurally from most other Gram+ pilin polymerases, which are sortases belonging to structural Class C (Kang *et al.*, 2011). It is, however, a surprise that the GAS polymerase belongs to sortase Class B, whose founding member is the *Staphylococcus aureus* sortase that anchors iron binding proteins to the cell wall, since in this organism this enzyme has no apparent role in pilus polymerization. However, the CWSS motif (NPQTN) recognized by *S. aureus* SrtB (Mazmanian *et al.*, 2002) also differs from the motifs present in the GAS pilin proteins, suggesting that different Class B sortases have different specificities. Further information about the mechanisms of substrate recognition by sortases and their mechanisms of action will be needed to understand these relationships.

The nomenclature for the pilin polymerase in GAS is confusing for historical reasons. Before it was clear that the T antigen is the pilus backbone protein, the sortase required specifically for surface localization of this protein in FCT-2 and FCT-3 strains was identified and named SrtC (Barnett *et al.*, 2004) because its sequence differs significantly from the SrtB gene previously identified as needed for detectable surface localization of T6 (FCT-1) (Barnett and Scott, 2002). However, when pilin polymerases were studied further and grouped into classes based on structure, it became clear that the pilin polymerase in GAS strains belongs to structural Class B and not Class C (Kang *et al.*, 2011). Perhaps it is time to rename these 'sortases', whose role and structure differ from the original *S. aureus* housekeeping sortase (Schneewind *et al.*, 1992) (a Class A sortase), based on their pilin polymerization function and on their structural class.

10.4.3 SipA2/LepA

The GAS pilus gene clusters FCT-2 through FCT-4 all contain a gene encoding a protein whose closest relative is the signal (or leader) peptidase, variously called *sipA* or *lepA*. In GAS, this gene is called *sipA* (Fig. 10.1). However, its amino acid similarity to signal peptidases is low and the active site serine required for peptidase activity is not present in the FCT protein. Although the FCT *sipA* gene is absolutely required for polymerization of pilins (Zahner and Scott, 2008), its function remains unclear. It was suggested that SipA might play the role of a chaperone for one or more individual pilin proteins and facilitate interaction of this protein with the pilin polymerase (Zahner and Scott, 2008). The lack of success of several excellent groups in development of an *in vitro* assay for the GAS pilus polymerase may indicate that an additional factor, possibly a chaperone, is required for transpeptidase activity of GAS pilin polymerases. Perhaps SipA plays this role.

10.4.4 Minor or accessory pilin proteins

Immunogold EM demonstrated that the other CWSS-containing proteins encoded in the

FCT-1, FCT-2 and FCT-3 loci are present in the pili found on the GAS surface (Mora *et al.*, 2005; Quigley *et al.*, 2009; Becherelli *et al.*, 2012). Because both EM and Western blots suggest that pili contain many fewer subunits of these two proteins than of the backbone protein, they are sometimes called minor pilin proteins. Since pili are polymerized and are visible on the cell surface in the absence of either of these proteins, they are also referred to as accessory proteins. In FCT-2 and FCT-3, the motif at the start of the CWSS for the first of these accessory proteins, AP1 (SPy125 in T1 and Cpa in T3) differs from the canonical LPXTG dramatically (VPTGV for FCT-2 and VPPTG for FCT-3), while AP2 of FCT-2 encodes the consensus sequence LPSTG and AP2 of FCT-3 encodes LPLAG, which is similar to the consensus. Although they are not required for pilus formation, both AP1 and AP2 are required for the normal adherence function of T1 (FCT-2) pili (see below; Abbot *et al.*, 2007), but their precise role has been difficult to establish.

Genetic analyses by Quigley *et al.* (2009) showed that the first FCT gene in the M3 FCT-3 locus, which had been named *cpa* for its ability to bind collagen (Podbielski *et al.*, 1999), can only be incorporated at the tip of pili. This was shown by determining that the lysine of the T3 shaft protein, K173, required for T3–T3 polymerization, is also required for addition of Cpa to T3 and this localization was confirmed by immunogold EM (Quigley *et al.*, 2009). This work also demonstrated that the CWSS motif VPPTG in Cpa is required for addition of Cpa to the T3 pilus, and thus identified the amino acids in each subunit that form the covalent bond.

Crystallographic analysis by the Banfield group of the protein that forms the tip of GAS FCT-2 pili identified another unusual type of internal covalent linkage (Pointon *et al.*, 2010). They found an internal thioester bond between the side chains of a cysteine and a glutamine residue and verified its presence by mass spectrometry. They also demonstrated by site-directed mutagenesis that removal of this bond compromises the ability of the GAS strain to adhere to model host cells (HaCaT). The authors suggested the possibility that during attachment to the host, the internal thioester bond in the pilus protein is broken and instead a similar bond forms with a host protein. This would establish a covalent linkage between the infecting bacterium and its host cell! It will be exciting to learn the importance of formation of this bond in colonization.

Because one of the accessory pilins in other Gram+ pili has often been identified as the anchor that is covalently bound to the cell-wall peptidoglycan, Smith *et al.* (2010) investigated the effect of deletion of each of the FCT-2 proteins on the distribution of pili in the culture supernatant versus cell wall. Using Western blots, they found that there is a small fraction of pili present in the culture supernatant of wild-type cells and of cells deleted for the gene encoding AP1, the adhesin. However, cells from which AP2 was deleted released most of their pili into the supernatant. This strongly implies that AP2 may serve as the pilus anchor protein. Further, the motif at the start of the CWSS in AP2 corresponds to the consensus LPXTG recognized by the 'housekeeping' sortase that anchors most cell-wall proteins in GAS. Thus, it seems likely that when AP2 is inserted into pili, it is recognized by the housekeeping sortase and attached to the peptidoglycan like other cell surface proteins. However, it should be remembered that in the absence of AP2, Mora *et al.* (2005) found pili attached to the cell wall and this attachment is considered to be covalent because it resists boiling in SDS. Furthermore, overexpression of the FCT-3 region without AP2 results in covalent linkage of the T3 pili to the cell wall (Barnett *et al.*, 2004; Quigley *et al.*, 2009). Therefore, it appears that although AP2 serves as the pilus anchor when it is present, it is not absolutely required for covalent linkage of pili to peptidoglycan. Experimental determination of which sortase anchors GAS pili has not yet been made. However, it should be remembered that although sortases have preferences for specific sequences in specific proteins, when overexpressed, sortases may act on substrates that are not preferred.

10.4.5 Pilus length determination

From the earliest work in the Schneewind lab, a major assay used for pili was formation of a high molecular weight 'ladder' on a Western blot following treatment of cell extracts with hot SDS. This indicates that pili of many lengths, differing by the addition of a single backbone subunit, appear to be covalently linked to the cell wall, so there is no 'ruler' protein that determines pilus length. Instead, it is likely that the relative amounts of the proteins involved determine when interaction of AP1 with the T protein and sortase leads to initiation of a new pilus, and when AP2 interacts with (presumably) SrtA to covalently bind pili to the growing cell wall to terminate pilus growth. This is difficult to address experimentally because most GAS strains grown in laboratory culture produce so few pili that overexpression of FCT genes is usually used to improve detection, and this distorts the ratio of pilin proteins to SrtA and cell-wall synthesis machinery.

10.4.6 Roles of FCT-1 genes in assembly of T6 pili

The FCT-1 pilus gene cluster differs significantly from FCT-2 and FCT-3 since it has only two pilin genes (encoding T6 and AP1, also called FctX) and lacks *sipA* (Fig. 10.1). The Podbielski group (Nakata *et al.*, 2011) used site-directed mutagenesis and immunogold EM to investigate assembly of T6 pili. The results suggest FCT-1 pili are assembled similarly to FCT-2 and FCT-3 pili. Nakata *et al.* (2011) demonstrated the necessity of the LPSTG motif in the CWSS of T6 for polymerization of T6, and showed that this motif is not required for formation of T6-AP1 heterodimers. They also showed the near consensus CWSS motif LPSSG in AP1 is required for formation of the heterodimer with T6, and they identified the lysine in T6 (K175) required for pilus polymerization. As in the case of T3 pili (Quigley *et al.*, 2009), K175 is also needed for addition of AP1 to T6, demonstrating that AP1 can only be present on the tip of pili, as is also suggested by immunogold EM (Nakata *et al.*, 2011). The

presence of a consensus CWSS motif in the backbone T6 protein, which is recognized by the 'housekeeping' sortase, SrtA, may obviate the need for a second ancillary pilin to serve as an anchor.

An unusual feature of the FCT-1 type pili is that deletion of the gene for AP1 reduces the polymerization of T6, leading to formation of shorter polymers (Nakata *et al.*, 2011; Becherelli *et al.*, 2012). The effect of lack of the accessory protein on polymerization of the backbone protein has been seen previously only in *Streptococcus suis* pili (Okura *et al.*, 2011) and not in other GAS pili. Perhaps this suggests that in FCT-1, which lacks a SipA protein, the accessory pilin is needed for correct presentation of the polymerizing backbone protein to the sortase that acts as a pilin polymerase. The authors suggest that SrtB is predominant in this polymerization, although SrtA also seems able to polymerize T6 when none of these proteins is being overproduced (Nakata *et al.*, 2011). The latter result is probably not surprising since the canonical LPSTG motif is present in T6.

10.5 Biological Roles of Pili

10.5.1 Adherence

Because initiation of infection depends on bacterial adherence to the host, bacteria have usually evolved many – often redundant – mechanisms for attachment. Since GAS can initiate infection from several different sites on the host, attachment of this pathogen is expected to involve a great variety of host receptor proteins, including MSCRAMMs as well as specific receptors. Therefore, the role of individual bacterial adhesins in virulence is very complex and many different surface proteins may be involved in adherence to the same and to different host receptors. It should also be remembered that all of these surface adhesins appear to be anchored by the same 'housekeeping' sortase enzyme (SrtA), so, if the amount of this enzyme is limiting, or if there is a limit to the number of cell-wall sites available for attachment of surface proteins, deletion of the gene for one surface protein from GAS may affect the amount of another on the GAS surface (Barnett *et al.*, 2004). Thus, the design of an experiment to test the role of specific GAS surface proteins in adherence is critical, and interpretation may be complex.

It is assumed that the primary role for pili on Gram positive bacteria is rapid binding to a specific receptor on the surface of a host cell to initiate the infection process. After specific binding occurs, bacterial–bacterial interactions mediated by other proteins, like the M protein, may increase the probability of a successful infection. Because GAS is a human-limited pathogen, animal models for studies of virulence can only mimic specific aspects of disease. In particular, since most receptor interactions are exquisitely specific, and animals are unlikely to express the human host-cell surface protein receptor(s) that are used by GAS for attachment, animal models are unlikely to accurately reflect colonization of humans, the first step in pathogenesis. Therefore, to analyse colonization, adherence to tissue culture cells derived from humans is generally used. However, different human cell types express different surface proteins and such proteins may be altered during passage of tissue cultures. For this reason, the Kehoe group (Abbot *et al.*, 2007) studied adherence of GAS to primary explants of relevant human tissue, i.e. tissue derived from sites at which GAS colonization is initiated. They used primary keratinocytes to model skin colonization and freshly isolated tonsillar tissue from several different people to model respiratory tract colonization. In this work, using an M1 strain, the most common type of GAS in the Western world, they demonstrated that the T1 pili are required for adherence to skin cells and greatly enhance adherence to tonsillar tissue. In contrast, these pili are not required for this strain to adhere to either HEp-2 or A549 cells, but are required for adherence to HaCaT cells. The authors also note that HEp-2 cells originate from a HeLa cell contamination and A549 originates from carcinomatous tissue of human alveoli, a site not normally colonized by GAS. HaCaT cells, in contrast, are derived from human keratinocytes. In addition, Detroit 562 cells, derived from a human pharyngeal carcinoma, appear to give results similar to HaCaT cells (Manetti *et al.*, 2007).

However, in contrast, for an FCT-1 M6 strain, which produce T6 pili with little homology to T1 pili, Becherelli *et al.* (2012) found that attachment to A549 cells does require pili, indicating that different pilus types have different receptors. It appears, therefore, that HaCaT cells constitute one of the most relevant tissue culture models for T1 pilus-mediated adherence of the FCT-2 strain. In contrast, the use of HEp-2 or A549 cells is unlikely to mimic colonization in humans, but this does not appear to be the case for an FCT-1 strain expressing T6 pili.

Different host proteins have been implicated in adherence of GAS, including collagen and fibronectin. However, Abbot *et al.* (2007) found that T1 pilus-mediated adhesion was not dependent on or inhibited by either fibronectin or type I or type IV collagen. This strongly suggests that none of these host proteins serves as the primary receptor for GAS binding, but instead that pili bind directly to a host protein that remains to be identified.

Pilus-mediated adherence has been found to display biphasic kinetics in two different studies: adherence to tonsillar cells mediated by T1 pili (Abbot *et al.*, 2007) and adherence of the T6 adhesin protein to A549 cells (Becherelli *et al.*, 2012). It is possible that other GAS surface proteins might be responsible for the second, faster binding, phase. Such proteins include the M protein, which mediates bacterial–bacterial interaction on tonsillar tissue, at least for an M6T6 strain (Caparon *et al.*, 1991). However, the Kehoe group also found that the purified adhesin protein of T1 pili undergoes spontaneous cleavage, leading to loss of the N terminal region (Smith *et al.*, 2010). The remaining shorter protein, or GAS cells engineered to produce this shorter protein, still binds to HaCaT cells and may be responsible for the steeper phase of the adherence curve. In this case, the delay in rapid adherence may result from the time required for proteolytic processing to produce the shorter form of the adhesin. It is also possible that the second phase of the curve represents the predicted reaction of the thioester bond in the adhesin with a host protein to form a covalent bond (Pointon *et al.*, 2010; see above). Although the details of adherence need clarification, the Kehoe group's work makes it clear that for tonsillar attachment, the primary adhesin of the M1 strain of GAS is the pilus. Similarly, they found pili to be the primary adhesin for keratinocytes and HaCaT cells and, in contrast to attachment to tonsillar tissue, the M1 protein has no apparent role in attachment to keratinocytes (in agreement with Caparon et al., 1991).

10.5.2 Biofilm formation

After rapid specific binding occurs, bacterial–bacterial interactions mediated by other proteins, including the M protein (which is the predominant surface protein on GAS), may increase the probability of a successful infection. Such specific protein interactions often mediate biofilm formation, which is also important for virulence. Using the same GAS M1 strain as the Kehoe group, the Novartis group demonstrated that pili have a critical role in formation of biofilms on both polystyrene plastic wells and on polylysine-coated coverslips (Manetti et al., 2007). Investigations of strains representing many different FCT types showed the importance of pili in biofilm formation and, as might be expected for a complex phenotype, found an effect of culture medium and growth conditions as well as differences for specific strains of the same FCT type (Cho and Caparon, 2005; Köller et al., 2010; Manetti et al., 2010; Becherelli et al., 2012; Kimura et al., 2012).

Pili on the surface of GAS, which are important for initiation of colonization and the subsequent formation of biofilms, may also have a negative effect on production of disseminated disease, both by complex interactions with the host (Crotty Alexander et al., 2010) and by tethering the growing bacterial mass at the initial site of infection. Even if pilus synthesis is 'turned off', if there are many pili covering the bacterial surface when bacteria grow in humans (as opposed to growing in pure culture), it will take many generations before non-piliated bacteria are produced. Regulation of pilus synthesis is discussed below, but whether mechanisms exist to remove pili from the bacterial surface,

like retraction of type 4 pili in Gram negative bacteria, has never been addressed for Gram positive bacteria. It is also possible that breakage of pili by stochastic processes will separate the bacteria from the tip-located adhesin and release the bacterial cells to invade new host sites.

10.5.3 Roles of specific pilins in adherence

To address the role of the minor pilins in attachment and colonization, Abbot et al. (2007) deleted the gene for each (AP1=SPy125 or AP2=SPy130) in turn in an FCT-2 strain producing T1 pili. They found that GAS cells expressing pili that lack either of the minor pilins are unable to adhere to tonsillar tissue, in contrast to the parental strain. Therefore, although these minor pilin proteins are not required for production of visible pili, pili lacking either of these proteins are functionally defective.

Purified recombinant AP1 or AP2, but not the major pilus backbone T1 protein, was found to bind to Detroit cells (Manetti et al., 2007) and to HaCaT cells as well as to human tonsillar epithelium (Smith et al., 2010). In addition, Smith et al. found that specific antiserum to AP2 or to T1 did not interfere with adhesion of this GAS strain to HaCaT cells or tonsillar epithelium, while antiserum specific for AP1 inhibited binding. This demonstrates that only AP1 acts as the adhesin in these relevant models of attachment. Further, the domain of AP1 required for adherence was also identified (Smith et al., 2010). AP1 was previously shown to be present exclusively on the tip of FCT-3 pili (Quigley et al., 2009) and this was also demonstrated for FCT-2 pili (Smith et al., 2010). In all GAS strains studied, the largest FCT protein (Fig. 10.1), usually AP1, appears to serve as the adhesin.

10.5.4 Biological roles of pilin proteins in the FCT-1 cluster

As mentioned above, FCT-1 pili contain only two pilin proteins, the backbone protein T6 and the tip protein AP1. Further, these proteins have less than 25% identity to the backbone protein and AP1 from other FCT types (Kratovac *et al.*, 2007; Falugi *et al.*, 2008). In the M6 FCT-1 gene cluster, the AP1 sequence contains a Cna_B domain, suggesting the presence of an intramolecular isopeptide bond (Kang *et al.*, 2007) and a possible Von-Willebrand factor type A domain, which is important for adherence in other streptococci (Becherelli *et al.*, 2012). Using a deletion mutant and transfer of the FCT-1 genes to *Lactococcus lactis*, the Novartis group demonstrated that AP1 is critical for adherence to A549 pulmonary epithelial cells (Becherelli *et al.*, 2012). However, it should be remembered that lack of AP1 reduces polymerization of pili, so there would be fewer pili on the bacterial cell surface and these will be shorter, which may account for reduced adherence of the AP1 deletion mutant. The Novartis group also found that AP1 is involved in formation of biofilms on polystyrene surfaces and that AP1 promotes inter-bacterial interactions, leading to spontaneous sedimentation of the GAS strain (Becherelli *et al.*, 2012). The latter appears to occur by interaction of the AP1 protein on one bacterial cell with the AP1 protein on another, and is specific for the homologous AP1 protein. Such spontaneous sedimentation of GAS cells had previously been associated only with the M protein, so it would be interesting to determine whether the M protein is required for the interactions identified in this work. Surprisingly, this work also showed that the GAS M6 strain deleted for AP1 is unable to grow in human blood and that it has a small growth disadvantage in a murine model following intraperitoneal inoculation. This led Becherelli *et al.* (2012) to suggest that pili, or at least the AP1 pilus protein, may have a role late in infection in addition to their role of initiation of colonization.

10.6 Regulation of Pilus Expression

Considering the importance of pili in pathogenesis of GAS and the notion that pili may be important also for the tissue affinity that determines disease type, remarkably little is known about regulation of pilus expression. The gene at the 5′ end of all FCT regions (Fig. 10.1), which is transcribed in the opposite direction from the rest of the pilus gene cluster, is a transcriptional regulator with homology either to RofA (regulation of protein F: (Fogg *et al.*, 1994)) or Nra (negative regulator in group A streptococcus: (Podbielski *et al.*, 1999)), which are about 63% homologous to each other. The former was first identified and studied in Caparon's lab in an FCT-1 strain, where it was shown to be a positive trans-acting regulator of AP1 responsive to oxygen level and redox potential (VanHeyningen *et al.*, 1993; Fogg *et al.*, 1994). This group also found that transcription of *rofA* is auto-activated and that the degree of *rofA* transcription in response to oxygen tension is determined by the host strain background (Fogg and Caparon, 1997). The latter implies that additional undefined regulators exist in this system.

The *nra* gene, identified by Podbielski's group in an M49 serotype strain with the FCT-3 pilus locus pattern, was shown to repress transcription of many genes, including itself and the global regulator *mga*. Thus, it may repress the FCT genes indirectly (Podbielski *et al.*, 1999). This seems likely since Luo *et al.* (2008) found that Nra acts as a positive regulator in an M53 serotype strain which also has an FCT-3 type pilus gene cluster (Table 10.1). In further contrast to the M49 strain results of Podbielski's group, Luo *et al.* found no effect of Nra on M protein or SpeB, both of which are regulated by Mga. However, they demonstrated that *nra* is important for pathogenesis of the M53 strain since its inactivation reduces virulence in a humanized mouse model of superficial skin infection (Luo *et al.*, 2008).

The gene downstream of the FCT-3 and FCT-4 locus, *msmR*, is homologous to the AraC transcriptional regulator (Fig. 10.1). In an M49 strain (FCT-3), MsmR activates transcription of the pilin operon and many other genes as well (Nakata *et al.*, 2005). Thus, in this M49 strain, MsmR and Nra have opposite effects on pilus operon transcription. MsmR was shown to bind to several chromosomal regions that it activates, including the pilus operon, suggesting a direct effect, but not precluding an indirect effect as well (Nakata *et al.*, 2005).

Recently, Manetti *et al.* (2010) observed an effect of pH (6.4 versus 7.5) on biofilm formation, which they found, at least in an FCT-3 type strain producing T3 pili, is dependent on the presence of pili. For this strain, and for an FCT-2 strain producing T1 pili, they demonstrated an increase in transcription of AP1 at the lower pH and showed that this could be due to more mRNA of the positive regulators RofA (for FCT-2) and MsmR (for FCT-3). No such pH effect on pilus production or transcription was seen, however, for the T6 FCT-1 strain tested. Further investigation of this and other stress effects on pilus production is needed.

Most recently, the Sumby group found that a small non-coding RNA, FasX, negatively regulates expression of the FCT-2 pili of an M1 strain (Liu *et al.*, 2012). They demonstrated that the FCT-2 pilus gene cluster is co-transcribed, and that FasX base-pairs with the extreme 5′ end of this mRNA to reduce its stability. FasX also reduces translation of AP1 (Cpa), the first protein in the pilus cluster. Since this protein is the adhesin, if the other pilus proteins are still produced, FasX regulation could lead to 'Trojan Horse' pili capable of interacting with anti-pilus antibodies but not able to adhere to the host receptor. FasX also activates expression of streptokinase (Kreikemeyer *et al.*, 2001; Ramirez-Pena *et al.*, 2010), so one might expect expression of pili, which serve to attach GAS to host cells, and expression of streptokinase, which promotes spreading through host tissue, to be alternative states. Because regulation of expression of FasX itself has not yet been studied, the mechanisms by which environmental stimuli that affect pilus production feed into the FasX system remain to be determined.

In conclusion, the only thing that is really clear about regulation of pilus expression in GAS is that the host-cell background has a large influence, as do the environmental conditions under which the strain is grown. Both suggest additional regulatory factors have not yet been identified. Obviously, this area cries out for further study.

10.7 Conclusions

I hope this review highlights the need for much additional work before an understanding of GAS pili can be claimed. So far, we know that there are two unusual aspects to pilus assembly in GAS: first, the pilus polymerase is exceptional in belonging to sortase Class B as opposed to Class C, and second, many GAS pilus gene clusters include a *sip/lep* gene, which is essential for pilus polymerization. The implications of these differences from other Gram+ pili have not yet been elucidated. Definition of the sortase usually used for attachment of GAS pili to the cell wall and the cell wall moiety to which it is attached also await experimental investigation. However, perhaps the largest gap in our knowledge of GAS and other Gram positive pili concerns regulation of pilus synthesis. The observations that most GAS strains are poorly piliated when grown in the lab although pili are probably critical to initiation of disease strongly implies that regulation is a fertile area for study. Nevertheless, considering that GAS pili were only discovered in 2005, we have come a long way in understanding their assembly and function. Given their importance in virulence and in vaccine development, we can hope for similar progress in learning more about these important structures in the next decade.

Acknowledgement

I am grateful to Dr I. Margarit y Ros for sending me an early version of the figure in Falugi *et al.* (2008).

References

Abbot, E.L., Smith, W.D., Siou, G.P., Chiriboga, C., Smith, R.J., Wilson, J.A., Hirst, B.H. and Kehoe, M.A. (2007) Pili mediate specific adhesion of *Streptococcus pyogenes* to human tonsil and skin. *Cellular Microbiology* 9, 1822–1833.

Alegre-Cebollada, J., Badilla, C.L. and Fernandez, J.M. (2010) Isopeptide bonds block the mechanical extension of pili in pathogenic *Streptococcus pyogenes*. *Journal of Chemical Biology* 285, 11235–11242.

Barnett, T.C. and Scott, J.R. (2002) Differential recognition of surface proteins in *Streptococcus pyogenes* by two sortase gene homologs. *Journal of Bacteriology* 184, 2181–2191.

Barnett, T.C., Patel, A.R. and Scott, J.R. (2004) A novel sortase, SrtC2, from *Streptococcus pyogenes* anchors a surface protein containing a QVPTGV motif to the cell wall. *Journal of Bacteriology* 186, 5865–5875.

Becherelli, M., Manetti, A.G., Buccato, S., Viciani, E., Ciucchi, L., Mollica, G., Grandi, G. and Margarit, I. (2012) The ancillary protein 1 of *Streptococcus pyogenes* FCT-1 pili mediates cell adhesion and biofilm formation through heterophilic as well as homophilic interactions. *Molecular Microbiology* 83, 1035–1047.

Bessen, D.E. and Kalia, A. (2002) Genomic localization of a T serotype locus to a recombinatorial zone encoding extracellular matrix-binding proteins in *Streptococcus pyogenes*. *Infection and Immunity* 70, 1159–1167.

Caparon, M.G., Stephens, D.S., Olsén, A. and Scott, J.R. (1991) Role of M protein in adherence of group A streptococci. *Infection and Immunity* 59, 1811–1817.

Cho, K.H. and Caparon, M.G. (2005) Patterns of virulence gene expression differ between biofilm and tissue communities of *Streptococcus pyogenes*. *Molecular Microbiology* 57, 1545–1556.

Crotty Alexander, L.E., Maisey, H.C., Timmer, A.M., Rooijakkers, S.H., Gallo, R.L., Von Kockritz-Blickwede, M. and Nizet, V. (2010) M1T1 group A streptococcal pili promote epithelial colonization but diminish systemic virulence through neutrophil extracellular entrapment. *Journal of Molecular Medicine (Berlin)* 88, 371–381.

Dochez, A.R., Avery, O.T. and Lancefield, R.C. (1919) Studies on the biology of *Streptococcus*: I. antigenic relationships between strains of *Streptococcus haemolyticus*. *Journal of Experimental Medicine* 30, 179–213.

Falugi, F., Zingaretti, C., Pinto, V., Mariani, M., Amodeo, L., Manetti, A.G., Capo, S., Musser, J.M., Orefici, G., Margarit, I., Telford, J.L., Grandi, G. and Mora, M. (2008) Sequence variation in group A *Streptococcus* pili and association of pilus backbone types with lancefield T serotypes. *Journal of Infectious Diseases* 198, 1834–1841.

Fogg, G.C. and Caparon, M.G. (1997) Constitutive expression of fibronectin binding in *Streptococcus pyogenes* as a result of anaerobic activation of rofA. *Journal of Bacteriology* 179, 6172–6180.

Fogg, G.C., Gibson, C.M. and Caparon, M.G. (1994) The identification of rofA, a positive-acting regulatory component of prtF expression: use of an m gamma delta-based shuttle mutagenesis strategy in *Streptococcus pyogenes*. *Molecular Microbiology* 11, 671–684.

Kang, H.J. and Baker, E.N. (2009) Intramolecular isopeptide bonds give thermodynamic and proteolytic stability to the major pilin protein of *Streptococcus pyogenes*. *Journal of Chemical Biology* 284, 20729–20737.

Kang, H.J., Coulibaly, F., Clow, F., Proft, T. and Baker, E.N. (2007) Stabilizing isopeptide bonds revealed in gram-positive bacterial pilus structure. *Science* 318, 1625–1628.

Kang, H.J., Coulibaly, F., Proft, T. and Baker, E.N. (2011) Crystal structure of Spy0129, a *Streptococcus pyogenes* class B sortase involved in pilus assembly. *PLoS One* 6, e15969.

Kimura, K.R., Nakata, M., Sumitomo, T., Kreikemeyer, B., Podbielski, A., Terao, Y. and Kawabata, S. (2012) Involvement of T6 pili in biofilm formation by serotype M6 *Streptococcus pyogenes*. *Journal of Bacteriology* 194, 804–812.

Köller, T., Manetti, A.G., Kreikemeyer, B., Lembke, C., Margarit, I., Grandi, G. and Podbielski, A. (2010) Typing of the pilus-protein-encoding FCT region and biofilm formation as novel parameters in epidemiological investigations of *Streptococcus pyogenes* isolates from various infection sites. *Journal of Medical Microbiology* 59, 442–452.

Kratovac, Z., Manoharan, A., Luo, F., Lizano, S. and Bessen, D.E. (2007) Population genetics and linkage analysis of loci within the FCT region of *Streptococcus pyogenes*. *Journal of Bacteriology* 189, 1299–1310.

Kreikemeyer, B., Boyle, M.D., Buttaro, B.A., Heinemann, M. and Podbielski, A. (2001) Group A streptococcal growth phase-associated virulence factor regulation by a novel operon (Fas) with homologies to two-component-type regulators requires a small RNA molecule. *Molecular Microbiology* 39, 392–406.

Lancefield, R.C. (1928) The antigenic complex of *Streptococcus haemolyticus*: I. Demonstration of a type-specific substance in extracts of *Streptococcus haemolyticus*. *Journal of Experimental Medicine* 47, 91–103.

Lancefield, R.C. (1940) Type-specific antigens, M and T, of matt and glossy variants of group a hemolytic streptococci. *Journal of Experimental Medicine* 71, 521–537.

Lauer, P., Rinaudo, C.D., Soriani, M., Margarit, I., Maione, D., Rosini, R., Taddei, A.R., Mora, M., Rappuoli, R., Grandi, G. and Telford, J.L. (2005)

Genome analysis reveals pili in Group B *Streptococcus*. *Science* 309, 105.

Liu, Z., Trevino, J., Ramirez-Pena, E. and Sumby, P. (2012) The small regulatory RNA FasX controls pilus expression and adherence in the human bacterial pathogen group A *Streptococcus*. *Molecular Microbiology* 86, 140–154.

Luo, F., Lizano, S. and Bessen, D.E. (2008) Heterogeneity in the polarity of Nra regulatory effects on streptococcal pilus gene transcription and virulence. *Infection and Immunity* 76, 2490–2497.

Maione, D., Margarit, I., Rinaudo, C.D., Masignani, V., Mora, M., Scarselli, M., Tettelin, H., Brettoni, C., Iacobini, E.T., Rosini, R., D'Agostino, N., Miorin, L., Buccato, S., Mariani, M., Galli, G., Nogarotto, R., Nardi Dei, V., Vegni, F., Fraser, C., Mancuso, G., Teti, G., Madoff, L.C., Paoletti, L.C., Rappuoli, R., Kasper, D.L., Telford, J.L. and Grandi, G. (2005) Identification of a universal Group B *Streptococcus* vaccine by multiple genome screen. *Science* 309, 148–150.

Manetti, A.G., Zingaretti, C., Falugi, F., Capo, S., Bombaci, M., Bagnoli, F., Gambellini, G., Bensi, G., Mora, M., Edwards, A.M., Musser, J.M., Graviss, E.A., Telford, J.L., Grandi, G. and Margarit, I. (2007) *Streptococcus pyogenes* pili promote pharyngeal cell adhesion and biofilm formation. *Molecular Microbiology* 64, 968–983.

Manetti, A.G., Koller, T., Becherelli, M., Buccato, S., Kreikemeyer, B., Podbielski, A., Grandi, G. and Margarit, I. (2010) Environmental acidification drives *S. pyogenes* pilus expression and microcolony formation on epithelial cells in a FCT-dependent manner. *PLoS One* 5, e13864.

Mazmanian, S.K., Ton-That, H., Su, K. and Schneewind, O. (2002) An iron-regulated sortase anchors a class of surface protein during *Staphylococcus aureus* pathogenesis. *Proceedings of the National Academy of Sciences of the USA* 99, 2293–2298.

Mora, M., Bensi, G., Capo, S., Falugi, F., Zingaretti, C., Manetti, A.G., Maggi, T., Taddei, A.R., Grandi, G. and Telford, J.L. (2005) Group A *Streptococcus* produce pilus-like structures containing protective antigens and Lancefield T antigens. *Proceedings of the National Academy of Sciences of the USA* 102, 15641–15646.

Nakata, M., Podbielski, A. and Kreikemeyer, B. (2005) MsmR, a specific positive regulator of the *Streptococcus pyogenes* FCT pathogenicity region and cytolysin-mediated translocation system genes. *Molecular Microbiology* 57, 786–803.

Nakata, M., Kimura, K.R., Sumitomo, T., Wada, S., Sugauchi, A., Oiki, E., Higashino, M.,

Kreikemeyer, B., Podbielski, A., Okahashi, N., Hamada, S., Isoda, R., Terao, Y. and Kawabata, S. (2011) Assembly mechanism of FCT region type 1 pili in serotype M6 *Streptococcus pyogenes*. *Journal of Chemical Biology* 286, 37566–37577.

Okura, M., Osaki, M., Fittipaldi, N., Gottschalk, M., Sekizaki, T. and Takamatsu, D. (2011) The minor pilin subunit Sgp2 is necessary for assembly of the pilus encoded by the srtG cluster of *Streptococcus suis*. *Journal of Bacteriology* 193, 822–831.

Podbielski, A., Woischnik, M., Leonard, B.A. and Schmidt, K.H. (1999) Characterization of nra, a global negative regulator gene in group A streptococci. *Molecular Microbiology* 31, 1051–1064.

Pointon, J.A., Smith, W.D., Saalbach, G., Crow, A., Kehoe, M.A. and Banfield, M.J. (2010) A highly unusual thioester bond in a pilus adhesin is required for efficient host cell interaction. *Journal of Chemical Biology* 285, 33858–33866.

Quigley, B.R., Zahner, D., Hatkoff, M., Thanassi, D.G. and Scott, J.R. (2009) Linkage of T3 and Cpa pilins in the *Streptococcus pyogenes* M3 pilus. *Molecular Microbiology* 72, 1379–1394.

Ramirez-Pena, E., Trevino, J., Liu, Z., Perez, N. and Sumby, P. (2010) The group A *Streptococcus* small regulatory RNA FasX enhances streptokinase activity by increasing the stability of the ska mRNA transcript. *Molecular Microbiology* 78, 1332–1347.

Schneewind, O., Model, P. and Fischetti, V.A. (1992) Sorting of protein A to the staphylococcal cell wall. *Cell* 70, 267–281.

Smith, W.D., Pointon, J.A., Abbot, E., Kang, H.J., Baker, E.N., Hirst, B.H., Wilson, J.A., Banfield, M.J. and Kehoe, M.A. (2010) Roles of minor pilin subunits Spy0125 and Spy0130 in the serotype M1 *Streptococcus pyogenes* strain SF370. *Journal of Bacteriology* 192, 4651–4659.

Spirig, T., Weiner, E.M. and Clubb, R.T. (2011) Sortase enzymes in Gram-positive bacteria. *Molecular Microbiology* 82, 1044–1059.

Ton-That, H. and Schneewind, O. (2004) Assembly of pili in Gram-positive bacteria. *Trends in Microbiology* 12, 228–234.

VanHeyningen, T., Fogg, G., Yates, D., Hanski, E. and Caparon, M. (1993) Adherence and fibronectin binding are environmentally regulated in the group A streptococci. *Molecular Microbiology* 9, 1213–1222.

Zahner, D. and Scott, J.R. (2008) SipA is required for pilus formation in *Streptococcus pyogenes* serotype M3. *Journal of Bacteriology* 190, 527–535.

11 The Role of Pili in the Formation of Biofilm and Bacterial Communities

Andrea G.O. Manetti and Tiziana Spadafina
Novartis Vaccines and Diagnostics, Siena, Italy

11.1 Introduction

Biofilms are mono- or multispecies complex microbial communities living in symbiotic relationship, and represent a protected mode of growth that allows bacteria to survive and proliferate in a hostile environment. The structure of the biofilm, made up of microcolonies of bacteria encased by an exopolysaccharidic matrix and crossed by a complex network of channels, allows sufficient nutrients to sustain growth and, at the same time, protects the bacteria from toxic compounds and reagents present in the environment, including those elicited by the innate and adaptive immune responses of the host (Costerton *et al.*, 1999; Lewis, 2001). The pattern of biofilm development involves bacterial attachment to a solid surface, the formation of microcolonies and the differentiation into exopolysaccharide (EPS) encased, mature biofilm (Hall-Stoodley *et al.*, 2004). EPS composition varies greatly, depending on the biofilm ecosystem, but typically contains a mixture of polysaccharides, as well as proteins, extracellular DNA (eDNA) due to cell lysis, and lipids, which are responsible for matrix strength and biofilm viscoelastic properties by means of weak physicochemical interactions (Rice *et al.*, 2007; Flemming and Wingender, 2010). Within the matrix, molecular gradients of nutrients, oxygen and signalling molecules

develop (Stewart and Franklin, 2008). The matrix can also act as a 'protective shield' against the diffusion and action of antibiotics and antimicrobials in the bulk of the biofilm, also protecting from host defence and phagocytosis (Fux *et al.*, 2005; Epstein *et al.*, 2011), thus posing a serious problem for treatment of biofilm-related infections and for microbial elimination. Microbial biofilms constitute the major lifestyle alternative to planktonic growth, and, according to National Institutes of Health estimations, they are responsible for over 80% of human microbial infections (http://grants.nih.gov/grants/guide/pa-files/PA-03-047.html; Houry *et al.*, 2012).

Biofilms are dynamic structures expressing fitness determinants, and may be subject to population shifts in response to microbial composition and environmental conditions. Single cells or cell aggregates can detach from a mature biofilm and disseminate via the bloodstream, causing metastatic infection and/or development of sepsis (Costerton *et al.*, 1999; Lewis, 2001). Within the biofilm community, bacteria communicate with each other by using chemical signal molecules in response to population density in a process that is called quorum sensing. Various physiological activities are regulated via quorum sensing in Gram-positive bacteria, including biofilm formation in staphylococci, streptococci and enterococci, expression of virulence factors in staphylococci, development of competence in streptococci, sporulation in Bacillus, and antibiotic biosynthesis in *Lactococcus lactis* (Waters and Bassler, 2005). The majority of microorganisms are able to develop biofilms and many different mechanisms induce biofilm formation in different species. The causative agents of biofilm-associated bacterial infections are different Gram-positive species of *Staphylococcus, Streptococcus, Enterococcus* and *Actinobacillus* as well as Gram-negative bacteria, such as *Pseudomonas aeruginosa* and *Escherichia coli*.

In order to establish an infection, pathogenic bacteria have to attach to host cells. This process is needed to colonize the host tissues and it is mediated by an array of different molecules, which have been termed MSCRAMMs (microbial surface components recognizing adhesive matrix molecules) and are able to bind specific receptors and/or structural elements of the membrane, such as for example collagen and fibronectin. Over many years, scientists have described appendages of different length and function protruding from the surface of both Gram-negative and -positive bacteria (Allen *et al.*, 2012; Kang and Baker, 2012; Schneewind and Missiakas, 2012), and several surface-associated proteins have been shown to be involved in bacterial attachment to cells and in microcolony formation in different pathogens (Costerton *et al.*, 1999). A general problem for bacterial adhesion to host tissues is the net repulsive force caused by the negative charges of both bacteria and host cells. This problem can be overcome by a cell surface structure in which the adhesin is located at the tip of hair-like, peritrichous, non-flagellar, filamentous surface appendages known as pili (Latin for hair) or fimbriae (Latin for thread or fibre). Pili are important virulence factors for several diseases, such as infections of the urinary, genital and gastrointestinal tracts, as well as important targets for vaccine development (Proft and Baker, 2009).

Besides mediating binding to host-cell receptors, pili have also been shown to promote co-aggregation, which is involved in the initial steps of biofilm formation. Many Gram-negative bacteria exploit hair-like structures exposed on the cell surface to promote attachment and aggregation to human cells (Helaine *et al.*, 2005), microcolony formation and biofilm formation (Di Martino *et al.*, 2003; Klausen *et al.*, 2003; Orndorff *et al.*, 2004; Paranjpye and Strom, 2005) and finally colonization of host tissue (Kirn *et al.*, 2000). Among them we can list three kinds of pili that have been shown to be involved in biofilm production: type I and type IV pili and curli. Type I pili of *E. coli* are required for initial surface attachment, a process that can be inhibited by mannose (Pratt and Kolter, 1998). Moreover, Orndorff *et al.* (2004) have shown that aggregation of *E. coli* by secretory IgA (SIgA) depend on type I pili. Type IV pili were found in a large variety of Gram-negative bacteria, including enteropathogenic *E. coli* (EPEC), EHEC, *Salmonella enterica, P. aeruginosa, Legionella pneumophila, Neisseria gonorrhoeae, N. meningitidis* and *Vibrio cholerae* (Strom and Lory, 1993; Craig *et al.*, 2004; Varga *et al.*, 2006; Craig and Li, 2008). Furthermore, type IV pili possess a number of unique features, including DNA uptake during transformation, phage transduction and a flagella-independent form of movement known as twitching motility. Finally, curli were first described in 1989 as proteinaceous coiled fibres found on enteric bacteria such as *E. coli* and *Salmonella* spp. Both *E. coli* and *Salmonella* spp. curli are important for biofilm development (Olsen *et al.*, 1989).

Recently pili structures have been described also in major Gram-positive human pathogens, such as *Streptococcus pyogenes* (Group A *Streptococcus*, GAS), *S. agalactiae* (Group B *Streptococcus*, GBS) and *S. pneumoniae* (pneumococcus). Moreover, pilus structures were described also in *Enterococcus faecalis* and *Enterococcus faecium*, which are Gram-positive human commensal organisms able to cause life-threatening diseases in immune-compromised patients. Biofilm formation is a characteristic lifestyle feature of the species discussed here, and several studies have suggested that pili promote biofilm formation (Nallapareddy *et al.*, 2006; Manetti *et al.*, 2007; Munoz-Elias *et al.*, 2008; Schluter *et al.*, 2009; Konto-Ghiorghi *et al.*, 2009; Rinaudo *et al.*, 2010).

Here we discussed the key role of pili expressed by Gram-positive bacterial pathogens in biofilm formation and the influence of environmental factors on their expression.

11.2 *Streptococcus pyogenes*

Group A *Streptococcus* is a Gram-positive pathogenic bacterium that exclusively infects humans. Primary infections usually take place at the level of the nasopharynx and skin and can result in either relatively mild diseases, such as pharyngitis, impetigo and cellulitis, or severe, life-threatening diseases, including myositis and necrotizing fasciitis. The ability of this pathogen to colonize and persist in distinct host sites, and to trigger infections with diverse clinical manifestations, relies on a wide range of virulence factors produced by the different isolates, and on complex regulatory networks that modulate gene expression in response to fluctuating environmental conditions (Cunningham, 2000; Musser and Shelburne, 2009). As with many other bacteria, initiation of GAS infections requires the capacity of the pathogen to adhere to host tissues and assemble in cell aggregates. Moreover, a role for biofilms in GAS pathogenesis has been proposed and experimentally supported in a number of publications (Neely *et al.*, 2002; Akiyama *et al.*, 2003; Hidalgo-Grass *et al.*, 2004; Cho and Caparon, 2005; Baldassarri *et al.*, 2006).

Several GAS components such as M protein, the *hasA* gene responsible for the production of hyaluronic capsule, the transcription regulatory proteins CovR and Mga and the putative quorum-sensing regulatory peptide SilC enable the pathogen to adhere to surfaces coated with matrix molecules and to eukaryotic cells and successively to produce biofilm (Frick *et al.*, 2000; Courtney *et al.*, 2002; Cho and Caparon, 2005; Lembke *et al.*, 2006). However, it has been postulated that the MSCRAMMs, some of which are encoded by genes located in the fibronectin-collagen-T-antigen (FCT) region of the GAS genome, could be other potential candidate proteins involved in biofilm formation (Lembke *et al.*, 2006).

The FCT pathogenicity island promotes the synthesis and assembly of pilus-like structures (Mora *et al.*, 2005). Nine major variants of the FCT region have been described on the basis of gene content and organization, each of which contains genes coding for pilus structural subunits (the backbone protein and ancillary proteins) and for the sortase enzymes required for pilus assembly (see Fig. 11.1). Interestingly, all GAS isolates so far analysed carry and express the FCT pilus locus (Kratovac *et al.*, 2007; Falugi *et al.*, 2008), suggestive of a key role of pili in streptococcal interaction with the human host.

Wild-type GAS SF370 adheres to human epithelial cells, forms aggregates in liquid culture and microcolonies on the cells and is capable of generating mature biofilm, producing the typical three-dimensional layer with bacterial microcolonies embedded in a carbohydrate polymeric matrix. All these properties are severely impaired when pili expression is abrogated by deleting either the gene coding for the pilus backbone structural protein (*spy_0128*) or the gene encoding the sortase C1 (*spy_0129*), indispensable for pilus assembly (Manetti *et al.*, 2007). In confirmation of the key role of pili, the complemented mutants had an adhesion and aggregation phenotype similar to the wild-type strain and were able to partially restore the capacity to produce biofilm (Manetti *et al.*, 2007) (see Fig. 11.2).

In particular, it has been shown that the adhesion of mutant strains to pharyngeal cells was remarkably reduced at short incubation times, indicating a role for pili in promoting the rapid attachment of GAS to host tissues. In addition, Abbot and coworkers, still using SF370 wild-type and pilus mutants, demonstrated the role of pili in adhesion to tonsil, primary keratinocyte and HaCaT host cells (Abbot *et al.*, 2007). Furthermore, bacterial aggregation, a precondition of biofilm formation, was clearly impaired in pilus-negative mutants both in liquid growth and during cell infection. Moreover, when grown on solid surfaces, the mutant strains did not develop the typical three-dimensional structures observed in the wild-type SF370 strain. Finally, the observation that

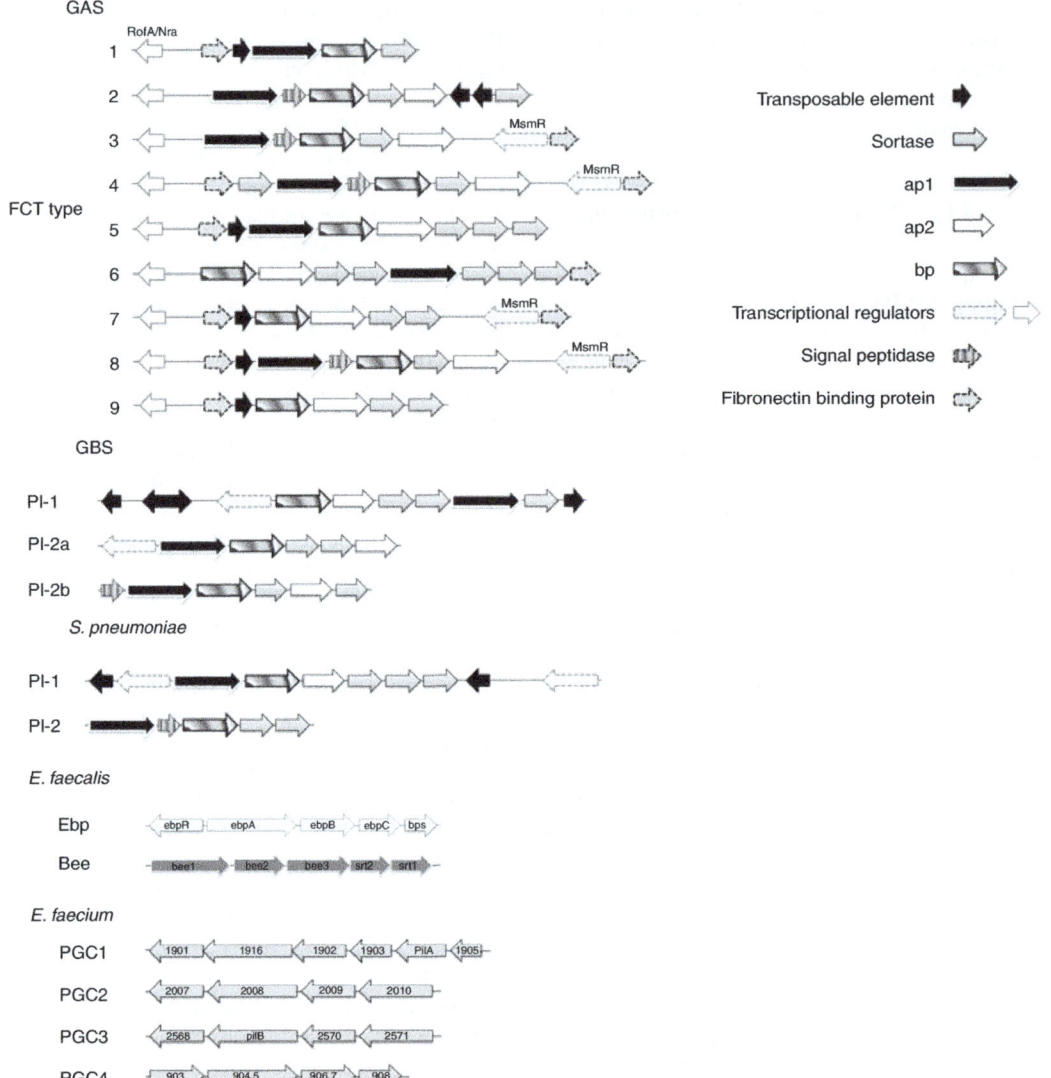

Fig. 11.1. Schematic representation of streptococcal and enterococcal pathogenic pilus regions. The nine different FCT types of *S. pyogenes*, the three different PI of *S. agalactiae*, the two different pilus operons of *S. pneumoniae*, the two different pilus operons of *E. faecalis* and the four different PGCs of *E. faecium* are reported. Position and orientation of the genes are indicated by arrows. Abbreviations: Ap: ancillary pilus protein; bp: backbone protein; FCT: fibronectin collagen T antigen; PI: pilus island.

Concanavalin A stained material is more abundant in the cellular aggregates formed on solid surface by wild-type and complemented strains further supports the evidence that pili are involved in the maturation of *S. pyogenes* EPS-containing biofilms. These data support the role of pili in GAS adherence and colonization and suggest a general role of pili in all pathogenic streptococci.

In another set of experiments, Becherelli and colleagues characterized the pilus ancillary protein 1 (AP1_M6) from GAS isolates expressing the FCT-1 pilus variant, known to be strong biofilm formers. Cell

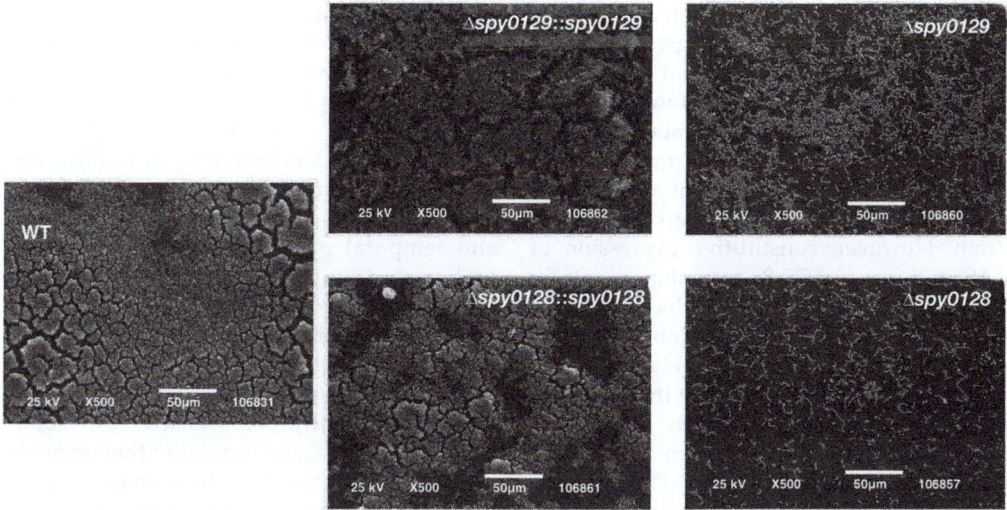

Fig. 11.2. Scanning electron microscopy (SEM) analysis of mature biofilm (72 hours) formed by GAS SF370 M1 wild type (Wt) strain, its backbone negative (ΔSpy0128) and sortase C1 negative (ΔSpy_0129) mutants and their complemented strains (ΔSpy0128::Spy_0128 and ΔSpy_0129::Spy_019). Wild-type strain forms a thick mature biofilm on an abiotic surface, whereas in pilus defective mutants the capacity to form biofilm is impaired, and single chains of bacteria are visible. However, complemented strains, correctly expressing pilus-like structures, produce biofilms similar to the wild-type strain. The SEM analysis was performed in collaboration with Dr Gabriella Gambellini from Centro Interdipartimentale di Microscopia Elettronica, Università della Tuscia, Viterbo, Italy.

binding and biofilm formation assays using in frame deletion mutants, *Lactococcus*-expressing heterologous FCT-1 pili and purified recombinant AP1_M6, indicated this pilin as a strong adhesin, also involved in bacterial biofilm formation. In fact, antibodies raised against AP1_M6 were capable of blocking biofilm formation, suggesting that vaccines targeting the pilus proteins could also prevent colonization of GAS target tissues, such as the skin and the pharyngeal epithelia. Moreover, it was shown that AP1_M6 establishes homophilic interactions that mediate inter-bacterial contacts, possibly promoting bacterial colonization of target epithelial cells in the form of three-dimensional microcolonies (Becherelli *et al.*, 2012).

In another study, Manetti *et al.* (2010) undertook a global analysis of GAS isolates representing the majority of FCT variants to investigate the effect of environmental growth conditions on their capacity to form multicellular communities. It was observed that all FCT-1 strains formed biofilm independently of the presence of glucose in the growth medium, whereas the capacity to form biofilm of strains belonging to other FCT variants required the presence of glucose or other sugars in the medium. This phenotype was found to be a consequence of the environmental acidification resulting from sugar consumption during bacterial growth, as a similar FCT dependent biofilm phenotype could be achieved by lowering the starting pH of the medium without any sugar supplementation.

Bacteria belonging to the pH-dependent FCT-type 3 pre-inoculated at a starting pH of 6.4 and co-incubated with epithelial cell mono layers, formed large three-dimensional microcolonies, suggesting that modulation of pilus expression and biofilm formation in response to pH could play an important role during the initial cell adhesion steps of GAS infection.

The fact that biofilm formation was pH dependent in some but not all GAS FCT types was associated with a difference in the expression of FCT-encoded genes. Higher amounts of pilus proteins and fibronectin binding adhesins (F2) were found in pH-dependent bacteria grown at a starting pH of 6.4 as compared with bacteria grown at pH 7.5. In contrast, pH-independent pilus expression was observed in the case of an FCT-1 strain. However, constitutive expression of GAS pili in a FCT-3 recombinant strain resulted in pH-independent biofilm formation, in contrast to the results obtained with a recombinant strain constitutively expressing the F2 protein, thus confirming that GAS pili and not F2 proteins are directly involved in pH-dependent auto aggregation and biofilm formation.

All FCT genomic regions contain genes coding for standalone transcription regulators, such as the RofA/Nra homologues and MsmR, which coordinately control the expression of the FCT open reading frames and are involved in the response to changes of environmental parameters like oxygen pressure (Fogg and Caparon, 1997; Podbielski et al., 1999; Granok et al., 2000; Nakata et al., 2005) and temperature (Nakata et al., 2009). The expression of the FCT transcription regulators is affected by pH. In fact, in a FCT-3 M3 strain, msmR transcript amounts increased at lower pH, in parallel to the enhanced expression of pili and the F2 protein, suggesting that in this strain MsmR exerts a positive effect on pilus transcription. Despite previous observations indicating that the Nra regulator is under the control of MsmR (Nakata et al., 2005), only a mild difference was detected in the amount of nra transcripts at the two investigated pH conditions.

Increased transcript amounts at acidic pH conditions were also detected for the FCT-2 pilus-positive regulator RofA. However, despite the fact that FCT-1 and FCT-2 rofA genes and their intergenic regions share 98% identity, the FCT-1 variant was highly and constitutively transcribed irrespective of environmental pH, probably due to independent pathways in addition to RofA (Granok et al., 2000). In conclusion, it appears that most GAS FCT types sense environmental pH as a signal to build pilus appendages on their surface, and this process leads to the formation of large cell-adhering multicellular communities that may assist the initial steps of GAS infection (Manetti et al., 2010).

A role for pH in host colonization and infection by Gram-positive pathogens, including GAS, has been proposed. First, the diverse environments encountered by these microorganisms are characterized by local and temporal pH fluctuations, such as the oral cavity from 5 to 7 (Dong et al., 1999), the naso-pharynx from 6.4 to 6.9 (Moscoso et al., 2006) and the skin from 4.2 to 5.9 (Ehlers et al., 2001; Rippke et al., 2002). In addition, bacterial growth can cause local pH alterations, which can subsequently be sensed by some organisms as a signal to modify their growth mode. For instance, it has been shown that in the oral cavity, a pH change associated with sugar metabolism by dental pathogens results in a switch from a healthy status to disease, which parallels the formation of mature multispecies biofilms (Donlan and Costerton, 2002; Beenken et al., 2004; Marsh, 2006). Furthermore, many Gram-positive pathogens modulate gene expression in response to pH changes. For instance, microarray analysis revealed a differential expression of approximately 10–15% of the total number of genes in Staphylococcus aureus (Weinrick et al., 2004), S. pyogenes (Loughman and Caparon, 2006) and S. agalactiae (Santi et al., 2009) under mildly acidic conditions compared with neutral pH.

11.3 Streptococus agalactiae

Streptococcus agalactiae is a common colonizer of the gastrointestinal and urogenital tracts of up to 30% of healthy individuals, and can infect newborns during delivery, causing severe invasive neonatal infections such as pneumonia, septicaemia and meningitis (Gibbs et al., 2004; Hansen et al., 2004). GBS strains have been isolated, in association with other known biofilm-forming bacteria and from biofilms on intrauterine devices (Marrie and Costerton, 1983; Donlan and Costerton, 2002). As already mentioned, in pathogenic streptococci, biofilm formation is mediated by pili, which promotes bacterial aggregation and attachment to the host surface. In GBS two pilus island (PI) variants have been described: PI-1 and PI-2 (differentiated in types 2a and 2b), each encoding three structural proteins (two ancillary proteins AP-1 and AP-2 and a backbone protein BP, and two dedicated subfamily C sortases (SrtC) involved in covalent polymerization of the subunits (Rosini *et al.*, 2006) (see Fig. 11.1).

Rinaudo *et al.* (2010) showed that the serotype Ia strain 515, expressing only pilus type 2a, was able to form biofilm *in vitro* by adhering to polystyrene surfaces. Also in the case of GBS the capacity to form biofilm was influenced by environmental conditions. In fact, only in the presence of glucose (or sucrose) concentrations ≥0.4% was surface adhesion and proliferation by GBS strain 515 observed. Under these experimental conditions, a high percentage of the analysed GBS isolates produced biofilm, and the capacity to form biofilms seemed to be associated with the presence of type 2a pili on the bacterial surface. In fact, using a series of mutant and complemented strains it was demonstrated that only pilus type 2a plays a role in biofilm formation. Furthermore, the presence of a polymerized pilus, correctly anchored to the cell surface, was necessary for efficient biofilm formation. In fact, any mutation that abrogated pilus expression (deletion of the backbone protein BP or deletion of the two sortase proteins SrtC-1 and SrtC-2), or impaired pilus anchoring to the cell surface (deletion of AP-2, SrtC-1 or SrtA proteins), prevented biofilm formation by GBS cells harbouring pilus type 2a. All these phenotypes were further confirmed by confocal laser scanning microscopy observations (Rinaudo *et al.*, 2010).

This findings are partially in contrast with those reported on GBS strain NEM316 by Konto-Ghiorghi *et al.* (2009), who observed that deletions of the PilA and PilB genes caused the loss of the auto aggregative and of the biofilm formation phenotypes, while deletion of PilC (homologous of AP2) did not affect biofilm formation. In addition, they found that the von Willebrand domain, involved in adherence to epithelial cells, was not required for biofilm formation.

Finally, it was observed that anti-BP antibodies significantly inhibited the ability of GBS wild-type strain 515 to form biofilm in a dose-dependent manner. Interestingly, antibodies specific for the backbone protein of pilus type 1 and 2b did not inhibit the capacity to form biofilm in strains expressing pilus 1 or pilus 2b even at very high levels. These data represent further evidence that only pilus 2a, but not the other two pili, is involved in biofilm formation *in vitro* (Rinaudo *et al.*, 2010).

11.4 *Streptococcus pneumoniae*

Streptococcus pneumoniae is a Gram-positive commensal of the nasopharyngeal tract of both children and healthy adults. However, *S. pneumoniae* is also a leading cause of morbidity and mortality worldwide, being responsible for non-invasive and invasive diseases, such as acute otitis media, pneumonia, sepsis and meningitis (Fletcher and Fritzell, 2007; Pelton and Leibovitz, 2009; van der Poll and Opal, 2009; Kim, 2010). The ability to form biofilms has been proposed to play an important role in otitis media with effusion and recurrent otitis media, which commonly involve *Haemophilus influenzae* and *S. pneumoniae* (Ehrlich *et al.*, 2002; Donlan *et al.*, 2004).

S. pneumoniae pilus 1 is encoded by the genetic islet PI-1, which is present in 30–50% of the pneumococcal strains and is implicated in adhesion to epithelial cells, lung infection and virulence (for review see Harfouche *et al.*, 2012). Pilus 1 is composed of the backbone subunit RrgB, the adhesin RrgA and the minor component RrgC (Hilleringmann *et al.*, 2008, 2009) (see Fig. 11.1). So far, three different variants of the pilus backbone RrgB (RrgB clade I, RrgB clade II and RrgB clade III), which have a degree of protein homology of 48–60% have been described, whereas RrgB proteins expressed by strains belonging to the same clade share 99% sequence identity (Moschioni *et al.*, 2008).

Munoz-Elias *et al.* (2008) showed that in noncapsulated strains *rrgA* mutants, but not *rrgB* or *rrgC* mutants, were impaired in their capacity to form biofilm. In fact, the surface expression of RrgA in a pilus structure or even on the surface itself in the absence of the pilus backbone was sufficient to impair biofilm formation. This finding was also supported by the fact that RrgA mediates adhesion to epithelial cells *ex vivo* independently of RrgB (Nelson *et al.*, 2007). As for *S. pyogenes* ancillary protein 1 of the FCT-1 pilus (Becherelli *et al.*, 2012), RrgA is likely either to bind a host molecule(s), the nature of which remains to be defined, or acts as a self-recognition adhesin, similar also to antigen 43 in *E. coli* (Schembri *et al.*, 2004). RrgA is found in clusters along the main shaft of the pilus (LeMieux *et al.*, 2006), favouring a possible interaction between pili from the same bacterium or adjacent bacteria. Similar to the relationship described between RrgA and capsule expression in *S. pneumoniae*, also in *E. coli*, the capsule shields the self-aggregation and biofilm formation function of the adhesin antigen 43 (Schembri *et al.*, 2004).

11.5 Other Non-Streptococcal Gram-Positive Organisms

11.5.1 *Enterococcus faecalis* and *faecium*

In the past Enterococci were considered harmless members of the mammalian gastrointestinal flora. However, two species, *E. faecalis* and *E. faecium*, have rapidly emerged as multi-resistant opportunistic pathogens and are currently considered the third source of hospital-associated infections, such as urinary tract infections, surgical site infections, bacteraemia and infective endocarditis (Werner *et al.*, 2008).

E. *faecalis* assembles two types of heterotrimeric pilus structures located in two pilin gene clusters (PGCs), which are designated as the endocarditis- and biofilm-associated pili (*ebp*) operon, and the biofilm-associated pili (*bee*) locus (Tendolkar *et al.*, 2004, 2006; Nallapareddy *et al.*, 2006) (see Fig. 11.1). The *bee* locus, placed on a conjugative plasmid, is only sporadically detected, whereas the *ebp* locus is found ubiquitously (Cobo Molinos *et al.*, 2008) and is also present in other species such as *E. gallinarum* and *E. casseliflavus* (Nallapareddy *et al.*, 2011). The assembly of these two loci is known to be

sortase-mediated and Srt-C-dependent, whereas expression of all PGCs is dependent on the EbpR transcriptional regulator (Hendrickx *et al.*, 2008; Gao *et al.*, 2010). In contrast, *E. faecium* assembles three heterotrimeric and one heterodimeric pilus (the PilA pilus) encoded by four PGCs, named PGC1 to 4, which are predominantly present in human hospital-acquired isolates (Hendrickx *et al.*, 2008) (see Fig. 11.1).

E. *faecalis* pili have been demonstrated to be directly involved in biofilm development. Mutations of *ebpA*, *ebpB*, *ebpC* and *srtC* genes showed a strong defect in biofilm formation and in the initial adherence to the host tissue (Nallapareddy *et al.*, 2006; Hendrickx *et al.*, 2009). Sinlampaa *et al.* (2010) showed that EbpC, the major putative pilus subunit protein of the *ebpABC* operon is also expressed by *E. faecium* strains (PilB expressed in PGC 3) and that this type of pilus strongly affected the initial adherence and the biofilm formation phenotype of this species.

In addition, the impairment of the pili encoded by the *bee* locus in *E. faecalis* was shown to affect biofilm formation (Schluter *et al.*, 2009). This study, based on a spontaneously derived mutation in the *srt-1* gene, suggested the involvement of the Bee proteins in the pilus formation and a possible role of the Srt-1 in the cross linking of the Bee protein, indicating a specific linkage of Bee-2 by Srt-1 within this Bee complex.

The environmental conditions affecting the expression of the ebpR-ebpABC locus has been recently investigated. The *ebp* locus expression was enhanced by the presence of 0.1 M sodium bicarbonate, but not by CO_2, resulting in an increasing number of cells producing pili. The molecular basis of the bicarbonate effect was not dissected; however, the authors reported that the pathway was independent of the Fsr system, an homologous of the Agr system of *Staphylococcus aureus* (Bourgogne *et al.*, 2010).

11.5.2 *Actinomyces*

The development of the complex oral biofilm referred to as dental plaque is due to interbacterial interactions between oral streptococci and actinomyces and their adherence to tooth surface and to the associated host cells. So-called pioneer bacteria form an adhesive platform to promote the colonization of bridging bacteria, which in turn attract the late colonizers of the biofilm (Rickard *et al.*, 2003; Kolenbrander *et al.*, 2006). These interactions depend largely on a lectin-like activity associated with the *Actinomyces oris* type 2 fimbriae, a surface structure assembled by sortase SrtC2-dependent polymerization of the shaft and tip fibrillins, FimA and FimB, respectively (Mishra *et al.*, 2011).

The genetic dissection of the fimbrial components showed that FimA, rather than FimB, is the multivalent adhesin that mediates bacterial co-aggregation with oral streptococci and biofilm formation (Mishra *et al.*, 2011). The fact that FimA was essential for auto aggregation as well as for the adherence of *Actinomyces* to the oral streptococci and host cells (Mishra *et al.*, 2011) indicated that it might contain distinct structural domains able to mediate these diverse interactions. A combination of x-ray crystallography and modelling revealed that FimA has three IgG-like domains, N1, N2 and N3, each one possessing an intra-molecular isopeptide bond. *Actinomyces* cells expressing mutated FimA failed to form monospecies biofilm in the presence of sugar (sucrose, fructose or glucose), while mutations of the residues required for intra-molecular isopeptide bond did not abrogate biofilm formation, indicating that these intra-molecular linkages are not involved in cell-to-cell interactions (Mishra *et al.*, 2011).

11.6 Conclusions

Microbial biofilms constitute the major lifestyle alternative to planktonic growth, representing a protected mode of growth, which allows bacteria to survive and proliferate in a hostile environment. In fact, according to the National Institutes of Health, they are responsible for over 80% of human microbial infections, and the matrix in which they are encased acts as a 'protective shield' against the diffusion and action of antibiotics

and antimicrobials, also protecting bacteria from host defence and thus posing a serious problem for treatment of biofilm-related infections and for microbial elimination. In the first decade of the new century, some of the most important bacterial human pathogens, such as GAS, GBS and pneumococcus, were found to harbour pilus-like structures protruding from their surface. Very interestingly, common commensal organisms of the intestinal flora such as *E. faecalis* and *faecium* were also found to express pili in relation to severe human diseases related to biofilms, such as urinary tract infections and infective endocarditis. Gram-negative pili have been implicated in auto aggregation and in biofilm formation, being part of a vast array of proteins located on the bacterial surface and assigned to the attachment to specific receptors or to basal membrane components. However, pili in Gram-positive bacteria were discovered more recently, and their involvement in biofilm formation has been demonstrated only in recent times. Although Gram-positive bacterial biofilm differs from Gram-negative bacterial biofilm, it is now accepted that streptococci and enterococci are also able to live in communities, and the role of pili in establishing these community is, as discussed in this chapter, pivotal. In fact, pilus-like structures protruding from the bacterial surface help the pathogen to begin the process of colonization by adhering to host cells and promote the enlargement of the community by favouring auto-aggregation through self-binding of specific pilins. Interestingly enough, the expression of pilin proteins of certain streptococcal strains is enhanced in mild acidic conditions, leading to the production of biofilm exclusively at mild acidic pH. Of note, diverse niches of the human body habitually colonized by bacterial pathogens, such as the oral cavity, nasopharynx and skin, have a local fluctuating pH, ranging around mild acidity. The same situation was also shown for the expression of pilus genes in enterococci, grown in medium supplemented with bicarbonate. In conclusion, pili are the means for bacteria to primarily adhere to the cells of the host tissue. In addition, pili promote bacterial self-aggregation, leading to the formation of bacterial communities and ultimately mature biofilms. As shown, antibodies raised against pilus components are able to impair biofilm formation, thus marking pilins as putative vaccine candidates to prevent biofilm-related diseases.

References

Abbot, E.L., Smith, W.D., Siou, G.P., Chiriboga, C., Smith, R.J., Wilson, J.A., Hirst, B.H. and Kehoe, M.A. (2007) Pili mediate specific adhesion of *Streptococcus pyogenes* to human tonsil and skin. *Cellular Microbiology* 9, 1822–1833.

Akiyama, H., Morizane, S., Yamasaki, O., Oono, T. and Iwatsuki, K. (2003) Assessment of *Streptococcus pyogenes* microcolony formation in infected skin by confocal laser scanning microscopy. *Journal of Dermatological Science* 32, 193–199.

Allen, W.J., Phan, G. and Waksman, G. (2012) Pilus biogenesis at the outer membrane of Gram-negative bacterial pathogens. *Current Opinion in Structural Biology* 22, 500–506.

Baldassarri, L., Creti, R., Recchia, S., Imperi, M., Facinelli, B., Giovanetti, E., Pataracchia, M., Alfarone, G. and Orefici, G. (2006) Therapeutic failures of antibiotics used to treat macrolide-susceptible *Streptococcus pyogenes* infections may be due to biofilm formation. *Journal of Clinical Microbiology* 44, 2721–2727.

Becherelli, M., Manetti, A.G., Buccato, S., Viciani, E., Ciucchi, L., Mollica, G., Grandi, G. and Margarit, I. (2012) The ancillary protein 1 of *Streptococcus pyogenes* FCT-1 pili mediates cell adhesion as well as biofilm formation through heterophilic as well as homophilic interactions. *Molecular Microbiology* 83, 1035–1047.

Beenken, K.E., Dunman, P.M., McAleese, F., Macapagal, D., Murphy, E., Projan, S.J., Blevins, J.S. and Smeltzer, M.S. (2004) Global gene expression in *Staphylococcus aureus* biofilms. *Journal of Bacteriology* 186, 4665–4684.

Bourgogne, A., Thomson, L.C. and Murray, B.E. (2010) Bicarbonate enhances expression of the endocarditis and biofilm associated pilus locus, ebpR-ebpABC, in *Enterococcus faecalis*. *BMC Microbiology* 10, 17.

Cho, K.H. and Caparon, M.G. (2005) Patterns of virulence gene expression differ between biofilm and tissue communities of *Streptococcus pyogenes*. *Molecular Microbiology* 57, 1545–1556.

Cobo Molinos, A., Abriouel, H., Omar, N.B., Lopez,

R.L. and Galvez, A. (2008) Detection of ebp (endocarditis- and biofilm-associated pilus) genes in enterococcal isolates from clinical and non-clinical origin. *International Journal of Food Microbiology* 126, 123–126.

Costerton, J.W., Stewart, P.S. and Greenberg, E.P. (1999) Bacterial biofilms: a common cause of persistent infections. *Science* 284, 1318–1322.

Courtney, H.S., Hasty, D.L. and Dale, J.B. (2002) Molecular mechanisms of adhesion, colonization, and invasion of group A streptococci. *Annals of Medicine* 34, 77–87.

Craig, L. and Li, J. (2008) Type IV pili: paradoxes in form and function. *Current Opinion in Structural Biology* 18, 267–277.

Craig, L., Pique, M.E. and Tainer, J.A. (2004) Type IV pilus structure and bacterial pathogenicity. *Nature Reviews. Microbiology* 2, 363–378.

Cunningham, M.W. (2000) Pathogenesis of group A streptococcal infections. *Clinical Microbiology Reviews* 13, 470–511.

Di Martino, P., Cafferini, N., Joly, B. and Darfeuille-Michaud, A. (2003) *Klebsiella pneumoniae* type 3 pili facilitate adherence and biofilm formation on abiotic surfaces. *Research in Microbiology* 154, 9–16.

Dong, Y.M., Pearce, E.I., Yue, L., Larsen, M.J., Gao, X.J. and Wang, J.D. (1999) Plaque pH and associated parameters in relation to caries. *Caries Research* 33, 428–436.

Donlan, R.M. and Costerton, J.W. (2002) Biofilms: survival mechanisms of clinically relevant microorganisms. *Clinical Microbiology Reviews* 15, 167–193.

Donlan, R.M., Piede, J.A., Heyes, C.D., Sanii, L., Murga, R., Edmonds, P., El-Sayed, I. and El-Sayed, M.A. (2004) Model system for growing and quantifying *Streptococcus pneumoniae* biofilms in situ and in real time. *Applied and Environmental Microbiology* 70, 4980–4988.

Ehlers, C., Ivens, U.I., Moller, M.L., Senderovitz, T. and Serup, J. (2001) Females have lower skin surface pH than men. A study on the surface of gender, forearm site variation, right/left difference and time of the day on the skin surface pH. *Skin Research and Technology* 7, 90–94.

Ehrlich, G.D., Veeh, R., Wang, X., Costerton, J.W., Hayes, J.D., Hu, F.Z., Daigle, B.J., Ehrlich, M.D. and Post, J.C. (2002) Mucosal biofilm formation on middle-ear mucosa in the chinchilla model of otitis media. *JAMA: The Journal of the American Medical Association* 287, 1710–1715.

Epstein, A.K., Pokroy, B., Seminara, A. and Aizenberg, J. (2011) Bacterial biofilm shows persistent resistance to liquid wetting and gas penetration. *Proceedings of the National*

Academy of Sciences of the United States of America 108, 995–1000.

Falugi, F., Zingaretti, C., Pinto, V., Mariani, M., Amodeo, L., Manetti, A.G., Capo, S., Musser, J.M., Orefici, G., Margarit, I., Telford, J.L., Grandi, G. and Mora, M. (2008) Sequence variation in group A *Streptococcus* pili and association of pilus backbone types with lancefield T serotypes. *Journal of Infectious Diseases* 198, 1834–1841.

Flemming, H.C. and Wingender, J. (2010) The biofilm matrix. *Nature Reviews. Microbiology* 8, 623–633.

Fletcher, M.A. and Fritzell, B. (2007) Brief review of the clinical effectiveness of PREVENAR against otitis media. *Vaccine* 25, 2507–2512.

Fogg, G.C. and Caparon, M.G. (1997) Constitutive expression of fibronectin binding in *Streptococcus pyogenes* as a result of anaerobic activation of rofA. *Journal of Bacteriology* 179, 6172–6180.

Frick, I.M., Morgelin, M. and Bjorck, L. (2000) Virulent aggregates of *Streptococcus pyogenes* are generated by homophilic protein–protein interactions. *Molecular Microbiology* 37, 1232–1247.

Fux, C.A., Costerton, J.W., Stewart, P.S. and Stoodley, P. (2005) Survival strategies of infectious biofilms. *Trends in Microbiology* 13, 34–40.

Gao, P., Pinkston, K.L., Nallapareddy, S.R., van Hoof, A., Murray, B.E. and Harvey, B.R. (2010) *Enterococcus faecalis* rnjB is required for pilin gene expression and biofilm formation. *Journal of Bacteriology* 192, 5489–5498.

Gibbs, R.S., Schrag, S. and Schuchat, A. (2004) Perinatal infections due to group B streptococci. *Obstetrics and Gynecology* 104, 1062–1076.

Granok, A.B., Parsonage, D., Ross, R.P. and Caparon, M.G. (2000) The RofA binding site in *Streptococcus pyogenes* is utilized in multiple transcriptional pathways. *Journal of Bacteriology* 182, 1529–1540.

Hall-Stoodley, L., Costerton, J.W. and Stoodley, P. (2004) Bacterial biofilms: from the natural environment to infectious diseases. *Nature Reviews. Microbiology* 2, 95–108.

Hansen, S.M., Uldbjerg, N., Kilian, M. and Sorensen, U.B. (2004) Dynamics of *Streptococcus agalactiae* colonization in women during and after pregnancy and in their infants. *Journal of Clinical Microbiology* 42, 83–89.

Harfouche, C., Filippini, S., Gianfaldoni, C., Ruggiero, P., Moschioni, M., Maccari, S., Pancotto, L., Arcidiacono, L., Galletti, B., Censini, S., Mori, E., Giuliani, M., Facciotti, C.,

Cartocci, E., Savino, S., Doro, F., Pallaoro, M., Nocadello, S., Mancuso, G., Haston, M., Goldblatt, D., Barocchi, M.A., Pizza, M., Rappuoli, R. and Masignani, V. (2012) RrgB321, a fusion protein of the three variants of the pneumococcal pilus backbone RrgB, is protective *in vivo* and elicits opsonic antibodies. *Infection and Immunity* 80, 451–460.

Helaine, S., Carbonnelle, E., Prouvensier, L., Beretti, J.L., Nassif, X. and Pelicic, V. (2005) PilX, a pilus-associated protein essential for bacterial aggregation, is a key to pilus-facilitated attachment of *Neisseria meningitidis* to human cells. *Molecular Microbiology* 55, 65–77.

Hendrickx, A.P., Bonten, M.J., van Luit-Asbroek, M., Schapendonk, C.M., Kragten, A.H. and Willems, R.J. (2008) Expression of two distinct types of pili by a hospital-acquired *Enterococcus faecium* isolate. *Microbiology* 154, 3212–3223.

Hendrickx, A.P., Willems, R.J., Bonten, M.J. and van Schaik, W. (2009) LPxTG surface proteins of enterococci. *Trends in Microbiology* 17, 423–430.

Hidalgo-Grass, C., Dan-Goor, M., Maly, A., Eran, Y., Kwinn, L.A., Nizet, V., Ravins, M., Jaffe, J., Peyser, A., Moses, A.E. and Hanski, E. (2004) Effect of a bacterial pheromone peptide on host chemokine degradation in group A streptococcal necrotising soft-tissue infections. *Lancet* 363, 696–703.

Hilleringmann, M., Giusti, F., Baudner, B.C., Masignani, V., Covacci, A., Rappuoli, R., Barocchi, M.A. and Ferlenghi, I. (2008) Pneumococcal pili are composed of protofilaments exposing adhesive clusters of Rrg A. *PLoS Pathogens* 4, e1000026.

Hilleringmann, M., Ringler, P., Muller, S.A., De Angelis, G., Rappuoli, R., Ferlenghi, I. and Engel, A. (2009) Molecular architecture of *Streptococcus pneumoniae* TIGR4 pili. *EMBO Journal* 28, 3921–3930.

Houry, A., Gohar, M., Deschamps, J., Tischenko, E., Aymerich, S., Gruss, A. and Briandet, R. (2012) Bacterial swimmers that infiltrate and take over the biofilm matrix. *Proceedings of the National Academy of Sciences of the United States of America* 109, 13088–13093.

Kang, H.J. and Baker, E.N. (2012) Structure and assembly of Gram-positive bacterial pili: unique covalent polymers. *Current Opinion in Structural Biology* 22, 200–207.

Kim, K.S. (2010) Acute bacterial meningitis in infants and children. *Lancet Infectious Diseases* 10, 32–42.

Kirn, T.J., Lafferty, M.J., Sandoe, C.M. and Taylor, R.K. (2000) Delineation of pilin domains required for bacterial association into microcolonies and intestinal colonization by *Vibrio cholerae. Molecular Microbiology* 35, 896–910.

Klausen, M., Heydorn, A., Ragas, P., Lambertsen, L., Aaes-Jorgensen, A., Molin, S. and Tolker-Nielsen, T. (2003) Biofilm formation by *Pseudomonas aeruginosa* wild type, flagella and type IV pili mutants. *Molecular Microbiology* 48, 1511–1524.

Kolenbrander, P.E., Palmer, R.J. Jr, Rickard, A.H., Jakubovics, N.S., Chalmers, N.I. and Diaz, P.I. (2006) Bacterial interactions and successions during plaque development. *Periodontology 2000* 42, 47–79.

Konto-Ghiorghi, Y., Mairey, E., Mallet, A., Dumenil, G., Caliot, E., Trieu-Cuot, P. and Dramsi, S. (2009) Dual role for pilus in adherence to epithelial cells and biofilm formation in *Streptococcus agalactiae. PLoS Pathogens* 5, e1000422.

Kratovac, Z., Manoharan, A., Luo, F., Lizano, S. and Bessen, D.E. (2007) Population genetics and linkage analysis of loci within the FCT region of *Streptococcus pyogenes. Journal of Bacteriology* 189, 1299–1310.

Lembke, C., Podbielski, A., Hidalgo-Grass, C., Jonas, L., Hanski, E. and Kreikemeyer, B. (2006) Characterization of biofilm formation by clinically relevant serotypes of group A streptococci. *Applied and Environmental Microbiology* 72, 2864–2875.

LeMieux, J., Hava, D.L., Basset, A. and Camilli, A. (2006) RrgA and RrgB are components of a multisubunit pilus encoded by the *Streptococcus pneumoniae* rlrA pathogenicity islet. *Infection and Immunity* 74, 2453–2456.

Lewis, K. (2001) Riddle of biofilm resistance. *Antimicrobial Agents and Chemotherapy* 45, 999–1007.

Loughman, J.A. and Caparon, M. (2006) Regulation of SpeB in *Streptococcus pyogenes* by pH and NaCl: a model for *in vivo* gene expression. *Journal of Bacteriology* 188, 399–408.

Manetti, A.G., Zingaretti, C., Falugi, F., Capo, S., Bombaci, M., Bagnoli, F., Gambellini, G., Bensi, G., Mora, M., Edwards, A.M., Musser, J.M., Graviss, E.A., Telford, J.L., Grandi, G. and Margarit, I. (2007) *Streptococcus pyogenes* pili promote pharyngeal cell adhesion and biofilm formation. *Molecular Microbiology* 64, 968–983.

Manetti, A.G., Koller, T., Becherelli, M., Buccato, S., Kreikemeyer, B., Podbielski, A., Grandi, G. and Margarit, I. (2010) Environmental acidification drives *S. pyogenes* pilus expression and microcolony formation on epithelial cells in a FCT-dependent manner. *PLoS One* 5, e13864.

Marrie, T.J. and Costerton, J.W. (1983) A scanning electron microscopic study of urine droppers and urine collecting systems. *Archives of Internal Medicine* 143, 1135–1141.

Marsh, P.D. (2006) Dental plaque as a biofilm and a microbial community – implications for health and disease. *BMC Oral Health* 6(Suppl 1), S14.

Mishra, A., Devarajan, B., Reardon, M.E., Dwivedi, P., Krishnan, V., Cisar, J.O., Das, A., Narayana, S.V. and Ton-That, H. (2011) Two autonomous structural modules in the fimbrial shaft adhesin FimA mediate *Actinomyces* interactions with streptococci and host cells during oral biofilm development. *Molecular Microbiology* 81, 1205–1220.

Mora, M., Bensi, G., Capo, S., Falugi, F., Zingaretti, C., Manetti, A.G., Maggi, T., Taddei, A.R., Grandi, G. and Telford, J.L. (2005) Group A *Streptococcus* produce pilus-like structures containing protective antigens and Lancefield T antigens. *Proceedings of the National Academy of Sciences of the United States of America* 102, 15641–15646.

Moschioni, M., Donati, C., Muzzi, A., Masignani, V., Censini, S., Hanage, W.P., Bishop, C.J., Reis, J.N., Normark, S., Henriques-Normark, B., Covacci, A., Rappuoli, R. and Barocchi, M.A. (2008) *Streptococcus pneumoniae* contains 3 rlrA pilus variants that are clonally related. *Journal of Infectious Diseases* 197, 888–896.

Moscoso, M., Garcia, E. and Lopez, R. (2006) Biofilm formation by *Streptococcus pneumoniae*: role of choline, extracellular DNA, and capsular polysaccharide in microbial accretion. *Journal of Bacteriology* 188, 7785–7795.

Munoz-Elias, E.J., Marcano, J. and Camilli, A. (2008) Isolation of *Streptococcus pneumoniae* biofilm mutants and their characterization during nasopharyngeal colonization. *Infection and Immunity* 76, 5049–5061.

Musser, J.M. and Shelburne, S.A. 3rd (2009) A decade of molecular pathogenomic analysis of group A *Streptococcus. Journal of Clinical Investigation* 119, 2455–2463.

Nakata, M., Podbielski, A. and Kreikemeyer, B. (2005) MsmR, a specific positive regulator of the *Streptococcus pyogenes* FCT pathogenicity region and cytolysin-mediated translocation system genes. *Molecular Microbiology* 57, 786–803.

Nakata, M., Koller, T., Moritz, K., Ribardo, D., Jonas, L., McIver, K.S., Sumitomo, T., Terao, Y., Kawabata, S., Podbielski, A. and Kreikemeyer, B. (2009) Mode of expression and functional characterization of FCT-3 pilus region-encoded proteins in *Streptococcus pyogenes* serotype M49. *Infection and Immunity* 77, 32–44.

Nallapareddy, S.R., Singh, K.V., Sillanpaa, J.,

Garsin, D.A., Hook, M., Erlandsen, S.L. and Murray, B.E. (2006) Endocarditis and biofilm-associated pili of *Enterococcus faecalis*. *Journal of Clinical Investigation* 116, 2799–2807.

Nallapareddy, S.R., Sillanpaa, J., Mitchell, J., Singh, K.V., Chowdhury, S.A., Weinstock, G.M., Sullam, P.M. and Murray, B.E. (2011) Conservation of Ebp-type pilus genes among Enterococci and demonstration of their role in adherence of *Enterococcus faecalis* to human platelets. *Infection and Immunity* 79, 2911–2920.

Neely, M.N., Pfeifer, J.D. and Caparon, M. (2002) *Streptococcus*-zebrafish model of bacterial pathogenesis. *Infection and Immunity* 70, 3904–3914.

Nelson, A.L., Ries, J., Bagnoli, F., Dahlberg, S., Falker, S., Rounioja, S., Tschop, J., Morfeldt, E., Ferlenghi, I., Hilleringmann, M., Holden, D.W., Rappuoli, R., Normark, S., Barocchi, M.A. and Henriques-Normark, B. (2007) RrgA is a pilus-associated adhesin in *Streptococcus pneumoniae*. *Molecular Microbiology* 66, 329–340.

Olsen, A., Jonsson, A. and Normark, S. (1989) Fibronectin binding mediated by a novel class of surface organelles on *Escherichia coli*. *Nature* 338, 652–655.

Orndorff, P.E., Devapali, A., Palestrant, S., Wyse, A., Everett, M.L., Bollinger, R.R. and Parker, W. (2004) Immunoglobulin-mediated agglutination of and biofilm formation by *Escherichia coli* K-12 require the type 1 pilus fiber. *Infection and Immunity* 72, 1929–1938.

Paranjpye, R.N. and Strom, M.S. (2005) A *Vibrio vulnificus* type IV pilin contributes to biofilm formation, adherence to epithelial cells, and virulence. *Infection and Immunity* 73, 1411–1422.

Pelton, S.I. and Leibovitz, E. (2009) Recent advances in otitis media. *Pediatric Infectious Disease Journal* 28, S133–137.

Podbielski, A., Woischnik, M., Leonard, B.A. and Schmidt, K.H. (1999) Characterization of nra, a global negative regulator gene in group A streptococci. *Molecular Microbiology* 31, 1051–1064.

Pratt, L.A. and Kolter, R. (1998) Genetic analysis of *Escherichia coli* biofilm formation: roles of flagella, motility, chemotaxis and type I pili. *Molecular Microbiology* 30, 285–293.

Proft, T. and Baker, E.N. (2009) Pili in Gram-negative and Gram-positive bacteria – structure, assembly and their role in disease. *Cellular and Molecular Life Sciences: CMLS* 66, 613–635.

Rice, K.C., Mann, E.E., Endres, J.L., Weiss, E.C., Cassat, J.E., Smeltzer, M.S. and Bayles, K.W. (2007) The cidA murein hydrolase regulator contributes to DNA release and biofilm development in *Staphylococcus aureus*. *Proceedings of the National Academy of Sciences of the United States of America* 104, 8113–8118.

Rickard, A.H., Gilbert, P., High, N.J., Kolenbrander, P.E. and Handley, P.S. (2003) Bacterial coaggregation: an integral process in the development of multi-species biofilms. *Trends in Microbiology* 11, 94–100.

Rinaudo, C.D., Rosini, R., Galeotti, C.L., Berti, F., Necchi, F., Reguzzi, V., Ghezzo, C., Telford, J.L., Grandi, G. and Maione, D. (2010) Specific involvement of pilus type 2a in biofilm formation in group B *Streptococcus*. *PloS One* 5, e9216.

Rippke, F., Schreiner, V. and Schwanitz, H.J. (2002) The acidic milieu of the horny layer: new findings on the physiology and pathophysiology of skin pH. *American Journal of Clinical Dermatology* 3, 261–272.

Rosini, R., Rinaudo, C.D., Soriani, M., Lauer, P., Mora, M., Maione, D., Taddei, A., Santi, I., Ghezzo, C., Brettoni, C., Buccato, S., Margarit, I., Grandi, G. and Telford, J.L. (2006) Identification of novel genomic islands coding for antigenic pilus-like structures in *Streptococcus agalactiae*. *Molecular Microbiology* 61, 126–141.

Santi, I., Grifantini, R., Jiang, S.M., Brettoni, C., Grandi, G., Wessels, M.R. and Soriani, M. (2009) CsrRS regulates group B *Streptococcus* virulence gene expression in response to environmental pH: a new perspective on vaccine development. *Journal of Bacteriology* 191, 5387–5397.

Schembri, M.A., Dalsgaard, D. and Klemm, P. (2004) Capsule shields the function of short bacterial adhesins. *Journal of Bacteriology* 186, 1249–1257.

Schluter, S., Franz, C.M., Gesellchen, F., Bertinetti, O., Herberg, F.W. and Schmidt, F.R. (2009) The high biofilm-encoding Bee locus: a second pilus gene cluster in *Enterococcus faecalis*? *Current Microbiology* 59, 206–211.

Schneewind, O. and Missiakas, D.M. (2012) Protein secretion and surface display in Gram-positive bacteria. *Philosophical Transactions of the Royal Society of London. Series B, Biological Sciences* 367, 1123–1139.

Sillanpaa, J., Nallapareddy, S.R., Singh, K.V., Prakash, V.P., Fothergill, T., Ton-That, H. and Murray, B.E. (2010) Characterization of the ebp(fm) pilus-encoding operon of *Enterococcus faecium* and its role in biofilm formation and virulence in a murine model of urinary tract infection. *Virulence* 1, 236–246.

Stewart, P.S. and Franklin, M.J. (2008) Physiological heterogeneity in biofilms. *Nature Reviews. Microbiology* 6, 199–210.

Strom, M.S. and Lory, S. (1993) Structure-function and biogenesis of the type IV pili. *Annual Review of Microbiology* 47, 565–596.

Tendolkar, P.M., Baghdayan, A.S., Gilmore, M.S. and Shankar, N. (2004) Enterococcal surface protein, Esp, enhances biofilm formation by *Enterococcus faecalis*. *Infection and Immunity* 72, 6032–6039.

Tendolkar, P.M., Baghdayan, A.S. and Shankar, N. (2006) Putative surface proteins encoded within a novel transferable locus confer a high-biofilm phenotype to *Enterococcus faecalis*. *Journal of Bacteriology* 188, 2063–2072.

van der Poll, T. and Opal, S.M. (2009) Pathogenesis, treatment, and prevention of pneumococcal pneumonia. *Lancet* 374, 1543–1556.

Varga, J.J., Nguyen, V., O'Brien, D.K., Rodgers, K., Walker, R.A. and Melville, S.B. (2006) Type IV pili-dependent gliding motility in the Gram-positive pathogen *Clostridium perfringens* and other Clostridia. *Molecular Microbiology* 62, 680–694.

Waters, C.M. and Bassler, B.L. (2005) Quorum sensing: cell-to-cell communication in bacteria. *Annual Review of Cell and Developmental Biology* 21, 319–346.

Weinrick, B., Dunman, P.M., McAleese, F., Murphy, E., Projan, S.J., Fang, Y. and Novick, R.P. (2004) Effect of mild acid on gene expression in *Staphylococcus aureus*. *Journal of Bacteriology* 186, 8407–8423.

Werner, G., Coque, T.M., Hammerum, A.M., Hope, R., Hryniewicz, W., Johnson, A., Klare, I., Kristinsson, K.G., Leclercq, R., Lester, C.H., Lillie, M., Novais, C., Olsson-Liljequist, B., Peixe, L.V., Sadowy, E., Simonsen, G.S., Top, J., Vuopio-Varkila, J., Willems, R.J., Witte, W. and Woodford, N. (2008) Emergence and spread of vancomycin resistance among enterococci in Europe. *Euro Surveillance: Bulletin Europeen Sur Les Maladies Transmissibles = European Communicable Disease Bulletin* 13.

12 Fimbriae/Pili from Oral Bacteria

Haley Echlin and Hui Wu

Departments of Pediatric Dentistry and Microbiology, University of Alabama at Birmingham, USA

12.1 Introduction

The main portal of entry into the human body is through the oral cavity. As such, it is the first site many microorganisms will come into contact with en route to the body. However, the oral cavity presents a challenge for microorganisms to adhere and colonize. The primary antagonists are the continual flow of saliva that washes out anything that cannot attach to the surface and the enzymes in saliva that target and prevent bacterial attachment. Some microorganisms may bind transiently to the oral cavity, but, without the appropriate machinery to firmly attach to the surface or to compete with those already present, are quickly detached. Thus, the majority of the microorganisms that enter the oral cavity are immediately displaced and swallowed to be passed through the rest of the body. Those that can successfully colonize the oral cavity have adapted to adhere to the oral surfaces and to utilize available nutrients. These successful colonizers are a large and diverse population of microbes including over 600 different species in 350 taxa and 37 genera; any one individual person may carry 100 or more different species in their mouth at any time (Kolenbrander and London, 1993; Dewhirst *et al.*, 2010). However, only a small proportion of these can actually adhere to the hard and soft tissues of the oral cavity. The majority of the colonizers will bind and interact with these primary colonizers. This cell–cell binding among microbes is thought to play a key role in building oral biofilms and complex interaction networks (Fig. 12.1). Interactions among microbes and between the microorganisms and their attached oral surfaces can occur in a great variety of ways; different species or strains are not equivalent in their capability to bind to hosts or bacterial ligands. At the simplest level, adhesion can involve basic non-covalent interactions, including electrostatic, ionic or hydrophobic interactions (Nobbs *et al.*, 2011). On a more advanced level, microbes can possess specific cell surface-associated adherence proteins that are responsible for adhesion and, in some instances, invasion. These adhesins recognize cognate receptors on oral surfaces and other cells; these receptors are diverse and can be proteins, glycoproteins or polysaccharides (Kolenbrander and London, 1993). They can include structures such as fimbriae, pili, antigen I/II, outer membrane proteins and auto-transporter adhesins. Here, we will focus on fimbriae and pili and the recent advances made in understanding their role in adhesion and invasion in the oral cavity. Because there are fundamental differences in the cell envelope structures between Gram-positive and Gram-negative bacteria, the adhesion mechanisms of the two groups can be quite distinct. The two primary groups of Gram-positive bacteria present in the oral cavity are Streptococci and Actinomyces. Gram-negative species often comprise the later colonizers of the oral cavity and the noticeable ones are often regarded as periodontal disease-associated pathogens, including *Aggregatibacter actinomycetemcomitans*, *Porphyromonas gingivalis* and *Fusobacterium nucleatum* (Table 12.1). In this chapter, we will use these bacteria as models to discuss bacterial surface structures pili and fimbriae and their contributions to bacterial adhesion and invasion.

12.2 Oral Streptococci

Streptococci species possess a multitude of adhesins on their cell surfaces that facilitate adherence to a wide range of substrates, including host-derived salivary glycoproteins, microbial products and other microbial cells (Morris and McBride, 1984; Jenkinson *et al.*, 1997; Wu *et al.*, 1998). Their ability to bind to a great number of substrates contributes, in part, to their high success rate in colonization of the oral and epithelial surfaces of humans.

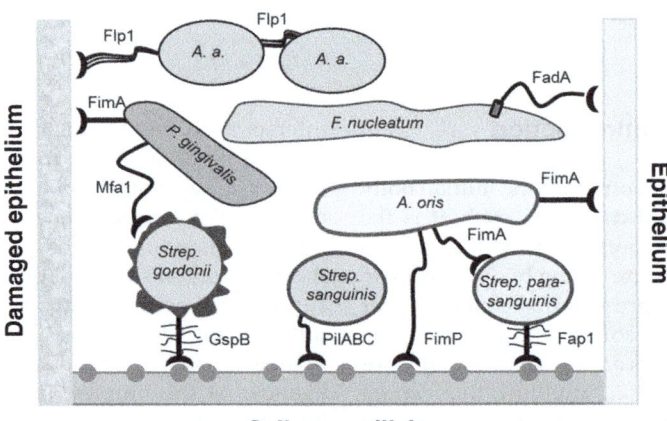

Salivary pellicle

Fig. 12.1. Diagrammatic representation of interactions occurring between oral microorganisms and other microorganisms/host tissues; these interactions contribute to the formation of microbial communities in the human oral cavity. Microbial adhesins are depicted as black structures extending from the bacterial cell surface and ending in a suction cup; the adhesins are labelled according to the nomenclature in Table 12.1. Pellicle receptors are represented as molecular spheres.

Table 12.1. Bacteria species and types, key characteristics and functions of adhesins.

Species	Adhesin	Key characteristic	Function/substrate
S. parasanguinis	Fap1	Glycoprotein	Salivary pellicle
S. gordonii	Hsa, GspB	Glycoprotein	Salivary pellicle (gp-340); fibronectin; host cells
S. sanguinis	PilA,B,C	Simple locus	Fibronectin; epithelial cells
A. oris	FimP/Q, FimA/B	Two distinct fimbriae	Salivary pellicle; host cells; co-aggregation (Strep. species)
A. actinomycetemcomitans	Flp1	Bundled pili; nonspecific adherence	Epithelial cells; salivary pellicle; auto-aggregation
P. gingivalis	FimA, Mfa1	Two distinct fimbriae; variety of substrates	Integrins; fibronectin; type I collagen; co-aggregation (*S. gordonii*)
F. nucleatum	FadA	Complex of pre-FadA and mature FadA	Epithelial/endothelial cells

Sanguis streptococci comprise a large proportion of oral bacterial species in dental plaque and are one of the first colonizers of the tooth surface (Wu et al., 1998; Palmer et al., 2003). Colonization of oral streptococci is dependent on the adhesion to salivary proteins, epithelial cells and tooth surfaces. Co-adherence partners include other early plaque colonizers, including Actinomyces species, and later colonizers, including Porphyromonas gingivalis, Tannerella forsythia and Treponema denticola, the three major causative agent of periodontal disease (Lamont et al., 1992; Yao et al., 1996; Whittaker et al., 1996; Jenkinson et al., 1997). Moreover, although the majority of streptococci are commensal, their colonization may be a precursor to tissue invasion and immune modulation, followed by disease development, when certain environmental conditions occur, thereby becoming opportunistic pathogens. Thus, modulation of environmental factors that impact bacterial colonization and understanding the mechanisms of bacterial adhesion may provide new insights into design and development of novel therapeutic targets against bacterial infections.

In the Sanguis streptococci group, Streptococcus parasanguinis and Streptococcus gordonii are primary colonizers of the tooth surface and, as such, play an important role in the formation of dental plaque and can act as a means to co-aggregate with other oral bacteria that are important for dental plaque formation. Moreover, S. parasanguinis and S. gordonii have been shown to be causes of infective endocarditis, an infection of the damaged heart valves (Burnette-Curley et al., 1995; Takahashi et al., 2002). S. parasanguinis has long peritrichous fimbriae that are required for adhesion, where mutants lacking fimbriae have a reduced adhesion to an in vitro tooth surface model, saliva-coated hydroxyapatite (SHA) (Fives-Taylor and Thompson, 1985; Stephenson et al., 2002). S. parasanguinis fimbriae consist of Fap1 (fimbriae-associated protein 1), a 200 kDa cell-wall-anchored serine-rich repeat glycoprotein (SRRP) (Fig. 12.2). Fap1 is required for fimbrial formation, bacterial adhesion (Wu et al., 1998; Stephenson et al., 2002) and biofilm formation (Froeliger and Fives-Taylor,

2001; Zhou and Wu, 2009). S. gordonii has two variants of SRRP similar to S. parasanguinis Fap1; these are Hsa and GspB. Hsa and GspB mediate adhesion to the salivary pellicle through recognition of sialic acid, including salivary mucin MG2 and salivary agglutinin (Takamatsu et al., 2006). Both GspB and Hsa are also able to bind to human platelets and are involved in pathogenesis of infective endocarditis; the binding is mediated by sialic acids on human platelets (Bensing et al., 2004b; Takamatsu et al., 2005a). Hsa is distinct from GspB in that it can bind both (2–3) sialyllactosamine [NeuAcα (2–3)Galβ (1–4) GlcNAc] and sialyl-T antigen (sT antigen) [NeuAcα (2–3)Galβ (1–3)GalNAc], whereas GspB binds only the sT antigen structure on GPIbα (Takamatsu et al., 2005a). All three SRRP have the characteristic regions of the SRRP family, including an N-terminal long signal peptide, an accessory Sec transport targeting domain, a short serine-rich region, a region that is rich in basic or acidic amino acid residues, a longer serine-rich region and a C-terminal LPXTG cell-wall-anchoring domain (Takahashi et al., 2002; Bensing et al., 2005; Takamatsu et al., 2006; Davies et al., 2009; Zhou and Wu, 2009; Bensing and Sullam, 2010).

The chromosomal region dedicated to SRRP biogenesis is quite large and highly conserved in many streptococci, staphylococci and even eubacteria, but the exact mechanism of SRRP biogenesis remains a mystery. Fap1 has an 11 gene cluster that is involved in its biogenesis and is separated into two regions: a core region that is conserved in every genome (secY2, gap1-3, secA2 and gtf1-2) and a variable region that includes four putative glycosyltransferases (gly, nss, galT1 and galT2) (Zhou and Wu, 2009). In the same way, GspB has a conserved locus dedicated to its biogenesis: a 14-kb chromosomal region just downstream of gspB which includes gly, nss, secA2, asp1, asp2, asp3, secY2, gtfA, gtfB, asp4 and asp5 (Fig. 12.2) (Takamatsu et al., 2004a). Fap1 is glycosylated with several monosaccharides, including glucose, N-acetyl glucosamine, N-acetyl galactosamine and rhamnose (Stephenson et al., 2002). Fap1 is glycosylated in the cytoplasm by Gtf1 and Gtf2 (glycosyltransferase 1 and 2), which

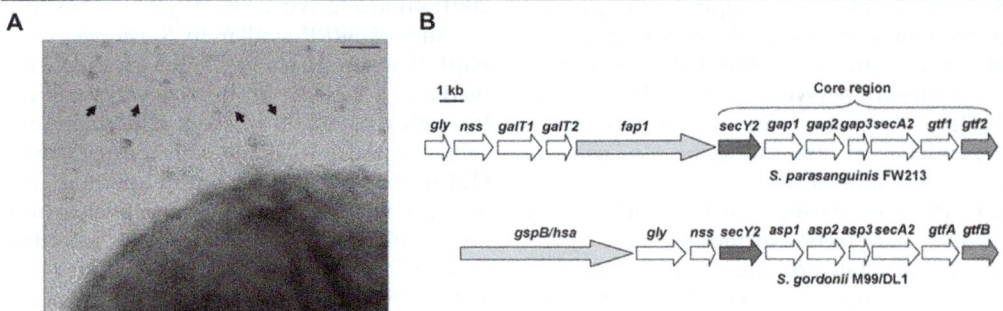

Fig. 12.2. *S. parasanguinis* fimbriae consist of Fap1. (A) Electron micrograph of the surface fimbriae of *S. parasanguinis*. Arrows point to the fimbriae. Bar 100 nm. (B) Diagram representation of loci arrangement of genes involved of SRRP biogenesis in *S. parasanguinis and S. gordonii* (Zhou and Wu, 2009).

interact together to add the first sugar (GlcNAc), followed by addition of the second sugar (glucose) via Gtf3 (glycosyltransferase 3) (Chen *et al.*, 2004; Wu *et al.*, 2007; Bu *et al.*, 2008; Zhou and Wu, 2009; Zhou *et al.*, 2010). Similarly, GspB is synthesized as a preprotein and is glycosylated intracellularly by Gly, Nss, GtfA and GtfB; glycosylation of GspB by GtfA and GtfB is important for the stability of GspB (Takamatsu *et al.*, 2004a, 2004b). To complete Fap1 biogenesis, accessory secretion components, including SecA2, SecY2, Gap1 and Gap3, are required to convert the immature form to the mature Fap1 (Chen *et al.*, 2004; Wu *et al.*, 2007; Zhou and Wu, 2009). Likewise, GspB is exported using a specialized accessory Sec system, which includes SecA2, SecY2, Asp1, Asp2, Asp3, Asp4 and Asp5 (Bensing and Sullam, 2002; Bensing *et al.*, 2004a; Takamatsu *et al.*, 2004b). Knock-out mutations of *secY2*, *gap1* or *gap3* produce an immature Fap1 in *S. parasanguinis* (Wu *et al.*, 2007; Li *et al.*, 2008; Peng *et al.*, 2008a, 2008b; Zhou *et al.*, 2008) and mutations of all *asps*, *secY2* and *secA2* result in lack of GspB export in *S. gordonii* (Bensing and Sullam, 2002; Takamatsu *et al.*, 2004b, 2005b), suggesting that these genes are involved in Fap1 and GspB biogenesis. SecA2 and SecY2 share homology with their counterparts in the canonical Sec system and most likely function in a similar manner; SecA2 has been shown to be an ATPase that shares structural features with SecA and SecY2 shows strong

similarity to the SecY transmembrane protein of the general protein secretion system (Bensing and Sullam, 2002, 2009). Moreover, SecA2 is required for the export of mature Fap1 to the cell-wall surface (Chen *et al.*, 2004). In *S. parasanguinis*, the interaction between Gap1 and Gap3 is required for Fap1 biogenesis (Li *et al.*, 2008); the Gap1/3 complex interacts with both SecA2 and SecA, indicating a link between accessory and canonical Sec system (Zhou *et al.*, 2011). In *S. gordonii*, Asp3 acts as a scaffold protein to coordinate the formation of the SecA2, Asp1, Asp2 and Asp3 complex for optimal export of GspB (Seepersaud *et al.*, 2010). Interestingly, both Asp2 and Asp3 can bind the un-glycosylated serine-rich repeat domains of GspB and these interactions are required for optimal GspB export. Glycosylation of the SRR domains prevent Asp binding, suggesting that binding of the Asps to the preprotein occurs prior to its full glycosylation and that Asps mediates GspB glycosylation (Yen *et al.*, 2011). However, these observations are not consistent with the findings that mature glycosylated GspB is still synthesized in the Asp mutants. Further studies are needed to clarify the discrepancy.

Like *S. parasanguinis* and *S. gordonii*, *S. sanguinis* is a primary colonizer of the oral cavity and is one of the most abundant species found in dental plaque (Kolenbrander and London, 1993; Nobbs *et al.*, 2009; Okahashi *et al.*, 2010). As such, *S. sanguinis* possesses a diverse array of adhesins. Other than the SRRP that is important for bacterial adhesion to human platelets (Plummer *et al.*, 2005), *S. sanguinis* have pili localized to the cell surface. Pilus structures on the surface of *S. sanguinis* are composed of polymers of three distinct subunits termed PilA, PilB and PilC. These subunits have homology with pilus proteins from more pathogenic strains, including *S. pneumoniae* (25–30% homology) (Okahashi *et al.*, 2010). However, in contrast to pathogenic strains, which often contain multiple genes involved in pilus formation, the pilus locus of *S. sanguinis* is more simply constructed with only three pilus subunits that contain an LPXTG motif and an 'E box' domain and a single sortase C (Kang *et al.*, 2007; Okahashi *et al.*, 2010). Although the pilus locus is simple, the pili of *S. sanguinis* are still functional and involved in bacterial adherence and invasion. Pilus subunit-deficient mutants have a greatly reduced ability to adhere and to invade epithelial cells. PilC has a strong ability to bind fibronectin, while PilA and PilB have no or weak binding (Okahashi *et al.*, 2010). None of the pilus subunits can bind collagen however, unlike those from several pathogenic species (Kreikemeyer *et al.*, 2005; Hilleringmann *et al.*, 2008; Okahashi *et al.*, 2010). This suggests functional differences among pilus structure, which may be due to the simplistic nature of

S. sanguinis pilus locus. Although PilA, B and C are involved in adhesion, there was not a complete loss of adhesion and invasion abilities in strains deficient in pili. This suggests that other bacterial surface components may be involved in adherence and invasion (McNab *et al.*, 1999; Okahashi *et al.*, 2010). Also, PilA, B and C are involved in *S. sanguinis* adhesion and colonization to saliva-coated tooth surfaces. PilB and PilC, but not PilA, bind to human whole saliva and PilC binds to multiple other salivary components, including salivary α-amylase. Moreover, pili are involved in biofilm formation, indicated by the inability of strains lacking pili to produce typical three-dimensional biofilm layers (Okahashi *et al.*, 2011). These studies further demonstrate binding tropism of adhesive fimbrial and fibrial structures from oral streptococci.

12.3 *Actinomyces oris*

Besides streptococcus species, one of the more prevalent species in the oral cavity belongs to the *Actinomyces* group, in particular the most abundant species in the oral cavity, *Actinomyces oris* (formerly *Actinomyces naeslundii*). *A. oris* produces two antigenically and functionally distinct types of fimbriae or pili (Fig. 12.3). Type 1 fimbriae promote

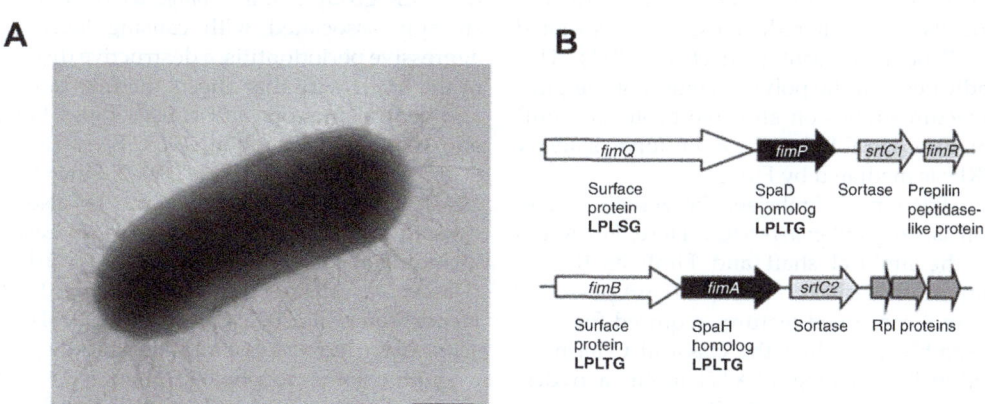

Fig. 12.3. *A. oris* displays two distinct fimbriae on its cell surface, Fim P/Q and Fim A/B. (A) Electron micrograph of the surface fimbriae of *A. oris*. Bar 200 nm. (B) Schematics of two gene clusters identified in the chromosome of *A. oris* MG-1 (Mishra *et al.*, 2007).

adhesion to salivary proline-rich proteins (PRPs) coating the tooth surface (Mishra *et al.*, 2007). This interaction was initially revealed by adhesion of *A. oris* to adsorbed acidic PRP1 (Gibbons and Hay, 1988), an un-glycosylated PRP, but subsequent studies (Clark *et al.*, 1989) have showed that type 1 fimbriae mediate adhesion to other PRPs, including acidic, basic and glycosylated proteins (Wu *et al.*, 2011). Type 2 fimbriae mediate interactions of *A. oris* with oral streptococci via a lectin activity by binding to GalNAc-β1-3-Gal and Gal-β-1-3-GalNAc (Stromberg and Karlsson, 1990; Cisar *et al.*, 1995). Type 2 fimbriae are also important for binding of *A. oris* to various host cells, including erythrocytes, epithelial cells and polymorphonuclear leukocytes (Mishra *et al.*, 2010, 2011). The genetic components for both types of fimbriae are arranged into distinct gene clusters (Mishra *et al.*, 2007), including genes for a fimbrial shaft, tip and a class C sortase (Fig. 12.3).

For type 1 fimbriae, the genetic cluster consists of *fimQ–fimP–srtC1*. FimP constitutes the fimbrial shaft and FimQ is the tip fimbrillin, which is the major adhesin that interacts with PRPs (Wu *et al.*, 2011). SrtC1 is the primary sortase used to assemble type 1 fimbriae to the cell surface (Wu *et al.*, 2011). A mutant lacking SrtC1 fails to produce FimP polymers but has only a 40% reduced ability to bind SHA. A strain deficient in FimQ displays a 75% reduction in adhesion to SHA and also results in reduced surface-associated FimP being present (Wu *et al.*, 2011). This indicates that the polymerization of the pilus structures relies on all three proteins (FimP, FimQ and SrtC1), but the adhesion to salivary PRPs is mediated by FimQ.

For type 2 fimbriae, the genetic cluster consists of *fimB–fimA–srtC2*. Here, FimA acts as the fimbrial shaft and FimB as the tip fimbrillin (Mishra *et al.*, 2007). FimA possesses several conserved features required for pilus assembly, including the classical C-terminal cell-wall signalling LPXTG motif, a hydro-phobic region and a positively charged tail (Navarre and Schneewind, 1999). FimA also contains a putative pin motif YPK, which is involved in the lysine-mediated trans-peptidation reaction necessary to cross-link

pilin into pilus polymers (Ton-That and Schneewind, 2004; Mishra *et al.*, 2011). Surprisingly, it is FimA, and not the tip FimB, that mediates *A. oris* co-aggregation with oral streptococci, adherence to red blood cells and biofilm formation (Mishra *et al.*, 2010). SrtC2 is a pilin-specific class sortase involved in type 2 fimbrial assembly and is required for *A. oris* to co-aggregate, adhere and form biofilms (Mishra *et al.*, 2010). More specifically, it is the C-terminal region of SrtC2, which contains a TM helix and a cytoplasmic domain, and the conserved catalytic resides Cys246 and His184 that are essential for pilus polymerization, bacterial co-aggregation and biofilm formation (Wu *et al.*, 2012). These studies have yielded novel and important information regarding fimbriae biogenesis in Gram-positive bacteria and provide a better understanding of molecular details of fimbriae-mediated bacterial adherence.

12.4 *Aggregatibacter actinomycetemcomitans*

Aggregatibacter (formerly *Actinobacillus*) *actinomycetemcomitans* is a facultative anaerobic coccobacillus that colonizes the human oral cavity and the upper respiratory tract. It is included in a clinically relevant group of microbes that includes *Haemophilus*, *Actinobacillus*, *Cardiobacter*, *Eikenella* and *Kingella* (HACEK group). *A. actinomycetemcomitans* is strongly associated with causing localized aggressive periodontitis, a destructive disease of the oral cavity that affects the first molars and central incisors and causes rapid bone and tooth loss (Zambon, 1985; Fives-Taylor *et al.*, 1999; Henderson *et al.*, 2002; Fine *et al.*, 2007). *A. actinomycetemcomitans* is also a causative agent of other human diseases, including abscess formation and endocarditis (Das *et al.*, 1997; Fives-Taylor *et al.*, 1999; Henderson *et al.*, 2002; Paturel *et al.*, 2004). *A. actinomycetemcomitans* clinical isolates display a rough colony morphology phenotype with star-like centres and adhere tenaciously to surfaces to form extremely strong biofilms (Kachlany *et al.*, 2001a; Perez *et al.*, 2006). A distinct feature of *A. actinomycetemcomitans* is that these microbes will autoaggregate and nonspecifically adhere on a variety of solid surfaces, including plastic, glass and hydroxyapatite (Kachlany *et al.*, 2000, 2001a); they can also adhere to the salivary pellicle, which suggests that *A. actinomycetemcomitans* may be considered an early colonizer (Fine *et al.*, 2010). The tenacious biofilm formed by *A. actinomycetemcomitans* is critical for colonization and persistence in the oral cavity and for pathogenesis in a rat model (Fine *et al.*, 1999; Schreiner *et al.*, 2003). Both autoaggregation and biofilm formation in *A. actinomycetemcomitans* is mediated by long bundled pili that resemble type IV pili and are composed of a glycosylated 6.5 kDa subunit protein, Flp-1 (Fig. 12.4) (Inoue *et al.*, 1998, 2000; Kachlany *et al.*, 2001a; Henderson *et al.*, 2002). It is unknown whether glycosylation of Flp-1 plays any role in pilus biogenesis and bacterial adhesion.

The secretion of Flp-1 is controlled by several proteins encoded by the tight adherence (*tad*) locus, which contains 14 genes: *flp-1-flp-2-tadV-rcpCAB-tadZABCDEFG* (Fig. 12.4) (Kachlany *et al.*, 2001b; Planet *et al.*, 2003). The *tad* locus is arranged as an operon and expressed as a single transcriptional unit from *flp-1* to at least *tadD* (Haase *et al.*, 2003; Perez *et al.*, 2006). Twelve of these genes (*flp-1*, *tadV*, *rcpA*, *rcpC* and *tadZABCDEFG*) are essential for pili formation, tight nonspecific adherence, autoaggregation and biofilm formation (Kachlany *et al.*, 2000, 2001a, 2001b; Planet *et al.*, 2003; Perez *et al.*, 2006). *flp-1* is translated into a type IVb prepillin protein

that is then post-translationally modified and then processed into a mature form prior to secretion and pilus assembly (Inoue *et al.*, 2000; Kachlany *et al.*, 2001b). TadA acts as an ATPase that is likely to energize or serve as a chaperone in the secretion and assembly of Flp-1 since it has ATPase activity and shares homology to secretion NTPases of type IV secretion systems (Bhattacharjee *et al.*, 2001; Perez *et al.*, 2006). Although it is very similar to Flp-1, Flp-2 seems to have no discernable activity as strains defective in Flp-2 have no observable phenotype, including pili formation, autoaggregation, colony morphology and biofilm formation in *A. actinomycetemcomitans* (Perez *et al.*, 2006). TadV acts as an aspartic acid prepillin peptidase required for processing the Flp-1 pilin subunit and for bacterial adherence and biofilm formation (Perez *et al.*, 2006; Tomich *et al.*, 2006). RcpB is a 20 kDa outer membrane component originally identified because it is exclusively produced by rough colony variants (Haase *et al.*, 1999), but its exact role remains unknown. To date, the complete *rcpB* has not been knocked-out in a wild-type strain, but has been in a spontaneous *tad*-deficient strain; this may suggest that *rcpB* is an essential gene in the context of a functional Tad secretion system (Perez *et al.*, 2006). RcpA is a member of the GspD/OutD/PulD secretin family of proteins (12, 19) and is predicted to form the outer membrane channel used for the secretion of Flp-1 (Haase *et al.*, 1999; Kachlany *et al.*, 2001a). To date, the remaining proteins

A   **B**

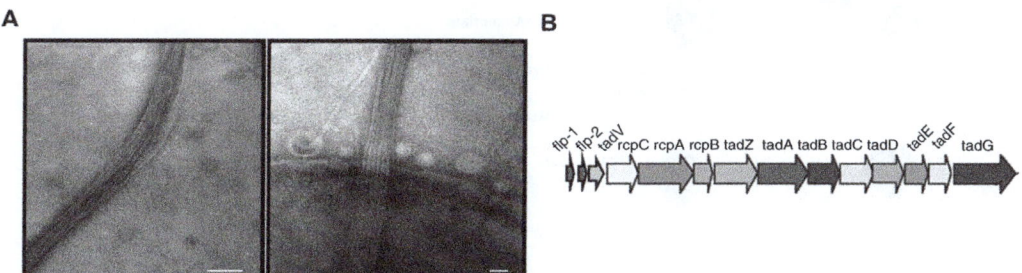

Fig. 12.4. The long bundled pili in *A. actinomycetemcomitans* are composed of a glycosylated 6.5 kDa subunit protein, Flp-1. (A, left) Electron micrograph of fibril bundles produced by the rough *A. actinomycetemcomitans* strain CU1000N. Bar 50 nm. (A, right) Array of fibrils passing over the edge of a bacterial cell. Bar 20 nm (Kachlany *et al.*, 2000). (B) Diagram of the *tad* locus of *A. actinomycetemcomitans* (Planet *et al.*, 2003).

encoded by the *tad* locus have no assigned functions and have no obvious homology to other known members of prokaryotic secretion systems. The *tad* locus is highly conserved in many Gram-negative, Gram-positive bacteria and even archaea (Tomich *et al.*, 2007), suggesting it plays an important role in bacterial fitness and virulence.

12.5 *Porphyromonas gingivalis*

Porphyromonas gingivalis is a black-pigmented obligate anaerobe that is strongly associated with periodontitis – a major cause of tooth loss in the adult population. In the oral cavity, this bacterium will settle in the gingival crevice or invade and persist within gingival epithelial cells (Lamont and Jenkinson, 1998; Socransky *et al.*, 1998; Andrian *et al.*, 2006). Moreover, *P. gingivalis* is implicated in other systemic inflammatory conditions, such as atherosclerosis (Seymour *et al.*, 2007). To colonize in the oral cavity, *P. gingivalis* interacts with the host cells, salivary and crevicular fluid components, and other periodontal pathogens, including *T. denticola* and *T. forsythia* (to form the red complex, a periodontal pathogen community) and many of the early plaque organisms such as oral streptococci (e.g. *S. gordonii* and *S. parasanguinis*) and *A. oris* (Lamont and Jenkinson,

1998; Socransky *et al.*, 1998; Holt and Ebersole, 2005). In order to interact with this multitude of cells and microbial and host components, *P. gingivalis* uses a wide variety of adhesion molecules, including fimbriae. *P. gingivalis* has two distinct types of fimbriae – long major fimbriae (0.3–1.5 μm) and shorter minor fimbriae (80–120 nm) (Fig. 12.5). The predominant protein subunits of these structures are FimA and Mfa1, respectively (Amano, 2010). Both types of fimbriae appear to be abundant in *Bacteroides* and represent evolutionarily unique groups since no homologues have been found in any other bacteria (Nishiyama *et al.*, 2007).

The major fimbriae binds to a number of eukaryotic proteins, including fibronectin, collagen, laminin, saliva-derived proline-rich protein, statherin, CXCchemokine receptor 4 on immune cells (which may provide a means of subversion of host immune function) and $\alpha_v\beta_3$ and $\alpha_5\beta_1$ integrin receptors on gingival epithelial cells (which mediate initial *P. gingivalis* attachment to the epithelial cells) (Yilmaz *et al.*, 2002; Nishiyama *et al.*, 2007; Hajishengallis *et al.*, 2008; Pierce *et al.*, 2009). FimA also binds prokaryotic proteins, including glyceraldehyde-3-phosphate dehydrogenase (GAPDH) of *Streptococcus oralis* (Maeda *et al.*, 2004). Six allelic forms of FimA have been identified (I, Ib, II, III, IV and V) and bind to integrin receptors with different

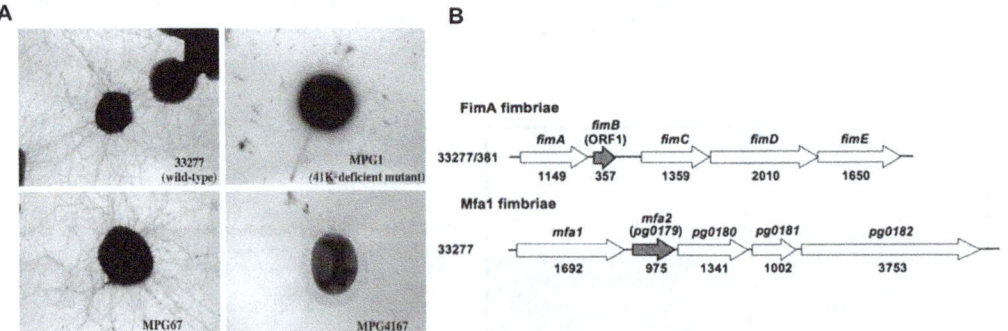

Fig. 12.5. *P. gingivalis* displays two distinct fimbriae on its cell surface, FimA (major) and Mfa1 (minor). (A) Electron micrograph of fimbriae present on *P. gingivalis* wild-type strain (33277) and *P. gingivalis* mutant strains lacking the major fimbriae (MPG1) or the minor fimbriae (MPG67). No fimbriae are observed in a *P. gingivalis* mutant strain deficient in both major and minor fimbriae (MPG4167) (Umemoto and Hamada, 2003). (B) Schematic diagrams of the genes downstream of *fimA* and *mfa1* in *P. gingivalis* strain 33277 (Yoshimura *et al.*, 2009).

affinities, which correlates with differences in pathogenicity among *P. gingivalis* strains (Amano *et al.*, 2000). Since FimA aids in the adherence and invasion of host cells, the major fimbriae plays a role in the virulence of *P. gingivalis* in rodent models of both periodontitis and atherosclerosis (Malek *et al.*, 1994; Gibson *et al.*, 2004). Albeit the function of the major fimbriae is critical, the biogenesis of the major fimbriae is less clear. FimA is translated as a prefimbrillin protein with a leader peptide and must be transported from the cytoplasm to the periplasm via the general secretion pathway. This is followed by cleavage by signal peptidase II at the cysteine residue of the N-terminus and transport to the outer face of the outer membrane where it is further processed by Rgp, a gingipain protease, to yield the mature form (Amano, 2010). Biosynthesis of FimA is controlled by FimR and FimS, a two-component regulatory system (Yoshimura *et al.*, 2009). Although FimA constitutes the main structural component of *P. gingivalis* major fimbriae, the native fimbrial structure contains additional protein components, encoded by genes immediately downstream of *fimA*- *fimB*, *fimC*, *fimD* and *fimE* (Fig. 12.5) (Nishiyama *et al.*, 2007). *fimB* is an unknown gene and presumably encodes an outer membrane protein (Yoshimura *et al.*, 2009). *fimC*, *fimD* and *fimE* encode for ancillary subunits that form a functional complex that mediates binding to ECM proteins fibronectin and type I collagen (Nishiyama *et al.*, 2007; Pierce *et al.*, 2009). Although FimCDE comprise less than 1% of the fimbrial protein content, they appear to contribute significantly to *P. gingivalis* virulence in that stains lacking FimCDE are dramatically less persistent and virulent in a mouse periodontitis model and express shorter fimbriae than the wild type (Wang *et al.*, 2007). Moreover, FimCDE are involved in bacterial autoaggregation and binding to several bacterial and eukaryotic proteins – including GAPDH of *S. oralis*, two extracellular matrix proteins, fibronectin and type I collagen (Nishiyama *et al.*, 2007). However, how FimCDE contribute to the assembly of the major fimbriae is unknown. It is suggested that FimCDE form the adhesive tip com-

ponent associated with the major fimbriae protein FimA (Amano, 2010). How the adhesive activities carried out by both FimCDE and FimA are differentiated requires further study.

Much research has been focused on the major fimbriae, but little is known regarding the minor fimbriae, Mfa1 fimbriae. Mfa1 fimbriae plays a role in adherence to host cells because it readily interacts with Toll-like receptor 2 and CD14 and induces the expression of various cytokines (Umemoto and Hamada, 2003; Amano, 2010). Further, Mfa1 is involved in *P. gingivalis* interactions with other bacteria, including interaction with the SspB region of *S. gordonii* (Nobbs *et al.*, 2011). A strain deficient in Mfa1 forms larger clumps of autoaggregated bacteria and a larger biofilm mass than wild-type strain, while a strain deficient in FimA forms smaller clumps and less biofilm mass, suggesting that FimA promotes auto-aggregation and biofilm formation while Mfa1 suppresses it (Umemoto and Hamada, 2003; Kuboniwa *et al.*, 2009; Amano, 2010). The exact mechanism underlying Mfa1 fimbriae biogenesis is unknown. Three proteins translated from genes downstream of *mfa1* may have accessory function to assist fimbrial assembly (Yoshimura *et al.*, 2009). Mfa2 – encoded by a gene that is located downstream of *mfa1* and co-transcribed with *mfa1* – has been shown to be involved in the biogenesis of Mfa1 fimbriae (Hasegawa *et al.*, 2009). Recently, the crystal structure of an Mfa2 homologue BT1062 from *Bacteroides thetaiotaomicron* has been solved and revealed a new conserved fimbrial fold. This fold shares some similarity with the minor pilin of GBS52 (PDB code 2pz4) and the major pilin Spy0128 (PDB code 3b2m) (Xu *et al.*, 2010), suggesting an evolutionary conservation among fimbrial structures from both Gram-negative and Gram-positive bacteria. Interaction between Mfa1 and SspB from *S. gordonii* has been explored to identify novel inhibitors to block bacterial virulence (Daep *et al.*, 2011). Further studies should focus on determining structures of these unique fimbriae and how they interact with SspB, thus revealing molecular details that are amenable to therapeutic discovery.

12.6 *Fusobacterium nucleatum*

One of the most abundant Gram-negative anaerobes colonizing the subgingival plaque is *Fusobacterium nucleatum* (Han *et al.*, 2005; Amano, 2010). This bacterium is a filamentous anaerobic commensal organism. However, *F. nucleatum* is considered an opportunistic pathogen because it has been implicated in various forms of periodontal disease and in infections and abscesses in other parts of the body, including, most notably, intra-uterine infections which can lead to pregnancy complications such as preterm birth and stillbirth (Moore and Moore, 1994; Amano, 2010; Han *et al.*, 2010; Fardini *et al.*, 2011). In the oral cavity during periodontal infection, *F. nucleatum* cell mass increases as much as 10,000-fold, making it one of the most abundant anaerobic species in the diseased site (Moore and Moore, 1994; Han *et al.*, 2005). Because *F. nucleatum* has an unusual promiscuity for co-aggregation with both early and late colonizers of the oral cavity, this organism is often thought of as a bridging bacterium in plaque by adhering to early colonizers and then providing a platform to facilitate the integration of later colonizers (Kolenbrander *et al.*, 2006; Amano, 2010; Nobbs *et al.*, 2011). Besides adherence to other microbial cells, *F. nucleatum* can also adhere to a variety of host cells – including epithelial and leukocytes, monocytes, erythrocytes, fibroblasts, and HeLa cells, salivary macromolecules, extracellular matrix proteins and human immunoglobulin G – and can invade host epithelial and endothelial cells (Han *et al.*, 2000, 2005). *F. nucleatum* also promotes the invasion and penetration of host cells by normally non-invasive bacteria (Edwards *et al.*, 2006; Fardini *et al.*, 2011).

F. nucleatum can adhere to other cells using a novel filamentous adhesin comprised of FadA. FadA has been shown to play a major role in the cell attachment process, invasion and development of uterine complications (Han *et al.*, 2005; Xu *et al.*, 2007; Ikegami *et al.*, 2009) and is conserved in oral fusobacteria, but not present in non-oral fusobacterial species (Han *et al.*, 2005). FadA exists in two forms: a non-secreted pre-FapA (129-amino-acid residues) and secreted, mature FadA (mFadA, 111-amino-acid residues). The secreted mFadA is exposed on the bacterial surface, while pre-FadA is anchored in the inner membrane and, together, they form a high molecular weight complex (FadAc). The complex, and not just mFadA, is required for attachment and invasion of the host cells (Fig. 12.6) (Han *et al.*, 2005; Xu *et al.*, 2007; Nithianantham *et al.*, 2009). The crystal structure of mFadA reveals two antiparallel alpha-helical arms linked by a non-alpha-helical hairpin loop. The mFadA subunits link in a head-to-tail pattern via a leucine chain structural motif to form elongated filaments. These filaments are stabilized by intermolecular hydrophobic interactions between the leucine residues and may intertwine to form thicker structures to support adhesion (Nithianantham *et al.*, 2009; Nobbs *et al.*, 2009). Proper oligomerization of FadA is essential for its function and the signal peptide (MKKFLLLAVLAVSASAFA) of mFadA determines the length and width of the filament, where short filaments and knots may be the active form (Temoin *et al.*, 2012).

12.7 Conclusions

Oral bacteria display a diverse array of filamental structures to meet the dynamic challenge they encounter in the oral cavity. The identification and characterization of key adhesins that promote oral bacterial adhesion and biofilm formation are important goals. Not only will these studies give us new understanding into the complex communities formed in the oral cavity, but they will allow us to better combat disease – not just in the oral cavity, but throughout the body. Within the oral cavity, many antibiotics have been rendered ineffective because of the ability of bacteria to form tenacious biofilms that form a physical barrier to those antibiotics. By understanding how the bacterial adhesins promote biofilm formation, new therapeutics can be developed to target the adhesins and the genes involved in their biogenesis. The late colonizers often associated with periodontal disease in this biofilm cannot adhere to the tooth surface directly and must interact with an initially layer of microbes. By

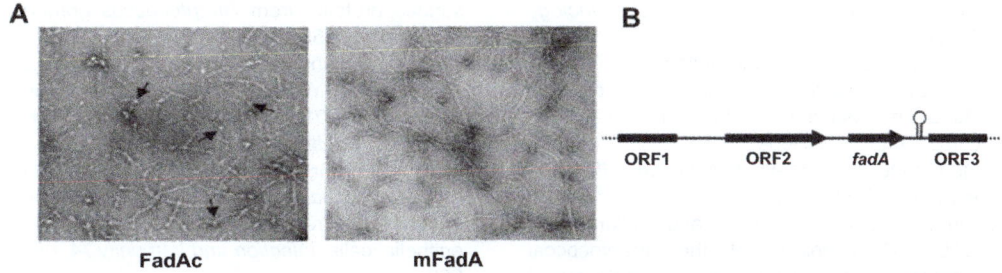

A

FadAc   mFadA

B

ORF1   ORF2   *fadA*   ORF3

Fig. 12.6. A novel *F. nucleatum* adhesin consists of a complex of pre-FadA and mature FadA. (A) Transmission electron micrographs of FadA proteins. Short filaments and knots (arrows) were present in the FadAc complex but absent in mature FadA alone (Temoin *et al.*, 2012). (B) Diagram demonstrating the location of *fadA* within the *F. nucleatum* chromosome. The hairpin indicates the location of a putative transcription terminator (Han *et al.*, 2005).

using our understanding of adhesion mechanisms and biofilm formation, we can control the population levels of both initial and late colonizers. Moreover, many of the types of adhesins discussed in this review are common to both commensal and pathogenic species of bacteria. For example, SRRPs have been identified in many streptococci, staphylococci and other Gram-positive bacteria and have been implicated in bacterial interactions with hosts and in pathogenesis (Zhou *et al.*, 2008) and pili of *S. sanguinis* and *A. oris* share components that have high homology to those in pathogenic strains (Okahashi *et al.*, 2010). Although we have determined many of the pathways of adhesin biogenesis, there is still much to be investigated and explored for potential drug discovery. For example, there are several genes involved in *A. actinomycetemcomitans* Flp-1 biogenesis and the minor fimbriae of *P. gingivalis* that are still uncharacterized. Also, while we have considerable information about co-aggregations among oral species, very little is known about the adhesins and receptors that are required for these co-aggregations. Finally, because the oral cavity is a constantly changing environment, more research should be done to understand how different environmental conditions will influence specific bacterial interactions. For example, how shear forces impact the formation of bacterial communities in the oral cavity. We have made much progress in understanding bacterial interactions and adhesins, but gaps in that knowledge still exist and must be filled through future research.

References

Amano, A. (2010) Bacterial adhesins to host components in periodontitis. *Periodontology 2000* 52, 12–37.

Amano, A., Kuboniwa, M., Nakagawa, I., Akiyama, S., Morisaki, I. and Hamada, S. (2000) Prevalence of specific genotypes of *Porphyromonas gingivalis* fimA and periodontal health status. *Journal of Dental Research* 79, 1664–1668.

Andrian, E., Grenier, D. and Rouabhia, M. (2006) *Porphyromonas gingivalis*–epithelial cell interactions in periodontitis. *Journal of Dental Research* 85, 392–403.

Bensing, B.A. and Sullam, P.M. (2002) An accessory sec locus of *Streptococcus gordonii* is required for export of the surface protein GspB and for normal levels of binding to human platelets. *Molecular Microbiology* 44, 1081–1094.

Bensing, B.A. and Sullam, P.M. (2009) Characterization of *Streptococcus gordonii* SecA2 as a paralogue of SecA. *Journal of Bacteriology* 191, 3482–3491.

Bensing, B.A. and Sullam, P.M. (2010) Transport of preproteins by the accessory Sec system requires a specific domain adjacent to the signal peptide. *Journal of Bacteriology* 192, 4223–4232.

Bensing, B.A., Gibson, B.W. and Sullam, P.M. (2004a) The *Streptococcus gordonii* platelet binding protein GspB undergoes glycosylation independently of export. *Journal of Bacteriology* 186, 638–645.

Bensing, B.A., Lopez, J.A. and Sullam, P.M. (2004b) The *Streptococcus gordonii* surface proteins GspB and Hsa mediate binding to sialylated carbohydrate epitopes on the platelet membrane glycoprotein Ibalpha. *Infection and Immunity* 72, 6528–6537.

Bensing, B.A., Takamatsu, D. and Sullam, P.M. (2005) Determinants of the streptococcal surface glycoprotein GspB that facilitate export by the accessory Sec system. *Molecular Microbiology* 58, 1468–1481.

Bhattacharjee, M.K., Kachlany, S.C., Fine, D.H. and Figurski, D.H. (2001) Nonspecific adherence and fibril biogenesis by *Actinobacillus actinomycetemcomitans*: TadA protein is an ATPase. *Journal of Bacteriology* 183, 5927–5936.

Bu, S., Li, Y., Zhou, M., Azadin, P., Zeng, M., Fives-Taylor, P. and Wu, H. (2008) Interaction between two putative glycosyltransferases is required for glycosylation of a serine-rich streptococcal adhesin. *Journal of Bacteriology* 190, 1256–1266.

Burnette-Curley, D., Wells, V., Viscount, H., Munro, C.L., Fenno, J.C., Fives-Taylor, P. and Macrina, F.L. (1995) FimA, a major virulence factor associated with *Streptococcus parasanguis* endocarditis. *Infection and Immunity* 63, 4669–4674.

Chen, Q., Wu, H. and Fives-Taylor, P.M. (2004) Investigating the role of secA2 in secretion and glycosylation of a fimbrial adhesin in *Streptococcus parasanguis* FW213. *Molecular Microbiology* 53, 843–856.

Cisar, J.O., Sandberg, A.L., Abeygunawardana, C., Reddy, G.P. and Bush, C.A. (1995) Lectin recognition of host-like saccharide motifs in streptococcal cell wall polysaccharides. *Glycobiology* 5, 655–662.

Clark, W.B., Beem, J.E., Nesbitt, W.E., Cisar, J.O., Tseng, C.C. and Levine, M.J. (1989) Pellicle receptors for *Actinomyces viscosus* type 1 fimbriae *in vitro*. *Infection and Immunity* 57, 3003–3008.

Daep, C.A., Novak, E.A., Lamont, R.J. and Demuth, D.R. (2011) Structural dissection and *in vivo* effectiveness of a peptide inhibitor of *Porphyromonas gingivalis* adherence to *Streptococcus gordonii*. *Infection and Immunity* 79, 67–74.

Das, M., Badley, A.D., Cockerill, F.R., Steckelberg, J.M. and Wilson, W.R. (1997) Infective endocarditis caused by HACEK microorganisms. *Annual Review of Medicine* 48, 25–33.

Davies, J.R., Svensater, G. and Herzberg, M.C. (2009) Identification of novel LPXTG-linked surface proteins from *Streptococcus gordonii*. *Microbiology* 155, 1977–1988.

Dewhirst, F.E., Chen, T., Izard, J., Paster, B.J., Tanner, A.C., Yu, W.H., Lakshmanan, A. and Wade, W.G. (2010) The human oral microbiome. *Journal of Bacteriology* 192, 5002–5017.

Edwards, A.M., Grossman, T.J. and Rudney, J.D. (2006) *Fusobacterium nucleatum* transports noninvasive *Streptococcus cristatus* into human epithelial cells. *Infection and Immunity* 74, 654–662.

Fardini, Y., Wang, X., Temoin, S., Nithianantham, S., Lee, D., Shoham, M. and Han, Y.W. (2011) *Fusobacterium nucleatum* adhesin FadA binds vascular endothelial cadherin and alters endothelial integrity. *Molecular Microbiology* 82, 1468–1480.

Fine, D.H., Furgang, D., Kaplan, J., Charlesworth, J. and Figurski, D.H. (1999) Tenacious adhesion of *Actinobacillus actinomycetemcomitans* strain CU1000 to salivary-coated hydroxyapatite. *Archives of Oral Biology* 44, 1063–1076.

Fine, D.H., Markowitz, K., Furgang, D., Fairlie, K., Ferrandiz, J., Nasri, C., McKiernan, M. and Gunsolley, J. (2007) *Aggregatibacter actinomycetemcomitans* and its relationship to initiation of localized aggressive periodontitis: longitudinal cohort study of initially healthy adolescents. *Journal of Clinical Microbiology* 45, 3859–3869.

Fine, D.H., Markowitz, K., Furgang, D. and Velliyagounder, K. (2010) *Aggregatibacter actinomycetemcomitans* as an early colonizer of oral tissues: epithelium as a reservoir? *Journal of Clinical Microbiology* 48, 4464–4473.

Fives-Taylor, P.M. and Thompson, D.W. (1985) Surface properties of *Streptococcus sanguis* FW213 mutants nonadherent to saliva-coated hydroxyapatite. *Infection and Immunity* 47, 752–759.

Fives-Taylor, P.M., Meyer, D.H., Mintz, K.P. and Brissette, C. (1999) Virulence factors of *Actinobacillus actinomycetemcomitans*. *Periodontol 2000* 20, 136–167.

Froeliger, E.H. and Fives-Taylor, P. (2001) *Streptococcus parasanguis* fimbria-associated adhesin fap1 is required for biofilm formation. *Infection and Immunity* 69, 2512–2519.

Gibbons, R.J. and Hay, D.I. (1988) Human salivary acidic proline-rich proteins and statherin promote the attachment of *Actinomyces viscosus* LY7 to apatitic surfaces. *Infection and Immunity* 56, 439–445.

Gibson, F.C. 3rd, Hong, C., Chou, H.H., Yumoto, H., Chen, J., Lien, E., Wong, J. and Genco, C.A. (2004) Innate immune recognition of invasive bacteria accelerates atherosclerosis in apolipoprotein E-deficient mice. *Circulation* 109, 2801–2806.

Haase, E.M., Zmuda, J.L. and Scannapieco, F.A. (1999) Identification and molecular analysis of rough-colony-specific outer membrane proteins of *Actinobacillus actinomycetemcomitans. Infection and Immunity* 67, 2901–2908.

Haase, E.M., Stream, J.O. and Scannapieco, F.A. (2003) Transcriptional analysis of the 5′ terminus of the flp fimbrial gene cluster from *Actinobacillus actinomycetemcomitans. Microbiology* 149, 205–215.

Hajishengallis, G., Wang, M., Liang, S., Triantafilou, M. and Triantafilou, K. (2008) Pathogen induction of CXCR4/TLR2 cross-talk impairs host defense function. *Proceedings of the National Academy of Sciences USA* 105, 13532–13537.

Han, Y.W., Shi, W., Huang, G.T., Kinder, Haake, S., Park, N.H., Kuramitsu, H, Genco, R.J. (2000) Interactions between periodontal bacteria and human oral epithelial cells: *Fusobacterium nucleatum* adheres to and invades epithelial cells. *Infection and Immunity* 68, 3140–3146.

Han, Y.W., Ikegami, A., Rajanna, C., Kawsar, H.I., Zhou, Y., Li, M., Sojar, H.T., Genco, R.J., Kuramitsu, H.K. and Deng, C.X. (2005) Identification and characterization of a novel adhesin unique to oral fusobacteria. *Journal of Bacteriology* 187, 5330–5340.

Han, Y.W., Fardini, Y., Chen, C., Iacampo, K.G., Peraino, V.A., Shamonki, J.M. and Redline, R.W. (2010) Term stillbirth caused by oral *Fusobacterium nucleatum. Obstetrics & Gynecology* 115, 442–445.

Hasegawa, Y., Iwami, J., Sato, K., Park, Y., Nishikawa, K., Atsumi, T., Moriguchi, K., Murakami, Y., Lamont, R.J., Nakamura, H., Ohno, N. and Yoshimura, F. (2009) Anchoring and length regulation of *Porphyromonas gingivalis* Mfa1 fimbriae by the downstream gene product Mfa2. *Microbiology* 155, 3333–3347.

Henderson, B., Wilson, M., Sharp, L. and Ward, J.M. (2002) *Actinobacillus actinomycetemcomitans. Journal of Medical Microbiology* 51, 1013–1020.

Hilleringmann, M., Giusti, F., Baudner, B.C., Masignani, V., Covacci, A., Rappuoli, R., Barocchi, M.A. and Ferlenghi, I. (2008) Pneumococcal pili are composed of protofilaments exposing adhesive clusters of Rrg A. *PLoS Pathogens* 4, e1000026.

Holt, S.C. and Ebersole, J.L. (2005) *Porphyromonas gingivalis, Treponema denticola,* and *Tannerella forsythia*: the 'red complex', a prototype polybacterial pathogenic consortium in periodontitis. *Periodontology 2000* 38, 72–122.

Ikegami, A., Chung, P. and Han, Y.W. (2009) Complementation of the fadA mutation in *Fusobacterium nucleatum* demonstrates that the surface-exposed adhesin promotes cellular invasion and placental colonization. *Infection and Immunity* 77, 3075–3079.

Inoue, T., Tanimoto, I., Ohta, H., Kato, K., Murayama, Y. and Fukui, K. (1998) Molecular characterization of low-molecular-weight component protein, Flp, in *Actinobacillus actinomycetemcomitans* fimbriae. *Microbiology and Immunology* 42, 253–258.

Inoue, T., Ohta, H., Tanimoto, I., Shingaki, R. and Fukui, K. (2000) Heterogeneous post-translational modification of *Actinobacillus actinomycetemcomitans* fimbrillin. *Microbiology and Immunology* 44, 715–718.

Jenkinson, H.F., McNab, R., Holmes, A.R., Loach, D.M. and Tannock, G.W. (1997) Function and immunogenicity of cell-wall-anchored polypeptide CshA in oral streptococci. *Advances in Experimental Medicine and Biology* 418, 703–705.

Kachlany, S.C., Planet, P.J., Bhattacharjee, M.K., Kollia, E., Desalle, R., Fine, D.H. and Figurski, D.H. (2000) Nonspecific adherence by *Actinobacillus actinomycetemcomitans* requires genes widespread in bacteria and archaea. *Journal of Bacteriology* 182, 6169–6176.

Kachlany, S.C., Planet, P.J., Desalle, R., Fine, D.H. and Figurski, D.H. (2001a) Genes for tight adherence of *Actinobacillus actinomycetemcomitans*: from plaque to plague to pond scum. *Trends in Microbiology* 9, 429–437.

Kachlany, S.C., Planet, P.J., Desalle, R., Fine, D.H., Figurski, D.H. and Kaplan, J.B. (2001b) flp-1, the first representative of a new pilin gene subfamily, is required for non-specific adherence of *Actinobacillus actinomycetemcomitans. Molecular Microbiology* 40, 542–554.

Kang, H.J., Coulibaly, F., Clow, F., Proft, T. and Baker, E.N. (2007) Stabilizing isopeptide bonds revealed in gram-positive bacterial pilus structure. *Science* 318, 1625–1628.

Kolenbrander, P.E. and London, J. (1993) Adhere today, here tomorrow: oral bacterial adherence. *Journal of Bacteriology* 175, 3247–3252.

Kolenbrander, P.E., Palmer, R.J. Jr, Rickard, A.H., Jakubovics, N.S., Chalmers, N.I. and Diaz, P.I. (2006) Bacterial interactions and successions during plaque development. *Periodontology 2000* 42, 47–79.

Kreikemeyer, B., Nakata, M., Oehmcke, S., Gschwendtner, C., Normann, J. and Podbielski, A. (2005) *Streptococcus pyogenes* collagen type I-binding Cpa surface protein. Expression profile, binding characteristics, biological functions, and potential clinical impact. *Journal of Biological Chemistry* 280, 33228–33239.

Kuboniwa, M., Amano, A., Hashino, E., Yamamoto, Y., Inaba, H., Hamada, N., Nakayama, K., Tribble, G.D., Lamont, R.J. and Shizukuishi, S. (2009) Distinct roles of long/short fimbriae and gingipains in homotypic biofilm development by *Porphyromonas gingivalis*. *BMC Microbiology* 9, 105.

Lamont, R.J. and Jenkinson, H.F. (1998) Life below the gum line: pathogenic mechanisms of *Porphyromonas gingivalis*. *Microbiology and Molecular Biology Reviews* 62, 1244–1263.

Lamont, R.J., Hersey, S.G. and Rosan, B. (1992) Characterization of the adherence of *Porphyromonas gingivalis* to oral streptococci. *Oral Microbiology and Immunology* 7, 193–197.

Li, Y., Chen, Y., Huang, X., Zhou, M., Wu, R., Dong, S., Pritchard, D.G., Fives-Taylor, P. and Wu, H. (2008) A conserved domain of previously unknown function in Gap1 mediates protein-protein interaction and is required for biogenesis of a serine-rich streptococcal adhesin. *Molecular Microbiology* 70, 1094–1104.

Maeda, K., Nagata, H., Yamamoto, Y., Tanaka, M., Tanaka, J., Minamino, N. and Shizukuishi, S. (2004) Glyceraldehyde-3-phosphate dehydrogenase of *Streptococcus oralis* functions as a coadhesin for *Porphyromonas gingivalis* major fimbriae. *Infection and Immunity* 72, 1341–1348.

Malek, R., Fisher, J.G., Caleca, A., Stinson, M., Van Oss, C.J., Lee, J.Y., Cho, M.I., Genco, R.J., Evans, R.T. and Dyer, D.W. (1994) Inactivation of the *Porphyromonas gingivalis* fimA gene blocks periodontal damage in gnotobiotic rats. *Journal of Bacteriology* 176, 1052–1059.

McNab, R., Forbes, H., Handley, P.S., Loach, D.M., Tannock, G.W. and Jenkinson, H.F. (1999) Cell wall-anchored CshA polypeptide (259 kilodaltons) in *Streptococcus gordonii* forms surface fibrils that confer hydrophobic and adhesive properties. *Journal of Bacteriology* 181, 3087–3095.

Mishra, A., Das, A., Cisar, J.O. and Ton-That, H. (2007) Sortase-catalyzed assembly of distinct heteromeric fimbriae in *Actinomyces naeslundii*. *Journal of Bacteriology* 189, 3156–3165.

Mishra, A., Wu, C., Yang, J., Cisar, J.O., Das, A. and Ton-That, H. (2010) The *Actinomyces oris* type 2 fimbrial shaft FimA mediates co-aggregation with oral streptococci, adherence to red blood cells and biofilm development. *Molecular Microbiology* 77, 841–854.

Mishra, A., Devarajan, B., Reardon, M.E., Dwivedi, P., Krishnan, V., Cisar, J.O., Das, A., Narayana, S.V. and Ton-That, H. (2011) Two autonomous

structural modules in the fimbrial shaft adhesin FimA mediate *Actinomyces* interactions with streptococci and host cells during oral biofilm development. *Molecular Microbiology* 81, 1205–1220.

Moore, W.E. and Moore, L.V. (1994) The bacteria of periodontal diseases. *Periodontology 2000* 5, 66–77.

Morris, E.J. and McBride, B.C. (1984) Adherence of *Streptococcus sanguis* to saliva-coated hydroxyapatite: evidence for two binding sites. *Infection and Immunity* 43, 656–663.

Navarre, W.W. and Schneewind, O. (1999) Surface proteins of gram-positive bacteria and mechanisms of their targeting to the cell wall envelope. *Microbiology and Molecular Biology Reviews* 63, 174–229.

Nishiyama, S., Murakami, Y., Nagata, H., Shizukuishi, S., Kawagishi, I. and Yoshimura, F. (2007) Involvement of minor components associated with the FimA fimbriae of *Porphyromonas gingivalis* in adhesive functions. *Microbiology* 153, 1916–1925.

Nithianantham, S., Xu, M., Yamada, M., Ikegami, A., Shoham, M. and Han, Y.W. (2009) Crystal structure of FadA adhesin from *Fusobacterium nucleatum* reveals a novel oligomerization motif, the leucine chain. *Journal of Biological Chemistry* 284, 3865–3872.

Nobbs, A.H., Lamont, R. and Jenkinson, H.F. (2009) Streptococcus adherence and colonization. *Microbiology and Molecular Biology Reviews* 73, 407–450.

Nobbs, A.H., Jenkinson, H.F. and Jakubovics, N.S. (2011) Stick to your gums: mechanisms of oral microbial adherence. *Journal of Dental Research* 90, 1271–1278.

Okahashi, N., Nakata, M., Sakurai, A., Terao, Y., Hoshino, T., Yamaguchi, M., Isoda, R., Sumitomo, T., Nakano, K., Kawabata, S. and Ooshima, T. (2010) Pili of oral *Streptococcus sanguinis* bind to fibronectin and contribute to cell adhesion. *Biochemical and Biophysical Research Communications* 391, 1192–1196.

Okahashi, N., Nakata, M., Terao, Y., Isoda, R., Sakurai, A., Sumitomo, T., Yamaguchi, M., Kimura, R.K., Oiki, E., Kawabata, S. and Ooshima, T. (2011) Pili of oral *Streptococcus sanguinis* bind to salivary amylase and promote the biofilm formation. *Microbial Pathogenesis* 50, 148–154.

Palmer, R.J. Jr, Gordon, S.M., Cisar, J.O. and Kolenbrander, P.E. (2003) Coaggregation-mediated interactions of streptococci and actinomyces detected in initial human dental plaque. *Journal of Bacteriology* 185, 3400–3409.

Paturel, L., Casalta, J.P., Habib, G., Nezri, M. and Raoult, D. (2004) *Actinobacillus actinomycetemcomitans* endocarditis. *Clinical Microbiology and Infection* 10, 98–118.

Peng, Z., Fives-Taylor, P., Ruiz, T., Zhou, M., Sun, B., Chen, Q. and Wu, H. (2008a) Identification of critical residues in Gap3 of *Streptococcus parasanguinis* involved in Fap1 glycosylation, fimbrial formation and *in vitro* adhesion. *BMC Microbiology* 8, 52.

Peng, Z., Wu, H., Ruiz, T., Chen, Q., Zhou, M., Sun, B. and Fives-Taylor, P. (2008b) Role of gap3 in Fap1 glycosylation, stability, *in vitro* adhesion, and fimbrial and biofilm formation of *Streptococcus parasanguinis*. *Oral Microbiology and Immunology* 23, 70–78.

Perez, B.A., Planet, P.J., Kachlany, S.C., Tomich, M., Fine, D.H. and Figurski, D.H. (2006) Genetic analysis of the requirement for flp-2, tadV, and rcpB in *Actinobacillus actinomycetemcomitans* biofilm formation. *Journal of Bacteriology* 188, 6361–6375.

Pierce, D.L., Nishiyama, S., Liang, S., Wang, M., Triantafilou, M., Triantafilou, K., Yoshimura, F., Demuth, D.R. and Hajishengallis, G. (2009) Host adhesive activities and virulence of novel fimbrial proteins of *Porphyromonas gingivalis*. *Infection and Immunity* 77, 3294–3301.

Planet, P.J., Kachlany, S.C., Fine, D.H., Desalle, R. and Figurski, D.H. (2003) The widespread colonization island of *Actinobacillus actinomycetemcomitans*. *Nature Genetics* 34, 193–198.

Plummer, C., Wu, H., Kerrigan, S.W., Meade, G., Cox, D. and Ian Douglas, C.W. (2005) A serine-rich glycoprotein of *Streptococcus sanguis* mediates adhesion to platelets via GPIb. *British Journal of Haematology* 129, 101–109.

Schreiner, H.C., Sinatra, K., Kaplan, J.B., Furgang, D., Kachlany, S.C., Planet, P.J., Perez, B.A., Figurski, D.H. and Fine, D.H. (2003) Tight-adherence genes of *Actinobacillus actinomycetemcomitans* are required for virulence in a rat model. *Proceedings of the National Academy of Sciences USA* 100, 7295–7300.

Seepersaud, R., Bensing, B.A., Yen, Y.T. and Sullam, P.M. (2010) Asp3 mediates multiple protein–protein interactions within the accessory Sec system of *Streptococcus gordonii*. *Molecular Microbiology* 78, 490–505.

Seymour, G.J., Ford, P.J., Cullinan, M.P., Leishman, S. and Yamazaki, K. (2007) Relationship between periodontal infections and systemic disease. *Clinical Microbiology and Infection* 13, 3–10.

Socransky, S.S., Haffajee, A.D., Cugini, M.A., Smith, C. and Kent, R.L. Jr (1998) Microbial complexes in subgingival plaque. *Journal of Clinical Periodontology* 25, 134–144.

Stephenson, A.E., Wu, H., Novak, J., Tomana, M., Mintz, K. and Fives-Taylor, P. (2002) The Fap1 fimbrial adhesin is a glycoprotein: antibodies specific for the glycan moiety block the adhesion of *Streptococcus parasanguis* in an *in vitro* tooth model. *Molecular Microbiology* 43, 147–157.

Stromberg, N. and Karlsson, K.A. (1990) Characterization of the binding of *Actinomyces naeslundii* (ATCC 12104) and *Actinomyces viscosus* (ATCC 19246) to glycosphingolipids, using a solid-phase overlay approach. *Journal of Biological Chemistry* 265, 11251–11258.

Takahashi, Y., Konishi, K., Cisar, J.O. and Yoshikawa, M. (2002) Identification and characterization of hsa, the gene encoding the sialic acid-binding adhesin of *Streptococcus gordonii* DL1. *Infection and Immunity* 70, 1209–1218.

Takamatsu, D., Bensing, B.A. and Sullam, P.M. (2004a) Four proteins encoded in the gspB-secY2A2 operon of *Streptococcus gordonii* mediate the intracellular glycosylation of the platelet-binding protein GspB. *Journal of Bacteriology* 186, 7100–7111.

Takamatsu, D., Bensing, B.A. and Sullam, P.M. (2004b) Genes in the accessory sec locus of *Streptococcus gordonii* have three functionally distinct effects on the expression of the platelet-binding protein GspB. *Molecular Microbiology* 52, 189–203.

Takamatsu, D., Bensing, B.A., Cheng, H., Jarvis, G.A., Siboo, I.R., Lopez, J.A., Griffiss, J.M. and Sullam, P.M. (2005a) Binding of the *Streptococcus gordonii* surface glycoproteins GspB and Hsa to specific carbohydrate structures on platelet membrane glycoprotein Ibalpha. *Molecular Microbiology* 58, 380–392.

Takamatsu, D., Bensing, B.A. and Sullam, P.M. (2005b) Two additional components of the accessory sec system mediating export of the *Streptococcus gordonii* platelet-binding protein GspB. *Journal of Bacteriology* 187, 3878–3883.

Takamatsu, D., Bensing, B.A., Prakobphol, A., Fisher, S.J. and Sullam, P.M. (2006) Binding of the streptococcal surface glycoproteins GspB and Hsa to human salivary proteins. *Infection and Immunity* 74, 1933–1940.

Temoin, S., Wu, K.L., Wu, V., Shoham, M. and Han, Y.W. (2012) Signal peptide of FadA adhesin from *Fusobacterium nucleatum* plays a novel structural role by modulating the filament's length and width. *FEBS Letters* 586, 1–6.

Tomich, M., Fine, D.H. and Figurski, D.H. (2006) The TadV protein of *Actinobacillus actinomycetemcomitans* is a novel aspartic acid prepilin peptidase required for maturation of the Flp1 pilin and TadE and TadF pseudopilins. *Journal of Bacteriology* 188, 6899–6914.

Tomich, M., Planet, P.J. and Figurski, D.H. (2007) The tad locus: postcards from the widespread colonization island. *Nature Reviews: Microbiology* 5, 363–375.

Ton-That, H. and Schneewind, O. (2004) Assembly of pili in Gram-positive bacteria. *Trends in Microbiology* 12, 228–234.

Umemoto, T. and Hamada, N. (2003) Characterization of biologically active cell surface components of a periodontal pathogen. The roles of major and minor fimbriae of *Porphyromonas gingivalis*. *Journal of Periodontology* 74, 119–122.

Wang, M., Shakhatreh, M.A., James, D., Liang, S., Nishiyama, S., Yoshimura, F., Demuth, D.R. and Hajishengallis, G. (2007) Fimbrial proteins of *Porphyromonas gingivalis* mediate *in vivo* virulence and exploit TLR2 and complement receptor 3 to persist in macrophages. *Journal of Immunology* 179, 2349–2358.

Whittaker, C.J., Klier, C.M. and Kolenbrander, P.E. (1996) Mechanisms of adhesion by oral bacteria. *Annual Review of Microbiology* 50, 513–552.

Wu, C., Mishra, A., Yang, J., Cisar, J.O., Das, A. and Ton-That, H. (2011) Dual function of a tip fimbrillin of *Actinomyces* in fimbrial assembly and receptor binding. *Journal of Bacteriology* 193, 3197–3206.

Wu, C., Mishra, A., Reardon, M.E., Huang, I.H., Counts, S.C., Das, A. and Ton-That, H. (2012) Structural determinants of *Actinomyces* sortase SrtC2 required for membrane localization and assembly of type 2 fimbriae for interbacterial coaggregation and oral biofilm formation. *Journal of Bacteriology* 194, 2531–2539.

Wu, H., Mintz, K.P., Ladha, M. and Fives-Taylor, P.M. (1998) Isolation and characterization of Fap1, a fimbriae-associated adhesin of *Streptococcus parasanguis* FW213. *Molecular Microbiology* 28, 487–500.

Wu, H., Bu, S., Newell, P., Chen, Q. and Fives-Taylor, P. (2007) Two gene determinants are differentially involved in the biogenesis of Fap1 precursors in *Streptococcus parasanguis*. *Journal of Bacteriology* 189, 1390–1398.

Xu, M., Yamada, M., Li, M., Liu, H., Chen, S.G. and Han, Y.W. (2007) FadA from *Fusobacterium nucleatum* utilizes both secreted and nonsecreted forms for functional oligomerization

for attachment and invasion of host cells. *Journal of Biological Chemistry* 282, 25000–25009.

Xu, Q., Abdubek, P., Astakhova, T., Axelrod, H.L., Bakolitsa, C., Cai, X., Carlton, D., Chen, C., Chiu, H.J., Chiu, M., Clayton, T., Das, D., Deller, M.C., Duan, L., Ellrott, K., Farr, C.L., Feuerhelm, J., Grant, J.C., Grzechnik, A., Han, G.W., Jaroszewski, L., Jin, K.K., Klock, H.E., Knuth, M.W., Kozbial, P., Krishna, S.S., Kumar, A., Marciano, D., Mcmullan, D., Miller, M.D., Morse, A.T., Nigoghossian, E., Nopakun, A., Okach, L., Puckett, C., Reyes, R., Sefcovic, N., Tien, H.J., Trame, C.B., Van Den Bedem, H., Weekes, D., Wooten, T., Yeh, A., Zhou, J., Hodgson, K.O., Wooley, J., Elsliger, M.A., Deacon, A.M., Godzik, A., Lesley, S.A. and Wilson, I.A. (2010) A conserved fold for fimbrial components revealed by the crystal structure of a putative fimbrial assembly protein (BT1062) from *Bacteroides thetaiotaomicron* at 2.2 Å resolution. *Acta Crystallographica Section F Structural Biology and Crystallization Communications* 66, 1281–1286.

Yao, E.S., Lamont, R.J., Leu, S.P. and Weinberg, A. (1996) Interbacterial binding among strains of pathogenic and commensal oral bacterial species. *Oral Microbiology and Immunology* 11, 35–41.

Yen, Y.T., Seepersaud, R., Bensing, B.A. and Sullam, P.M. (2011) Asp2 and Asp3 interact directly with GspB, the export substrate of the *Streptococcus gordonii* accessory Sec system. *Journal of Bacteriology* 193, 3165–3174.

Yilmaz, O., Watanabe, K. and Lamont, R.J. (2002) Involvement of integrins in fimbriae-mediated binding and invasion by *Porphyromonas gingivalis*. *Cellular Microbiology* 4, 305–314.

Yoshimura, F., Murakami, Y., Nishikawa, K., Hasegawa, Y. and Kawaminami, S. (2009) Surface components of *Porphyromonas gingivalis*. *Journal of Periodontal Research* 44, 1–12.

Zambon, J.J. (1985) *Actinobacillus actinomycetemcomitans* in human periodontal disease. *Journal of Clinical Periodontology* 12, 1–20.

Zhou, M. and Wu, H. (2009) Glycosylation and biogenesis of a family of serine-rich bacterial adhesins. *Microbiology* 155, 317–327.

Zhou, M., Peng, Z., Fives-Taylor, P. and Wu, H. (2008) A conserved C-terminal 13-amino-acid motif of Gap1 is required for Gap1 function and necessary for the biogenesis of a serine-rich glycoprotein of *Streptococcus parasanguinis*. *Infection and Immunity* 76, 5624–5631.

Zhou, M., Zhu, F., Dong, S., Pritchard, D.G. and

Wu, H. (2010) A novel glucosyltransferase is required for glycosylation of a serine-rich adhesin and biofilm formation by *Streptococcus parasanguinis*. *Journal of Biological Chemistry* 285, 12140–12148.

Zhou, M., Zhang, H., Zhu, F. and Wu, H. (2011) Canonical SecA associates with an accessory secretory protein complex involved in biogenesis of a streptococcal serine-rich repeat glycoprotein. *Journal of Bacteriology* 193, 6560–6566.

13 Pilus-based Vaccine Development in Streptococci: Variability, Diversity and Immunological Responses

Daniela Rinaudo and Monica Moschioni
Novartis Vaccines and Diagnostics, Siena, Italy

13.1 Introduction

Existing vaccines that target bacterial diseases are based either on whole organisms (inactivated or attenuated) or on specific bacterial subunits. In general, subunit vaccines consist of potentially conserved and immune-dominant components isolated from the pathogen (either proteins like bacterial toxins or capsular polysaccharides of encapsulated bacteria); they are preferred over whole cell vaccines because they have the advantage of being well defined and characterized as well as leading to fewer unpleasant side effects. However, isolated components often do not activate a response as broad or as robust as that of whole organism vaccines.

During recent decades, the application of recombinant DNA techniques and the accumulating genomic information have favoured the application of non-conventional approaches to vaccinology strategies and also the identification of new vaccine candidates. Indeed, vaccine components have to be well surface exposed (to be more accessible to antibodies), strongly immunogenic, con-served and expressed in a wide panel of strains and be able to confer protection from disease development (Rappuoli *et al.*, 2011a, 2011b).

With this respect, following the quite recent identification of pilus-like structures on the bacterial surface of Gram-positive bacteria, pilus-forming subunits have been evaluated in different pathogens (*Streptococcus pneumoniae*, *S. agalactiae*, *S. pyogenes* and *S. suis*) as possible vaccine candidates (Maione *et al.*, 2005; Mora *et al.*, 2005; Gianfaldoni *et al.*, 2007; Margarit *et al.*, 2009). Indeed, pilus components are well exposed on the bacterial surface and therefore accessible to the immune system, immunogenic and also implicated at different stages of bacterial pathogenesis (Soriani and Telford, 2010; Kreikemeyer *et al.*, 2011).

In Gram-positive bacteria, pili are heteropolymeric structures consisting of a backbone protein and either one or two ancillary proteins covalently assembled and linked to the cell wall by a series of sortase-mediated transpeptidase reactions (Ton-That and Schneewind, 2004; Kreikemeyer *et al.*, 2011). The genes coding for the pilus subunits and the specific sortases involved in their assembly are clustered in genomic islets inserted in specific loci of the genome. The pilus encoding islets in Gram-positive bacteria are all characterized by the presence of genes coding for LPXTG (where X is any amino acid) or LPXTG-like surface-anchored proteins (pilus subunits) and genes coding for the sortases. Conversely, genes coding for fibronectin binding proteins, transcriptional regulators, heat shock proteins, signal peptidases and transposable elements are only present in a fraction of the pilus islets (Kreikemeyer *et al.*, 2011).

This chapter will review genetic variability and molecular epidemiology data available on the pilus encoding islets of *S. pneumoniae*, *S. agalactiae*, *S. pyogenes* and *S. suis*. Immunological and preclinical protection data will also be presented, if available. Finally, similarities and differences identified among the different streptococci will be discussed.

13.2 *Streptococcus agalactiae* pili

13.2.1 Genomic organization of *S. agalactiae* pilus islands

Streptococcus agalactiae (also known as Group B *Streptococcus* or GBS) is the most common cause of sepsis and meningitis in neonates and is also the primary colonizer of the anogenital mucosa of healthy women (Johri *et al.*, 2006). In 2005, characterization studies of protective antigens, identified by a multiple genome approach aiming at the development of an effective vaccine against GBS infections, revealed for the first time in a streptococcal species the existence of high-molecular-weight (HMW) polymers, visible by electron microscopy as pilus-like structures extending out from the bacterial surface (Lauer *et al.*, 2005; Maione *et al.*, 2005; Dramsi *et al.*, 2006). Subsequently, comparative analysis of the eight published genome sequences have permitted the discovery in GBS of three genomic pilus islands (PIs), named PI-1, PI-2a and PI-2b (Rosini *et al.*, 2006). The overall organization of the three islands is similar to pilus gene clusters identified in other Gram-positive bacteria (Ton-That and Schneewind, 2003; Ton-That and Schneewind, 2004) (Fig. 13.1). Each of the GBS PIs encodes three structural pilus components, corresponding to the major pilus subunit (the backbone protein) forming the pilus shaft and the two ancillary proteins (AP1 and AP2). These structural subunits harbour a (L/I)PXTG sorting motif that is typical of cell-wall-anchored proteins. In addition, the pilus clusters contain at least two genes coding for pilus-associated class C sortase enzymes (SrtC1 and SrtC2) catalysing pilus protein polymerization (Dramsi *et al.*, 2006; Rosini *et al.*, 2006).

PI-1 is part of a genomic island of approximately 16 Kb, flanked by an 11 bp direct repeat. In addition to the pilus operon, the region contains a gene coding for an AraC-type transcriptional regulator and genes with homology to transposon-like genes. Genome sequence analysis in two GBS sequenced strains (A909 and CJB111) revealed an insertion of a 51.2 Kb prophage at one end of the island.

PI-2a and PI-2b represent two variants of pilus island 2 since they are alternatively present in the same genomic locus and define an approximately 11 Kb region flanked by identical conserved genes. In addition to the genes coding for the three pilus structural subunits and two sortases, and upstream of the *ap1* gene, the PI-2a region contains a gene coding for a *rogB* type transcriptional regulator (Gutekunst *et al.*, 2003), while PI-2b contains a gene coding for a LepA-type signal peptidase.

Genetic studies performed in the genomic PI-1 and PI-2a loci allowed the establishment of the relative contribution of SrtC1 and SrtC2 sortases in pilus assembly (Rosini *et al.*, 2006). SrtC1 and SrtC2 were found to be specific in terms of ancillary proteins incorporation, while they can both efficiently polymerize the backbone protein. In addition, unlike the other two islands, PI-1 carries an additional gene predicted to code for a third sortase C enzyme (SrtC3) of unknown function. However, the heterologous expression in *Lactococcus lactis* of a PI-1, carrying the genes coding for the three structural proteins and for SrtC1 and SrtC2 enzymes and lacking the SrtC3 gene resulted in pilus 1 formation, suggesting that SrtC3 is either redundant or not directly involved in pilus assembly (Buccato *et al.*, 2006).

Streptococcus agalactiae

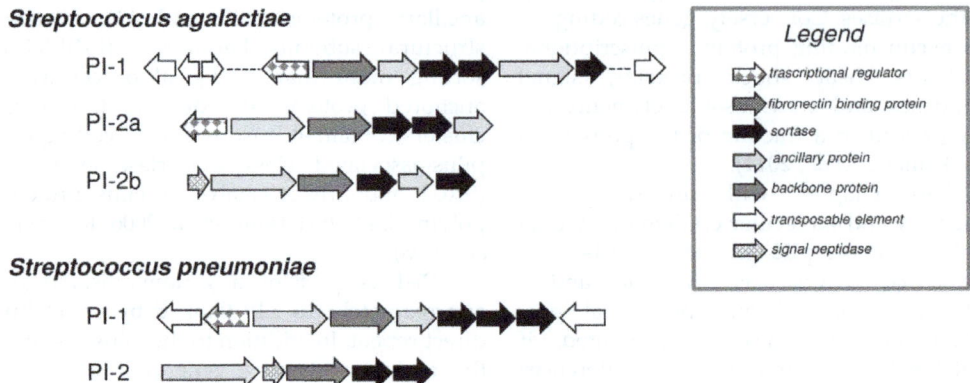

Streptococcus pneumoniae

Streptococcus pyogenes

FCT type

Streptococcus suis

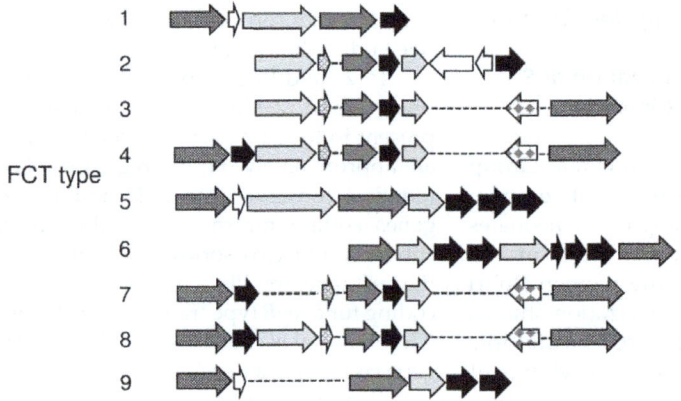

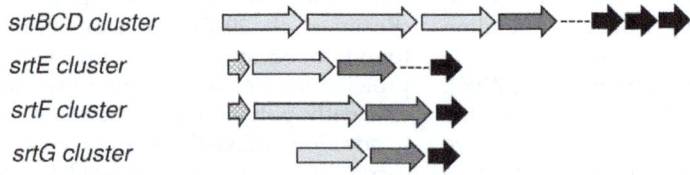

Fig. 13.1. Schematic representation of the genomic pilus islands' organization identified in the main streptococcal pathogens. Three pilus gene clusters in *S. agalactiae* (PI-1, pilus island 1; PI-2a, pilus island 2a and PI-2b, pilus island 2b); two pilus islets in *S. pneumoniae* (PI-1, pilus islet 1 and PI-2, pilus islet 2); nine FCT (fibronectin-binding, collagen-binding, T-antigen) types in *S. pyogenes*; and four putative pilus gene clusters, designated *srtBCD*, *srtE*, *srtF* and *srtG* according to class C sortase genes in the respective locus, in *S. suis*. Each cluster contains genes encoding a major or backbone protein (dark grey arrows), one or two ancillary protein(s) (light grey arrows) and pilus-specific sortase gene(s) (black arrows). Some clusters display genes encoding transcriptional regulators and putative signal peptidase-like proteins. Pilus cluster names are indicated on the right. Dashed black lines represent dimensions that are not to scale.

13.2.2 Global distribution of *S. agalactiae* pilus islands

By means of an extensive analysis of pili distribution and conservation in a collection of 289 clinical isolates from patients (infants and adults) with invasive GBS infection or asymptomatic colonization from three different geographical areas, Margarit and co-workers found that all isolates carried at least one of the three islands or a combination of two PIs. More specifically, all 289 strains carried a PI-2 locus; in particular, the PI-2a variant was present in 73% of the strains, whereas the PI-2b variant was found in the remaining 27% of the clinical isolates analysed (Margarit *et al.*, 2009). Interestingly, pilus type 2a was found alone in a large number of strains (27%) or in association with PI-1 (46% of strains). By contrast, PI-2b variant was almost always associated with PI-1; indeed, only 4 out of 289 strains (1.4%) carried PI-2b alone and all belonged to serotype IV isolates. Finally, the PI-1 locus was absent in 28% of the strains. Although, to date, there is not sufficient evidence to support any correlation between a specific pilus pattern and a particular pathological feature and/or the type of disease or carriage, Margarit *et al.* reported that the combination of PI-1 and PI-2b was relatively more frequent in strains from infants than in those from adults, who conversely showed a more frequent combination of PI-1 with PI-2a.

Interestingly, a correlation was observed between the presence of a particular combination of PIs and the capsule (CPS) type; indeed, the same pilus variant was shared by different capsular serotypes, thus suggesting that capsular serotypes have evolved after pilus divergence. Most serotype Ia isolates (91%) contained only the PI-2a island, whereas the majority of type Ib (85%), type V (96%) and non-typeable (71%) strains carried PI-1 and PI-2a. All serotype III isolates (the most epidemiologically relevant serotype and also the most frequent in the strain collection analysed) carried a combination of two pilus islands: 30% of the strains contained PI-1 and PI-2a, while 70% carried PI-1 and PI-2b.

In a recent study evaluating global human and bovine GBS, Sorensen *et al.* (2010) analysed the presence of pilus islands in 238 bovine and human isolates from nine countries on five continents. In accordance with Margarit *et al.* (2009), they demonstrated that all the isolates contained at least one and often two islands and, while all 238 isolates carried the PI-2 locus, PI-1 was missing in several lineages and specifically in 54% of the bovine isolates (59/110) and in 24% of the human isolates (31/128). In addition, the presence or absence of pilus islands and their allelic variants correlated with the genetic clusters identified as determined by MLST analysis. Interestingly, the human-associated and highly virulent sequence type 17 (ST17) seemed to be strongly associated with the PI-1 plus PI-2b pattern (Sorensen *et al.*, 2010).

The general presence of PI-2 genes irrespective of host association supports their significance in colonization of both the human and bovine host and underscores their potential as a component of a vaccine against GBS infections.

13.2.3 Sequence variability of *S. agalactiae* pilus components

To evaluate the sequence conservation of pilus subunits, Margarit *et al.* carried out sequence analyses of the backbone protein and the two ancillary proteins encoded by each island in a panel of 186 clinical isolates collected from two geographical areas (USA and Italy). The three genes coding for the structural proteins of PI-1 were well conserved (overall sequence conservation of ~99%), and their products differed by very few amino acids. Eight out of 126 PI-1-carrying strains, belonging to serotype II, carried a point mutation resulting in a frameshift and premature termination after Thr360. Similarly, in a subset of serotype III isolates, the major ancillary protein (AP1-1) contained only three amino acid replacements. A high degree of conservation was also observed for the backbone protein and the two ancillary proteins encoded by PI-2b. In 40 PI-2b-carrying isolates analysed, the backbone protein (BP-2b) shared 100% identity, whereas the AP1-2b protein clustered into two alleles that differed in seven amino acids and for the presence of a variable region of 181 amino acids showing 65% sequence identity between the two variants. In 146 clinical isolates carrying the pilus variant 2a, sequence analysis revealed that PI-2a was the most variable of the three islands. In particular, the backbone protein (BP-2a) shared the highest level of gene variability among all pilin subunits, and grouped into seven main immunologically different variants (named 515, CJB111, DK21, H36B, 2603 and CJB110, based on their reference strain) with sequence identities ranging from 48% to 98%. A similar clustering in 7 alleles was found when comparing AP1-2a sequences, with identities ranging from 87%–98%. Noteworthy, all the minor ancillary proteins AP2, coded by the three pilus islands showed highly conserved sequences. A correlation was observed between capsular serotypes, particular PI combinations, and specific PI-2a alleles. For instance, serotype Ia strains predominantly carried only BP-2a allele 515, while the BP-2a variant DK21 was restricted to serotype II strains and serotype III isolates nearly always contained PI-2a 2603 V/R variant. In conclusion, with the exception of the backbone protein in PI-2a, the sequences of the three pilus structural proteins were well conserved and 94% of the strains analysed expressed at least one pilus type on their surface as detected by FACS analysis on the entire strain collection using antibodies against the BP and AP1 subunits of the three islands. In general, whenever PI-2a and PI-2b islands were present, pili were expressed at high levels on the bacterial surface in more than 90% of the strains analysed. In contrast, only 31% of the strains containing PI-1 expressed the pilus antigens at high levels on the bacterial surface (Margarit et al., 2009).

13.2.4 GBS pili as promising vaccine candidates

Currently there is no vaccine against GBS. This microorganism is classified into ten capsular polysaccharide serotypes, each antigenically and structurally unique. Although multivalent capsular conjugate vaccines are in development, to overcome serotype-specific immunity, the increasing number of non-typeable isolates and problems of serotype replacement/switching, vaccines based on conserved protective proteins are under investigation. The above evidence showing that all GBS isolates carried at least one or a combination of two pilus islands confirms the key role played by pili in GBS virulence and pathogenesis. Such structures have also been implicated in promoting phagocyte resistance and systemic virulence in animal models (Maisey et al., 2008; Papasergi et al., 2011), in mediating attachment to human epithelial cells (Dramsi et al., 2006; Krishnan et al., 2007; Konto-Ghiorghi et al., 2009), in the adhesion and invasion of brain microvascular endothelial cells (Maisey et al., 2007), in promoting transepithelial migration (Pezzicoli et al., 2008) and in biofilm formation (Konto-Ghiorghi et al., 2009; Rinaudo et al., 2010). Maione et al. (2005) provided the first evidence that streptococcal pilus components, specifically the backbone protein of PI-1 (BP-1, also named GBS80), and the major ancillary proteins from PI-1 (AP1-1, also called GBS104) and from PI-2a (AP1-2a, also named GBS67 or PilA) were able to induce protective immunity in animal models of streptococcal diseases. These proteins, when tested in mice in the active maternal immunization–neonatal pup challenge model in combination, and together with a fourth conserved protein, GBS322 also known as Sip protein (Martin et al., 2002), were able to provide broad protective activity against a large panel of GBS strains representative of all circulating serotypes. Interestingly, protection against any particular strain also correlated with antigen accessibility on the bacterial surface and with the induction of opsonophagocytic antibodies. Thus, despite the absence of universal antigens, appropriate combinations of protective antigens, each effective against overlapping populations of isolates, can confer broad serotype-independent protection.

Subsequently, following the identification in GBS of three different pilus variants,

Rosini *et al.* (2006) demonstrated that also the backbone protein of pilus variant 2a (BP-2a, also known as GBS59 or PilB) was able to confer protection in mice. Finally, Margarit *et al.* (2009) showed that, as in the case of PI-1 and PI-2a, two pilus components of the PI-2b variant (the backbone protein, named BP-2b, and the ancillary protein 1, named AP1-2b) were able to elicit protection against GBS challenge in mice. Therefore, they provided the first evidence that a pilus-based vaccine exclusively constituted by three pilus components, one from each pilus type (BP-1, AP1-2a and BP-2b), can be effective in preventing GBS infections and capable of providing broad protection (Margarit *et al.*, 2009). This evidence provides for the first time a rationale for the development of a universal pilus-based vaccine that is potentially capable of preventing disease by all GBS serotypes. Although BP-2a was capable of eliciting high opsonophagocitic titres, it was excluded from this vaccine formulation because of its high gene variability and variant-specific protection *in vivo*. Indeed, the protein was able to protect only pups challenged with strains carrying the allelic variant used to immunize their mothers, and protection was not observed against strains expressing a heterologous allele (Margarit *et al.*, 2009). The high variability of this protein could reflect a strategy to escape the immune system of the host, thus representing an additional confirmation of the important role of this protein in virulence.

Thus, considering that the backbone proteins, being the bulk of the pilus structures, tend to perform better than ancillary proteins in animal models, the 'ideal' vaccine should include all pilus backbone proteins. To develop an easily producible and efficacious backbone proteins-based vaccine, Nuccitelli *et al.* (2011) successfully applied a Structural Vaccinology technology for the rational design of an optimized BP-2a protein. By solving the crystallographic structure of BP-2a from GBS strain 515, the authors discerned four Ig-like domains (D1-D2-D3-D4), which could function as independent folding units. Individual domains were tested as subunit vaccines; however, only D3 elicited

protection similar to the full-length BP-2a. Since the other BP-2a variants showed a similar structural organization, Nuccitelli *et al.* cobbled together the D3 domains of six variant BP-2a molecules connected by short Gly-Ser-Gly-Ser spacers. Expression in *Escherichia coli* generated a stable hybrid product (named 6xD3) that provided protection against all six prototypic pilus variant GBS strains examined (Nuccitelli *et al.*, 2011). The use of this technology can likely be expanded well beyond the boundaries of GBS vaccines to develop vaccines against other bacterial pathogens.

In 2006, Buccato *et al.* showed that an entire pilus encoding island can be transferred from GBS to a nonpathogenic species (i.e. the food-grade microorganism *L. lactis*) and that recombinant bacteria acquire the capacity to express pilus-like structures on their surface. Furthermore, systemic and mucosal vaccination of mice with *L. lactis* engineered to express GBS pili (specifically an entire PI-1 or hybrid pili containing GBS pilus 1 and pilus 2a components) conferred protection against infections (Buccato *et al.*, 2006). These data indicate that fimbriae from different streptococcal isolates can be functionally expressed in nonpathogenic bacteria and open the way for the development of formulations of multivalent live vaccines able to provide broad intra- and inter-species protection against streptococcal diseases.

In conclusion, all three pili identified in GBS elicit protective immunity; moreover, it is noteworthy that antibodies raised against immunogenic pilin subunits of PI-2a inhibited bacterial adherence to both abiotic and biotic surfaces (Rinaudo *et al.*, 2010). These observations suggest that a pilus-based vaccine may work according to two different mechanisms of action, through the elicitation of opsonophagocytic antibodies capable of killing bacteria in a complement-dependent manner and via the induction of antibodies capable of preventing colonization in the host. This observation could be of relevance for other piliated Gram-positive pathogens, including *S. pyogenes* and *S. pneumoniae*, known to grow in biofilm on biological surfaces and implanted medical devices.

13.3 *Streptococcus pneumoniae* pili

13.3.1 Genomic organization of *S. pneumoniae* pilus-encoding islets

Streptococcus pneumoniae is a Gram-positive commensal of the nasopharyngeal tract of both children and healthy adults. However, *S. pneumoniae* is also a leading cause of morbidity and mortality worldwide, being responsible for non-invasive and invasive diseases such as acute otitis media, pneumonia, sepsis and meningitis (Fletcher and Fritzell, 2007; Pelton and Leibovitz, 2009; van der Poll and Opal, 2009; Kim, 2010). In *S. pneumoniae*, extensive investigation of the available genome sequences has led to the identification of two genomic elements containing features typical of Gram-positive pilus encoding islets, pilus islet 1 (PI-1, also known as *rlrA* pathogenicity islet) and pilus islet 2 (PI-2) (Barocchi *et al.*, 2006; LeMieux *et al.*, 2006; Bagnoli *et al.*, 2008) (Fig. 13.1).

PI-1, first identified in the genome of TIGR4 (a serotype 4 strain), is a chromosomal region of approximately 12 Kb comprising seven genes: *rlrA*, encoding a RofA-like transcriptional positive regulator, *rrgA*, *rrgB* and *rrgC*, which encode the pilus subunits, and *srtB*, *srtC* and *srtD* coding for three sortases. PI-1 is flanked by two insertion sequences (IS1167) containing inverted repeats characteristic of mobile genetic elements and is always inserted between the genes coding for the formate acetyltransferase (*pfl*) and a conserved hypothetical protein (Barocchi *et al.*, 2006; LeMieux *et al.*, 2006).

PI-2, first identified in the genome of the serotype 1 strain INV104 is a chromosomal region of approximately 6.5 Kb, comprising five genes: *pitA* and *pitB*, which encode the pilus subunits, *srtG1* and *srtG2* coding for two sortases and *sipA* (putative signal peptidase), coding for a protein that is essential for pilus assembly but whose role in polymerization has not yet been clearly determined. Interestingly, in most of the strains analysed, *pitA* and *srtG2* are pseudo-genes due to the presence in both of them of a stop-codon causing the premature termination of protein translation. PI-2 is always inserted between the conserved genes *pepT* (peptidase T) and *hemH* (ferrochetalase) (Bagnoli *et al.*, 2008).

13.3.2 Global distribution of pilus-encoding islets

Molecular epidemiology analyses performed on a variety of isolate collections have revealed that both PI-1 and PI-2 are present in a fraction of the clinical pneumococcal isolates. PI-1 is present in approximately 25–30% of the isolates (Basset *et al.*, 2007; Moschioni *et al.*, 2008; Vainio *et al.*, 2011), and PI-2 in about 16–21% (Barocchi *et al.*, 2006; Zahner *et al.*, 2010; Vainio *et al.*, 2011), depending on the strain collection analysed. Indeed, independent studies have demonstrated that the presence of the islets correlates with the genotype of the isolates (defined by multi locus sequence typing, MLST (Enright and McKenzie, 1997) and not with the serotype (Aguiar *et al.*, 2008; Moschioni *et al.*, 2008). Therefore, variation in PIs prevalence is mainly associated with different regional distributions of the PIs positive clones. In addition, PIs prevalence in invasive (bacteraemic pneumonia, sepsis, meningitis), carriage and acute otitis media isolates is similar regardless of the diseases tested (Basset *et al.*, 2007; Aguiar *et al.*, 2008; Moschioni *et al.*, 2008, 2009; Vainio *et al.*, 2011).

PI-1 has been found in isolates belonging to the international clones 156, 205, 490, 176, 320, 247 and 558 (Aguiar *et al.*, 2008; Moschioni *et al.*, 2008). These clones are mostly associated with serotypes 4, 6B, 9V, 14, 19A, 19F, 23F and 35B. Since the majority of these serotypes are included in the 7-valent conjugated poly-saccharide vaccine (Darkes and Plosker, 2002), the resulting prevalence of PI-1 positive strains among vaccine types (VT) is higher (45–50%). Probably due to this association, Regev-Yochay *et al.* (2010) found that in the USA, following conjugate pneumococcal vaccination with Prevnar-7® (2001–2004), the elimination of VT strains corresponded to a dramatic decrease of PI-1 prevalence. Interestingly, this initial decline has been followed in the years between 2004 and 2007 by an increase in PI-1 prevalence to rates exceeding the pre-conjugate vaccine era, primarily due to the emergence of PI-1 positive non-VT strains. This suggests that pilus-1 expression may confer an intrinsic advantage to PI-1 positive isolates. In this respect, a recent epidemiological study focused on the evaluation of PI-1 presence in a pneumococcal carriage collection has excluded this advantage to be correlated with longer carriage duration in infants or with higher transmissibility rates between mother and infants (Turner *et al.*, 2011). Besides, the clonal success of drug-resistant pneumococci is often accompanied by PI-1 presence, thus indicating that pilus-1 expression may act as an additional fitness factor favouring the expansion of antibiotic-resistant clones (Sjostrom *et al.*, 2007; Imai *et al.*, 2009, 2011; Sadowy *et al.*, 2010; Siira *et al.*, 2011). In line with this observation, different reports have highlighted that PI-1 prevalence among antibiotic- (penicillin, erythromycin or multidrug) resistant isolates is significantly higher than among antibiotic-susceptible isolates, suggesting that PI-1 positive clones are more likely to acquire antibiotic resistance (Moschioni *et al.*, 2009; Regev-Yochay *et al.*, 2010).

Noteworthy, while the majority of the PI-positive isolates contain either PI-1 or PI-2, those belonging to the MLST multi-drug-resistant international clone 320 bear both islets. In these isolates the two pili (pilus-1 and pilus-2) are both expressed and independently assembled (Bagnoli *et al.*, 2008).

PI-2 presence is associated with the international clones 62, 74, 171, 251, 320, 304 and 306. These clones are mostly associated with serotypes 1, 2, 7F, 11A, 19A and 19F. Unlike PI-1, with the exception of serotype 19F, PI-2 is associated only with non-VT (Bagnoli *et al.*, 2008; Moschioni *et al.*, 2009; Zahner *et al.*, 2010). In addition, excluding clone 320, all the clones associated with PI-2 are sensitive to antibiotics. Indeed, following vaccine introduction in the USA, PI-2 prevalence rose from 3.6% to 21% due to the expansion of the non-VTs 19A and 7F, serotype 19A being associated with the antibiotic-resistant clone 320 (Zahner *et al.*, 2010).

13.3.3 Nucleotide and protein variability of pneumococcal pilus components

Extensive PI-1 and PI-2 sequencing in different PI positive strains allowed for the identification of three PI-1 clades (named I, II and III) with an overall inter-clade nucleotide identity ranging from 88% to 92% and one single PI-2 clade (overall conservation of ~99% along the entire islet) (Moschioni *et al.*, 2008). Strains belonging to the same MLST clone are also homogeneous for PI-1 clade. Interestingly, the most variable genes within PI-1 are *rrgA* and *rrgB*, coding for the pilus-1 adhesin, RrgA, located at the pilus tip (Nelson *et al.*, 2007; Hilleringmann *et al.*, 2008), and the pilus backbone structural subunit RrgB (Barocchi *et al.*, 2006; LeMieux *et al.*, 2006). Notably, RrgC, which is believed to be the pilus-1 cell-wall anchor and is not well exposed on the bacterial surface (Hilleringmann *et al.*, 2009; De Angelis *et al.*, 2011), is 100% conserved. In detail, RrgA exists in two variants (I and II) with an overall protein homology of 83%, while RrgB has been classified in three variants (I, II and III) sharing an overall protein homology ranging from 49% to 67%. *rrgB* is the most variable gene, and the three RrgB variants correspond to the PI-1 clades. Interestingly, within PI-1, RrgA clade I is associated with either RrgB clade I or III, whereas RrgA clade II is associated only with RrgB clade II (Moschioni *et al.*, 2008).

Representative PI positive bacteria have been taken into account to evaluate pilus expression in different *in vitro* growth conditions. Immuno-gold electron micrography demonstrated that PI-1 positive bacteria are decorated by several pilus-1 structures of different length protruding outside the capsule, mainly composed of the backbone RrgB (Barocchi *et al.*, 2006); in PI-2 positive bacteria, instead, one single pilus-2 is assembled on the surface of each single diplococcus (Bagnoli *et al.*, 2008). However, while PI-2 positive bacteria are homogenous for pilus-2 expression, pilus-1 expression in all the strains tested is biphasic. Indeed, in PI-1 positive strains it is possible to identify bacteria with two distinct phenotypes, one expressing and one not expressing the pilus on the bacterial surface (Bagnoli *et al.*, 2008;

Basset *et al.*, 2011; De Angelis *et al.*, 2011). In all the strains tested the ratio between the pilus-positive and pilus-negative bacteria is independent of the serotype, genotype and growth conditions.

13.3.4 Pilus components' immunogenicity and protection

In order to explore the possibility of using pneumococcal pilus subunits as components of a multi-component protein-based vaccine targeting pneumococcal disease, immunogenicity and protection induced by the immunization performed with different pilus subunits or combinations thereof were evaluated.

In detail, pilus-2 PitB resulted in high immunogenicity, but, probably because of the presence of single pilus-2 structures on the bacterial surface, was not protective in colonization, pneumonia and sepsis animal models (unpublished data).

By contrast, immunization with both RrgA and RrgB pilus-1 recombinant subunits resulted in significant protection in sepsis mouse models of infection (Gianfaldoni *et al.*, 2007); however, the protective efficacy of RrgB immunization was the highest, probably because it is more abundant on the bacterial surface than RrgA. Noteworthy, passive immunization performed with RrgA and RrgB antisera reduced bacteraemia and mortality in mice challenged intraperitoneally with *S. pneumoniae*, indicating that antibody-mediated bacterial opsonization is an important mechanism in the protection exerted by pilus components (Gianfaldoni *et al.*, 2007). Indeed, antibodies to RrgB and RrgA, the latter to a lower extent, are also able to mediate a complement-dependent opsonophagocytic killing of PI-1positive strains *in vitro* at levels comparable with those obtained with antisera raised against pneumococcal glycoconjugates (Harfouche *et al.*, 2011).

Interestingly, the resolution of the RrgB crystal structure has allowed the rational design of RrgB recombinant portions, which have been used by Gentile *et al.* to molecularly dissect the position of the protective epitopes

within RrgB (Spraggon *et al.*, 2010; El Mortaji *et al.*, 2011; Gentile *et al.*, 2011; Paterson and Baker, 2011). The comparison of the protective efficacy exerted by the full-length RrgB with that obtained with the four single RrgB domains (D1, D2, D3 and D4), or a combination thereof, has allowed the determination that the best protective effects in terms of reduction of bacteraemia and increase of survival in infected animals is obtained with the entire molecule and the combination, followed by the single domains D1 and D4. Similar results have been obtained with passive immunization experiments (Gentile *et al.*, 2011). Hence, the protective effect induced by RrgB immunization is not due to a single antigenic domain but to protective epitopes dispersed throughout the molecule, thus supporting the use of the full-length protein rather than portions of it in a pilus-1-containing vaccine.

Considering the degree of protein variability displayed by the two pilus-1 protective antigens RrgA and RrgB, cross-reaction and cross-protection studies have been performed in order to estimate their possible coverage when included in a protein-based vaccine (Moschioni *et al.*, 2010; Harfouche *et al.*, 2011). These analyses have then been followed by antigen optimization studies with the aim of broadening the coverage as much as possible.

In this respect, Moschioni *et al.* demonstrated that antisera to RrgA clade I are cross-reactive towards clade II and vice-versa and that immunization with antisera against RrgA clade II are protective against the challenge performed with a strain expressing RrgA clade I (Moschioni *et al.*, 2010). In contrast, because protein similarity among the three RrgB variants is lower, the three proteins were neither cross-reactive nor cross-protective (Harfouche *et al.*, 2011). Since the highest protection levels are obtained only using the full-length protein (see above), the absence of RrgB cross-protection has been overcome by Harfouche *et al.* (2011) generating a fusion protein antigen containing the three RrgB variants, III, II and I (named RrgB321), in a head to tail organization. Indeed, the strategy to create a fusion protein to broaden the coverage of RrgB has proved successful. In fact, the immune-response raised by this fusion protein is directed against all the three variants, although variant I was the most immunogenic; and when animals, upon RrgB321 active or passive immunization, are challenged intravenously or intra-peritoneally with *S. pneumoniae* strains expressing a pilus-1 of whichever three clades, they are all significantly protected.

Since pneumococcal pilus-1 is not essential for bacterial growth and presents a biphasic pattern of expression, in the context of the evaluation of the coverage of a pilus vaccine candidate the expression level of the antigen has to be carefully taken into account. In this context, Moschioni *et al.* performed RrgB321 immunization studies challenging the animals with strains presenting either a low (L) or a high (H) proportion of the bacteria expressing pilus-1 at the moment of infection. Both resulted in significant protection in sepsis and pneumonia animal models, thus indicating that, in principle, even infections due to strains naturally expressing very low amounts of pilus-1 could be cleared upon RrgB321 immunization (Moschioni *et al.*, 2012).

Overall, based on the currently available molecular epidemiology data and the antigen optimization performed, a protein-vaccine solely composed of RrgB321 could, theoretically, reach coverage of about 30% in the pneumococcal population and up to 60% among antibiotic-resistant pneumococcal isolates.

13.4 *Streptococcus pyogenes* pili

13.4.1 Genomic organization of *S. pyogenes* pilus-encoding islets

Streptococcus pyogenes (group A *Streptococcus*, GAS) is a Gram-positive bacterium that colonizes the pharynx and skin of humans and, besides being the most common cause of bacterial pharyngitis in children, is responsible for more than a dozen different diseases in humans. In fact, GAS causes severe invasive disease, which can result in streptococcal toxic shock syndrome or necrotizing fasciitis; in addition, autoimmune

sequelae due to unresolved pharyngitis episodes may result in severe cardiac pathology and glomerulonephritis (Cunningham, 2000).

Strains of *S. pyogenes* have been traditionally classified based on serum recognition of a variable trypsin-sensitive surface protein, the M protein, and a variable trypsin-resistant antigen, the T antigen. More than 100 M proteins and 21 T antigens have been identified so far (Johnson and Kaplan, 1993). The M protein is an important *S. pyogenes* virulence factor and its genetic variability (through direct *emm* gene sequencing) has been extensively characterized (Facklam *et al.*, 1999; Tanaka *et al.*, 2002); in contrast, the genetic basis for T sero-typing (the method set up by Rebecca Lancefield; Lancefield and Dole, 1946) has been poorly understood until a few years ago, when the T antigens were shown to correspond with trypsin-resistant pilus proteins, encoded by genes included in the so-called FCT regions (Bessen and Kalia, 2002; Mora *et al.*, 2005).

Epidemiological studies on *S. pyogenes* have so far revealed the existence of nine different pilus-encoding islets (Fig. 13.1). These have been termed FCT types 1–9 based on the presence within the regions of fibronectin- and collagen-binding proteins as well as T antigens. FCT-type regions are present in all *S. pyogenes* isolates tested and are inserted exclusively between two highly conserved genes, *hsp33* and *spy0136* (Bessen and Kalia, 2002; Kratovac *et al.*, 2007). In addition, the FCT regions are extremely variable in their gene composition (number and order of the encoded genes) and, for this reason, range from 11 to 16 Kb in size.

In detail, based on PCR (FCT-types 7 and 8) and sequence data (FCT-types 1–6 and 9), the gene content of the FCT regions is as follows: the genes coding for the pilus backbone protein (BP, FctA) and the sortases as well as that encoding a transcriptional regulator (either *nra* or *rofA*) are present in all the nine FCT regions; *sfb1* coding for a protein F fibronectin-binding MSCRAMM is lacking in FCT-type 3 and FCT-type 6; the gene coding for the collagen binding MSCRAMM Cpa, also named AP-1 (ancillary protein 1) is present in all but FCT-type 9 and 7, while that coding for AP-2 (ancillary protein 2) also known as FctB lacks only in FCT-type 1; *sipA/lepA* coding for a putative signal peptidase is present in the FCT-type regions 2, 3, 4, 7, 8; an additional transcriptional regulator coded by an *araC/xylS*-type gene is exclusively present in the FCT-type regions 3, 4, 7, 8; and finally, a gene coding a second fibronectin-binding MSCRAMM protein F2 (existing in two variants FbaA and FbaB) is found only in the FCT-types regions 3, 4, 6, 7 and 8 (Kratovac *et al.*, 2007; Falugi *et al.*, 2008).

13.4.2 Sequence variability of S. pyogenes pilus components

To evaluate sequence diversity among the FCT regions, Falugi *et al.* carried out a protein variability analysis of the pilus components (backbone and ancillary proteins) in 57 *S. pyogenes* strains that were representative of 23 epidemiologically relevant M-types. The analysis was performed only on seven out of the nine FCT-type regions, since strains carrying the FCT types 7 and 8 were not present among those analysed (Falugi *et al.*, 2008).

In detail, the 57 backbone proteins of the seven FCT-type regions were classified into 15 variants sharing <90% amino acid identity (within each variant the amino-acid sequence identity was >97%). The highest percentage of similarity was found among variants present in the same FCT-type regions. The variants could then be grouped into six distinct clusters (corresponding to FCT1, 2, 5, 6 and 9), with the backbone protein variants of FCT-3 and 4 clustering together (51–81% identity). Interestingly, the amino-acid identity among FCT-type variants other than FCT-3 and 4 was less than 30%.

Sequence analysis of the pilus AP1 revealed the existence of 14 major variants sharing <90% identity. Noteworthy, the AP1 variants clustered similarly to the backbone proteins with each cluster corresponding to specific FCT-type region (intra-cluster protein identity <32%) with variants of FCT-3 and 4 clustering together.

Sequence analysis of the genes coding for the pilus AP2 identified the existence of

five variants, with high conservation within each FCT-type (one variant for each FCT region) and AP2 of FCT-3 and FCT-4 clustering together (>96% identity).

13.4.3 Distribution of the FCT regions

In order to evaluate the global distribution of the FCT regions, a strong effort has been made to correlate the FCT-type with M type and the T antigen classification.

In particular, given the correspondence between T types and 'pilus variants', an extensive FCT sequence analysis paired with T typing has found that the 15 backbone protein variants account for all but four of the 21 T serotypes (Falugi *et al.*, 2008). Indeed, the antibodies against the pilus backbone proteins were found to be the major players in promoting agglutination and determining T typing. Out of the four T sera that do not recognize any of the backbone protein variants, two (3/13/B3264 and 8/25/Imp19) induce unspecific agglutination, while the other two (T18 and T22) have not been tested so far. Based on these data, Falugi *et al.* (2008) proposed a multiplex PCR amplification of specific regions in the genes coding for the backbone proteins (*tee* typing) as an alternative to T typing by agglutination test with sera, for which different centres had reported discrepant results.

With respect to the correlation between FCT regions and M types, sequence analysis highlighted that one or more M types are associated with specific FCT-type regions, while strains belonging to the same M type share almost identical pilus protein variants (Falugi *et al.*, 2008).

Several global and population-based epidemiological studies have demonstrated that M types, and more precisely *emm* pattern genotypes (defined based on the number of *emm* genes coding for the M proteins, their ancestral lineages and their chromosomal arrangements), can serve as a reliable genetic marker for tissue site preferences among *S. pyogenes* causing infection. Specific *emm* patterns (A–C) show strong preference for throat infections, the *emm* pattern D is more associated with skin infections, while the *emm* pattern E strains, so called generalists, lack a strong tissue site preference. Thus, given the linkage among FCT regions and M types, Kratovac *et al.* (2007) tried to establish a correlation between FCT region diversity and *emm* patterns. Indeed, pili (which are among the FCT region gene products) are good candidates to have key roles in pathogenesis by mediating adherence to host tissues and to trigger, in this way, host-tissue specificity. They found that the *emm* pattern A–C strains (throat specialists) display the most diversity in FCT region genotypes; in fact the *emm* A–C strains correspond to six FCT types (1, 2, 3, 4, 5, 7), with no single FCT region form corresponding to more than 25% of the strains. In contrast, many of the *emm* pattern D strains (skin specialists) are associated to FCT3 and most pattern E strains are restricted to two FCT region forms (FCT-4 and FCT-5). If gene products of the entire FCT region are critical for infection at a specific tissue site, then *emm* A–C strains should rely upon a variety of strategies to cause throat infection, while *emm* pattern D strains may be more dependent on a single mechanism (Kratovac *et al.*, 2007).

13.4.4 Pilus components of *S. pyogenes* as protective antigens

The pili of *S. pyogenes* mediate cell adhesion and microcolony formation on the surface of host epithelia and indeed, sera from pharyngitis patients contain antibodies to pilus components (Manetti *et al.*, 2007, 2010; Bombaci *et al.*, 2009). Therefore, GAS pilins are expected to be promising vaccine candidates. However, the high variability of pilus components (see above) clearly represents the most evident drawback for their use in a GAS vaccine.

Probably due to the high number of protein variants, systematic studies considering all the pilus backbone and ancillary proteins for their immunogenicity and protective ability have not been reported so far. Indeed, the only report indicating that immunization of mice with a combination of recombinant pilus proteins confers significant protection against mucosal challenge with virulent *S. pyogenes* bacteria comes from Mora

et al. (2005). They show how immunization with the combination of the backbone and the two ancillary proteins is able to protect against homologous intranasal challenge with a M1-FCT-type 2 strain at levels comparable with immunization with the M1 protein.

The establishment of the correlation between M types and FCT-type regions allows, even based on only M-type classification, the prediction of the prevalence of pilus variants (FCT regions) in specific M-typed isolate collection, and ultimately to estimate the coverage of a vaccine based on pilus components. As an example, Falugi *et al.*, based on epidemiologic data accumulated in surveillance studies in Europe and the USA (Shulman *et al.*, 2004; Creti *et al.*, 2007), estimated that a vaccine comprising a combination of 12 pilus backbone protein variants (corresponding to 12 distinct T antigens) could theoretically protect against more than 90% of currently circulating GAS strains. Indeed, this vaccine could be effective against 24 out of 27 predominant M types in a global *S. pyogenes* isolate collection and potentially interfere with GAS colonization because of the role of GAS pili in promoting bacterial adhesion to host tissues (Falugi *et al.*, 2008).

13.5 *Streptococcus suis* pili

13.5.1 Genomic organization of *S. suis* pilus islands

The availability of growing numbers of complete genomes of several bacterial species has led to the identification of pilus islands in other streptococcal pathogens, including *S. suis*. This Gram-positive coccus is responsible for a wide range of diseases in pigs, including meningitis, septicaemia and endocarditis, and can also affect humans in close contact with diseased pigs or swine products (Staats *et al.*, 1997; Lee *et al.*, 2008; Takamatsu *et al.*, 2008). Recently, at least four distinct putative *S. suis* pilus gene clusters were identified in sequenced genomes and sequences obtained from several clinical isolates (Takamatsu *et al.*, 2009) (Fig. 13.1). All four clusters (named *srtBCD*, *srtE*, *srtF* and *srtG* clusters according to class C sortase genes in the respective locus) contain either one or three class C sortase genes as well as several genes encoding cell-wall anchor family proteins containing C-terminal cell-wall sorting signals (CWSSs). In addition to three sortase genes, the *srtBCD* cluster contains genes encoding for four putative pilin structural proteins termed *sbp1* (for srtB-associated protein1), *sbp2*, *sbp3*, *sbp4*. By contrast, the *srtE*, *srtF* and *srtG* clusters consist of only one sortase gene and two putative pilin subunit genes each, called *sep1*, *sep2*; *sfp1*, *sfp2*; and *sgp1*, *sgp2*, in the *srtE*, *srtF* and *srtG* clusters, respectively. In *srtE* and *srtF* clusters, genes encoding signal peptidase homologues (named *sipE* and *sipF* [for signal peptidase gene in the *srtE* and *srtF* clusters], respectively) were also found (Takamatsu *et al.*, 2009).

Genetic organizations of the *srtBCD* and *srtG* clusters were similar to those of the abovementioned *S. pneumoniae* PI-1 and *S. pyogenes* FCT-1 region, respectively (Telford *et al.*, 2006). On the other hand, genetic organizations of the *srtE* and *srtF* clusters were similar to that of PI-2b of *S. agalactiae* COH1 (Telford *et al.*, 2006; Garibaldi *et al.*, 2010). In support of the hypothesis that the identified pilus gene clusters in *S. suis* mediate the expression of pilus-like polymers on the cell surface, experimental evidence was recently produced on the *srtF* and *srtG* clusters (Fittipaldi *et al.*, 2010a; Okura *et al.*, 2011). By generating mutant strains for each gene coding for the putative pilin proteins and by immunoblot and immunogold electron microscopy analysis using antibodies against the pilin subunits, the authors demonstrated that *srtF* and *srtG* pilus types can be assembled on the cell surface of several highly virulent invasive *S. suis* isolates, according to the current mechanism of Gram-positive sortase-mediated pilus assembly (Fittipaldi *et al.*, 2010a; Okura *et al.*, 2011).

13.5.2 Prevalence of the putative pilus gene clusters among the *S. suis* population

The distribution of putative pilus gene clusters has been evaluated in the *S. suis* population by Takamatsu *et al.* (2009). A total of 108 *S. suis* isolates (21 from healthy pigs, 60 from diseased pigs and 27 from human patients) were PCR-analysed for the presence of pilus-associated genes belonging to the four putative pilus clusters. None of the isolates bore all pilus islands. The *srtF* cluster was the most prevalent among the isolates tested, present in nearly 90% of the strains from diseased pigs and humans, and 78.7% of the isolates were positive for all genes in the cluster. On the other hand, *srtBCD* and *srtG* clusters were moderately prevalent among the population, and 27.8% and 58.3% of the isolates were positive for all genes in the *srtBCD* and *srtG* clusters, respectively. In contrast, the intact *srtE* cluster was rarely detectable. Although *srtE* and *sipE* genes were amplified from 94.4% of the isolates tested, *sep1* and *sep2* were detected in only three isolates from healthy pigs (Takamatsu *et al.*, 2009). On the basis of the presence or absence of putative pilus-associated genes by PCR analysis, the authors classified the 108 *S. suis* isolates into 12 genotypes (genotypes A–L) and demonstrated a strong correlation between pilus-associated gene profiles and ST complexes (Takamatsu *et al.*, 2009).

13.5.3 Immunoprotective activity of *S. suis* pilus subunits

The first evidence showing the immuno-protective potential of *S. suis* pilin subunit came from a recent study from Garibaldi *et al.* (2010) demonstrating the protective activity of a pilin subunit encoded by *srtF* pilus cluster in a murine infection model. This protein was identified by a proteomic approach that selectively detected surface components and showed a high level of sequence similarity with the ancillary protein 1 of *S. agalactiae* PI-2b (AP1-2b). The authors demonstrated that immunization with recombinant fragments of this protein, designated as pilus ancillary protein of island 2b or PAPI-2b, markedly protected mice from systemic *S. suis* infection (Garibaldi *et al.*, 2010). Considering that, as described above (Takamatsu *et al.*, 2009), the *srtF* pilus cluster is present in nearly 90% of the strains analysed in that study, these data open the way to the potential use of recombinant pilus subunits for the development of a protein vaccine against this pathogen. However, further studies are required to verify the immunoprotective effects of pilus components of the other clusters, to assess the percentage of clinical strains in which pilus clusters actually mediate pilus formation and to determine the degree of sequence conservation of the pilus subunits. Moreover, additional investigations will be needed to understand the biological significance and/or advantage of pilus expression in *S. suis* pathogenesis.

13.6 Conclusions

The build-up over a very short period (from 2005 to date) of experimental evidence related to the presence and the characterization of pilus structures in different Gram-positive microorganisms represents an example of the amazing impact of genomics in accelerating the discovery of previously unknown functions. The demonstration that in the human streptococcal pathogens (*S. pneumoniae*, *S. pyogenes*, *S. agalactiae*) as well as in livestock-associated species (i.e. *S. suis*) pili are composed of protective antigens and therefore potentially useful as vaccine components strengthens the importance of understanding the role of these structures in bacterial virulence/pathogenesis (Table 13.1).

To date, all streptococcal pili have been ascribed with adhesive properties; however, the tropisms for different biological niches associated with each streptococcal species, as well as the diverse distribution and expression patterns observed for the pilus-encoding islets within and among the Streptococci, strongly suggest that these intriguing appendages could have different roles in the bacterial life cycle within the host. Although the last few years have seen huge progress in the understanding of structure, assembly and function of these long filamentous structures, several questions still remain open. For instance, the notion of pili being essential for streptococcal virulence has not yet

been systematically evaluated from an epidemiological perspective. In this respect, *S. pneumoniae* pili seem not to be critical for virulence in invasive isolates, because the protective pneumococcal pilus (pilus-1) has been found only in about 30% of invasive clinical isolates (Basset *et al.*, 2007; Aguiar *et al.*, 2008; Moschioni *et al.*, 2008). Similar evidence has been reported also for *S. suis* pili (Fittipaldi *et al.*, 2010b), while all *S. agalactiae* and *S. pyogenes* strains contain at least one islet coding for a pilus. The fact that all pilus components display a certain degree of protein diversity (in particular the backbone protein) supports the idea that this variability might be the result of positive selection due to immune-evasion mechanisms triggered by the high accessibility and immunogenicity of these structures.

Noteworthy, allowed by the increasing availability of whole genome sequences, genetic variability studies performed in Mitis-group Streptococci highlighted suggestive examples of inter-species horizontal gene transfer. In fact, Zähner *et al.* (2011) recently identified *S. pneumoniae* PI-2-like islets in the commensal oral streptococci *S. oralis*, *S. mitis* and *S. sanguinis*; in addition, they showed in representative strains that pili encoded by these islets are morphologically similar to the pneumococcal PI-2 pilus. Interestingly, in all Mitis-group strains analysed, the PI-2 islet was always inserted in the same genomic site between the *pepT-hemH* genes, as reported for *S. pneumoniae*. On the other hand, the analysis of *S. pyogenes* FCT regions revealed examples of both inter- and intra-species recombination. In fact, the two ancillary proteins and the two sortases of the FCT-6 region share 92–93% and 88–90% identity, respectively, with those of group B *Streptococcus* pilus island PI-1. In addition, although the FCT-type regions are extremely variable in their gene composition, in terms of number and order of the encoded genes, all nine variants identified are inserted exclusively between two highly conserved genes, *hsp33* and *spy0136*. A similar observation can be made for the PI-2 locus in *S. agalactiae*. In fact, the two PI-2a and PI-2b variants are alternatively present in the same genomic site and a PI-2 pilus-encoding island

Table 13.1. Protective efficacy of *Streptococcus* pilus antigen.

Antigen name	Specifications	Immunogenic characteristics	Limitations	References
Streptococcus pneumoniae				
RrgB321	Fusion protein of the three variants (I, II, III) of RrgB, PI-1 backbone subunit	Protective efficacy demonstrated against all PI-1-positive strains in murine sepsis and pneumonia models; induces opsonophagocytic antibodies	Coverage (30–50%)	Gianfaldoni *et al.* (2007); Harfouche *et al.* (2011); Moschioni *et al.* (2012)
RrgA	PI-1 major ancillary protein	Cross-protection demonstrated in sepsis mouse models between the two RrgA variants; induces opsonophagocytic antibodies	Coverage (30–50%)	Gianfaldoni *et al.* (2007); Moschioni *et al.* (2010)
Streptococcus agalactiae				
GBS 80 (BP-1)	PI-1 backbone protein	Protective efficacy demonstrated against PI-1-expressing strains in active maternal mouse immunization/neonatal pup challenge model; induces opsonophagocytic antibodies	Coverage (30%)	Maione *et al.* (2005)
GBS 104 (AP1-1)	PI-1 major ancillary protein	Induces protection against AP1-1-expressing strains in active maternal mouse model	Coverage (30%)	Maione *et al.* (2005)
GBS 67 (AP1-2a)	PI-2a major ancillary protein	Cross-protection between the two AP1-2a variants demonstrated against AP1-2a-expressing strains in animal models; induces opsonophagocytic antibodies	Coverage (70%)	Maione *et al.* (2005); Margarit *et al.* (2009)
GBS 59 (BP-2a)	PI-2a backbone protein	Protective efficacy demonstrated against PI-2a-expressing strains in mouse models; elicits high bactericidal antibody titres	Coverage due to the high variability and variant-specific protection	Rosini *et al.* (2006); Margarit *et al.* (2009)
BP-2b	PI-2b backbone protein	Induces protection against PI-2b-expressing strains in active maternal mouse model	Coverage (<30%)	Margarit *et al.* (2009)

Antigen name	Specifications	Immunogenic characteristics	Limitations	References
AP1-2b	PI-2b major ancillary protein	Induces protection against AP1-2b-expressing strains in active maternal mouse model	Coverage (<30%)	Margarit *et al.* (2009)
BP-1+BP-2b+AP1-2a	Combination of the backbone proteins of PI-1 and PI-2a and major ancillary protein of PI-2a	Confers broad protection in animal models; estimated coverage >94%		Margarit *et al.* (2009)
BP-2a 6xD3	Fusion protein of D3 domain of the six variants of BP-2a, PI-2a backbone subunit	Protective efficacy demonstrated against strains expressing all BP-2a variants in murine models; induces opsonophagocytic antibodies	Coverage (70%), due to PI-2a-specific protection	Nuccitelli *et al.* (2011)
Streptococcus pyogenes				
Cpa+M1_128+ M1_130	Combination of the backbone (128) and major (Cpa) and minor (130) ancillary proteins of a FCT-2 pilus (M1-type strain)	Protective efficacy comparable to the homologous M1 protein in a intranasal challenge mouse model	Coverage due to the high variability and the number of FCT regions	Mora *et al.* (2005)
Streptococcus suis				
PAPI-2b	Ancillary pilin subunit encoded by the srtF pilus cluster	Immunoprotective activity demonstrated in a murine infection model	Coverage still to be determined (srtF pilus cluster distribution and sequence conservation assessment)	Garibaldi *et al.* (2009)

has been detected in all GBS strains analysed so far.

A further confirmation of the wide distribution of pilus structures in Gram-positive bacteria is represented by the recent identification of pilus appendages on the surface of strains of *S. gallolyticus*, a causative agent of infective endocarditis, associated with colon cancer. In *S. gallolyticus*, genome analysis revealed the existence of three pilus loci (named *pil1*, *pil2* and *pil3*). The experimental evidence that Pil1 is involved in the development of endocarditis identifies pili as the first virulence factor documented in this emerging pathogen (Danne *et al.*, 2011).

In conclusion, broader epidemiological as well as further functional characterization studies will be necessary to shed light on the effective contribution of pili in bacterial virulence in Gram-positive bacteria, thus supporting the choice of pilus components in the development of vaccines against these pathogenic microorganisms.

References

Aguiar, S.I., Serrano, I., Pinto, F.R., Melo-Cristino, J. and Ramirez, M. (2008) The presence of the pilus locus is a clonal property among pneumococcal invasive isolates. *BMC Microbiology* 8, 41.

Bagnoli, F., Moschioni, M., Donati, C., Dimitrovska, V., Ferlenghi, I., Facciotti, C., Muzzi, A., Giusti, F., Emolo, C., Sinisi, A., Hilleringmann, M., Pansegrau, W., Censini, S., Rappuoli, R., Covacci, A., Masignani, V. and Barocchi, M.A. (2008) A second pilus type in *Streptococcus pneumoniae* is prevalent in emerging serotypes and mediates adhesion to host cells. *Journal of Bacteriology* 190, 5480–5492.

Barocchi, M.A., Ries, J., Zogaj, X., Hemsley, C., Albiger, B., Kanth, A., Dahlberg, S., Fernebro, J., Moschioni, M., Masignani, V., Hultenby, K., Taddei, A.R., Beiter, K., Wartha, F., von Euler, A., Covacci, A., Holden, D.W., Normark, S., Rappuoli, R. and Henriques-Normark, B. (2006) A pneumococcal pilus influences virulence and host inflammatory responses. *Proceedings of the National Academy of Sciences of the USA* 103, 2857–2862.

Basset, A., Trzcinski, K., Hermos, C., O'Brien, K.L., Reid, R., Santosham, M., McAdam, A.J., Lipsitch, M. and Malley, R. (2007) Association of the pneumococcal pilus with certain capsular serotypes but not with increased virulence. *Journal of Clinical Microbiology* 45, 1684–1689.

Basset, A., Turner, K.H., Boush, E., Sayeed, S., Dove, S.L. and Malley, R. (2011) Expression of the type 1 pneumococcal pilus is bistable and negatively regulated by the structural component RrgA. *Infection and Immunity* 79, 2974–2983.

Bessen, D.E. and Kalia, A. (2002) Genomic localization of a T serotype locus to a recombinatorial zone encoding extracellular matrix-binding proteins in *Streptococcus pyogenes*. *Infection and Immunity* 70, 1159–1167.

Bombaci, M., Grifantini, R., Mora, M., Reguzzi, V., Petracca, R., Meoni, E., Balloni, S., Zingaretti, C., Falugi, F., Manetti, A.G., Margarit, I., Musser, J.M., Cardona, F., Orefici, G., Grandi, G. and Bensi, G. (2009) Protein array profiling of tic patient sera reveals a broad range and enhanced immune response against group A *Streptococcus* antigens. *PLoS ONE* 4, e6332.

Buccato, S., Maione, D., Rinaudo, C.D., Volpini, G., Taddei, A.R., Rosini, R., Telford, J.L., Grandi, G. and Margarit, I. (2006) Use of *Lactococcus lactis* expressing pili from group B *Streptococcus* as a broad-coverage vaccine against strepto-coccal disease. *Journal of Infectious Diseases* 194, 331–340.

Creti, R., Imperi, M., Baldassarri, L., Pataracchia, M., Recchia, S., Alfarone, G. and Orefici, G. (2007) *emm* Types, virulence factors, and antibiotic resistance of invasive *Streptococcus pyogenes* isolates from Italy: what has changed in 11 years? *Journal of Clinical Microbiology* 45, 2249–2256.

Cunningham, M.W. (2000) Pathogenesis of group A streptococcal infections. *Clinical Microbiology Reviews* 13, 470–511.

Danne, C., Entenza, J.M., Mallet, A., Briandet, R., Debarbouille, M., Nato, F., Glaser, P., Jouvion, G., Moreillon, P., Trieu-Cuot, P. and Dramsi, S. (2011) Molecular characterization of a *Streptococcus gallolyticus* genomic island encoding a pilus involved in endocarditis. *Journal of Infectious Diseases* 204, 1960–1970.

Darkes, M.J. and Plosker, G.L. (2002) Pneumococcal conjugate vaccine (Prevnar; PNCRM7): a review of its use in the prevention of *Streptococcus pneumoniae* infection. *Paediatric Drugs* 4, 609–630.

De Angelis, G., Moschioni, M., Muzzi, A., Pezzicoli, A., Censini, S., Delany, I., Lo, S.M., Sinisi, A., Donati, C., Masignani, V. and Barocchi, M.A. (2011) The *Streptococcus pneumoniae* pilus-1 displays a biphasic expression pattern. *PLoS ONE* 6, e21269.

Dramsi, S., Caliot, E., Bonne, I., Guadagnini, S., Prevost, M.C., Kojadinovic, M., Lalioui, L., Poyart, C. and Trieu-Cuot, P. (2006) Assembly and role of pili in group B streptococci. *Molecular Microbiology* 60, 1401–1413.

El Mortaji, L., Contreras-Martel, C., Moschioni, M., Ferlenghi, I., Manzano, C., Vernet, T., Dessen, A. and Di Guilmi, A.M. (2011) The full-length *Streptococcus pneumoniae* major pilin RrgB crystallizes in a fiber-like structure, which presents the D1 isopeptide bond and provides details on the mechanism of pilus polymerization. *Biochemical Journal* 441, 833–841.

Enright, M.C. and McKenzie, H. (1997) *Moraxella (Branhamella) catarrhalis* – clinical and molecular aspects of a rediscovered pathogen. *Journal of Medical Microbiology* 46, 360–371.

Facklam, R., Beall, B., Efstratiou, A., Fischetti, V., Johnson, D., Kaplan, E., Kriz, P., Lovgren, M., Martin, D., Schwartz, B., Totolian, A., Bessen, D., Hollingshead, S., Rubin, F., Scott, J. and Tyrrell, G. (1999) *emm* typing and validation of provisional M types for group A streptococci. *Emerging Infectious Diseases* 5, 247–253.

Falugi, F., Zingaretti, C., Pinto, V., Mariani, M., Amodeo, L., Manetti, A.G., Capo, S., Musser, J.M., Orefici, G., Margarit, I., Telford, J.L.,

Grandi, G. and Mora, M. (2008) Sequence variation in group A *Streptococcus* pili and association of pilus backbone types with lancefield T serotypes. *Journal of Infectious Diseases* 198, 1834–1841.

Fittipaldi, N., Takamatsu, D., de la Cruz Dominguez-Punaro, M., Lecours, M.P., Montpetit, D., Osaki, M., Sekizaki, T. and Gottschalk, M. (2010a) Mutations in the gene encoding the ancillary pilin subunit of the *Streptococcus suis* srtF cluster result in pili formed by the major subunit only. *PLoS ONE* 5, e8426.

Fittipaldi, N., Takamatsu, D., de la Cruz Dominguez-Punaro, M., Lecours, M.P., Montpetit, D., Osaki, M., Sekizaki, T. and Gottschalk, M. (2010b) Mutations in the gene encoding the ancillary pilin subunit of the *Streptococcus suis* srtF cluster result in pili formed by the major subunit only. *PLoS ONE* 5, e8426.

Fletcher, M.A. and Fritzell, B. (2007) Brief review of the clinical effectiveness of PREVENAR against otitis media. *Vaccine* 25, 2507–2512.

Garibaldi, M., Rodriguez-Ortega, M.J., Mandanici, F., Cardaci, A., Midiri, A., Papasergi, S., Gambadoro, O., Cavallari, V., Teti, G. and Beninati, C. (2010) Immunoprotective activities of a *Streptococcus suis* pilus subunit in murine models of infection. *Vaccine* 28, 3609–3616.

Gentile, M.A., Melchiorre, S., Emolo, C., Moschioni, M., Gianfaldoni, C., Pancotto, L., Ferlenghi, I., Scarselli, M., Pansegrau, W., Veggi, D., Merola, M., Cantini, F., Ruggiero, P., Banci, L. and Masignani, V. (2011) Structural and functional characterization of the *Streptococcus pneumoniae* RrgB pilus backbone D1 domain. *Journal of Biological Chemistry* 286, 14588–14597.

Gianfaldoni, C., Censini, S., Hilleringmann, M., Moschioni, M., Facciotti, C., Pansegrau, W., Masignani, V., Covacci, A., Rappuoli, R., Barocchi, M.A. and Ruggiero, P. (2007) *Streptococcus pneumoniae* pilus subunits protect mice against lethal challenge. *Infection and Immunity* 75, 1059–1062.

Gutekunst, H., Eikmanns, B.J. and Reinscheid, D.J. (2003) Analysis of RogB-controlled virulence mechanisms and gene repression in *Streptococcus agalactiae*. *Infection and Immunity* 71, 5056–5064.

Harfouche, C., Filippini, S., Gianfaldoni, C., Ruggiero, P., Moschioni, M., Maccari, S., Pancotto, L., Arcidiacono, L., Galletti, B., Censini, S., Mori, E., Giuliani, M., Facciotti, C., Cartocci, E., Savino, S., Doro, F., Pallaoro, M., Nocadello, S., Mancuso, G., Haston, M., Goldblatt, D., Barocchi, M.A., Pizza, M., Rappuoli, R. and Masignani, V. (2011) RrgB321, a fusion protein of the three variants of the pneumococcal pilus backbone RrgB, is protective *in vivo* and elicits opsonic antibodies. *Infection and Immunity* 80, 451–460.

Hilleringmann, M., Giusti, F., Baudner, B.C., Masignani, V., Covacci, A., Rappuoli, R., Barocchi, M.A. and Ferlenghi, I. (2008) Pneumococcal pili are composed of protofilaments exposing adhesive clusters of Rrg A. *PLoS Pathogens* 4, e1000026.

Hilleringmann, M., Ringler, P., Muller, S.A., De Angelis, G., Rappuoli, R., Ferlenghi, I. and Engel, A. (2009) Molecular architecture of *Streptococcus pneumoniae* TIGR4 pili. *EMBO Journal* 28, 3921–3930.

Imai, S., Ito, Y., Ishida, T., Hirai, T., Ito, I., Maekawa, K., Takakura, S., Iinuma, Y., Ichiyama, S. and Mishima, M. (2009) High prevalence of multidrug-resistant Pneumococcal molecular epidemiology network clones among *Streptococcus pneumoniae* isolates from adult patients with community-acquired pneumonia in Japan. *Clinical Microbiology and Infection* 15, 1039–1045.

Imai, S., Ito, Y., Ishida, T., Hirai, T., Ito, I., Yoshimura, K., Maekawa, K., Takakura, S., Niimi, A., Iinuma, Y., Ichiyama, S. and Mishima, M. (2011) Distribution and clonal relationship of cell surface virulence genes among *Streptococcus pneumoniae* isolates in Japan. *Clinical Microbiology and Infection* 17, 1409–1414.

Johnson, D.R. and Kaplan, E.L. (1993) A review of the correlation of T-agglutination patterns and M-protein typing and opacity factor production in the identification of group A streptococci. *Journal of Medical Microbiology* 38, 311–315.

Johri, A.K., Paoletti, L.C., Glaser, P., Dua, M., Sharma, P.K., Grandi, G. and Rappuoli, R. (2006) Group B *Streptococcus*: global incidence and vaccine development. *Nature Reviews Microbiology* 4, 932–942.

Kim, K.S. (2010) Acute bacterial meningitis in infants and children. *Lancet Infectious Diseases* 10, 32–42.

Konto-Ghiorghi, Y., Mairey, E., Mallet, A., Dumenil, G., Caliot, E., Trieu-Cuot, P. and Dramsi, S. (2009) Dual role for pilus in adherence to epithelial cells and biofilm formation in *Streptococcus agalactiae*. *PLoS Pathogens* 5, e1000422.

Kratovac, Z., Manoharan, A., Luo, F., Lizano, S. and Bessen, D.E. (2007) Population genetics and linkage analysis of loci within the FCT region of *Streptococcus pyogenes*. *Journal of Bacteriology* 189, 1299–1310.

Kreikemeyer, B., Gamez, G., Margarit, I., Giard, J.C., Hammerschmidt, S., Hartke, A. and Podbielski, A. (2011) Genomic organization, structure, regulation and pathogenic role of pilus constituents in major pathogenic Streptococci and Enterococci. *International Journal of Medical Microbiology* 301, 240–251.

Krishnan, V., Gaspar, A.H., Ye, N., Mandlik, A., Ton-That, H. and Narayana, S.V. (2007) An IgG-like domain in the minor pilin GBS52 of *Streptococcus agalactiae* mediates lung epithelial cell adhesion. *Structure* 15, 893–903.

Lancefield, R.C. and Dole, V.P. (1946) The properties of T antigens extracted from group A hemolytic Streptococci. *Journal of Experimental Medicine* 84, 449–471.

Lauer, P., Rinaudo, C.D., Soriani, M., Margarit, I., Maione, D., Rosini, R., Taddei, A.R., Mora, M., Rappuoli, R., Grandi, G. and Telford, J.L. (2005) Genome analysis reveals pili in Group B *Streptococcus*. *Science* 309, 105.

Lee, G.T., Chiu, C.Y., Haller, B.L., Denn, P.M., Hall, C.S. and Gerberding, J.L. (2008) *Streptococcus suis* meningitis, United States. *Emerging Infectious Diseases* 14, 183–185.

LeMieux, J., Hava, D.L., Basset, A. and Camilli, A. (2006) RrgA and RrgB are components of a multisubunit pilus encoded by the *Streptococcus pneumoniae* rlrA pathogenicity islet. *Infection and Immunity* 74, 2453–2456.

Maione, D., Margarit, I., Rinaudo, C.D., Masignani, V., Mora, M., Scarselli, M., Tettelin, H., Brettoni, C., Iacobini, E.T., Rosini, R., D'Agostino, N., Miorin, L., Buccato, S., Mariani, M., Galli, G., Nogarotto, R., Nardi Dei, V, Vegni, F., Fraser, C., Mancuso, G., Teti, G., Madoff, L.C., Paoletti, L.C., Rappuoli, R., Kasper, D.L., Telford, J.L. and Grandi, G. (2005) Identification of a universal Group B streptococcus vaccine by multiple genome screen. *Science* 309, 148–150.

Maisey, H.C., Hensler, M., Nizet, V. and Doran, K.S. (2007) Group B streptococcal pilus proteins contribute to adherence to and invasion of brain microvascular endothelial cells. *Journal of Bacteriology* 189, 1464–1467.

Maisey, H.C., Quach, D., Hensler, M.E., Liu, G.Y., Gallo, R.L., Nizet, V. and Doran, K.S. (2008) A group B streptococcal pilus protein promotes phagocyte resistance and systemic virulence. *FASEB Journal* 22, 1715–1724.

Manetti, A.G., Zingaretti, C., Falugi, F., Capo, S., Bombaci, M., Bagnoli, F., Gambellini, G., Bensi, G., Mora, M., Edwards, A.M., Musser, J.M., Graviss, E.A., Telford, J.L., Grandi, G. and Margarit, I. (2007) *Streptococcus pyogenes* pili promote pharyngeal cell adhesion and biofilm formation. *Molecular Microbiology* 64, 968–983.

Manetti, A.G., Koller, T., Becherelli, M., Buccato, S., Kreikemeyer, B., Podbielski, A., Grandi, G. and

Margarit, I. (2010) Environmental acidification drives *S. pyogenes* pilus expression and microcolony formation on epithelial cells in a FCT-dependent manner. *PLoS ONE* 5, e13864.

Margarit, I., Rinaudo, C.D., Galeotti, C.L., Maione, D., Ghezzo, C., Buttazzoni, E., Rosini, R., Runci, Y., Mora, M., Buccato, S., Pagani, M., Tresoldi, E., Berardi, A., Creti, R., Baker, C.J., Telford, J.L. and Grandi, G. (2009) Preventing bacterial infections with pilus-based vaccines: the group B streptococcus paradigm. *Journal of Infectious Diseases* 199, 108–115.

Martin, D., Rioux, S., Gagnon, E., Boyer, M., Hamel, J., Charland, N. and Brodeur, B.R. (2002) Protection from group B streptococcal infection in neonatal mice by maternal immunization with recombinant Sip protein. *Infection and Immunity* 70, 4897–4901.

Mora, M., Bensi, G., Capo, S., Falugi, F., Zingaretti, C., Manetti, A.G., Maggi, T., Taddei, A.R., Grandi, G. and Telford, J.L. (2005) Group A *Streptococcus* produce pilus-like structures containing protective antigens and Lancefield T antigens. *Proceedings of the National Academy of Sciences of the USA* 102, 15641–15646.

Moschioni, M., Donati, C., Muzzi, A., Masignani, V., Censini, S., Hanage, W.P., Bishop, C.J., Reis, J.N., Normark, S., Henriques-Normark, B., Covacci, A., Rappuoli, R. and Barocchi, M.A. (2008) *Streptococcus pneumoniae* contains 3 rlrA pilus variants that are clonally related. *Journal of Infectious Diseases* 197, 888–896.

Moschioni, M., De Angelis, G., Melchiorre, S., Masignani, V., Leibovitz, E., Barocchi, M.A. and Dagan, R. (2009) Prevalence of pilus encoding islets among acute otitis media *Streptococcus pneumoniae* isolates from Israel. *Clinical Microbiology and Infection* 16, 1501–1504.

Moschioni, M., Emolo, C., Biagini, M., Maccari, S., Pansegrau, W., Donati, C., Hilleringmann, M., Ferlenghi, I., Ruggiero, P., Sinisi, A., Pizza, M., Norais, N., Barocchi, M.A. and Masignani, V. (2010) The two variants of *Streptococcus pneumoniae* pilus-1 RrgA adhesin retain the same function and elicit cross-protection *in vivo*. *Infection and Immunity* 78, 5033–5042.

Moschioni, M., De Angelis, G., Harfouche, C., Bizzarri, E., Filippini, S., Mori, E., Mancuso, G., Doro, F., Barocchi, M.A., Ruggiero, P. and Masignani, V. (2012) Immunization with the RrgB321 fusion protein protects mice against both high and low pilus-expressing *Streptococcus pneumoniae* populations. *Vaccine* 30, 1349–1356.

Nelson, A.L., Ries, J., Bagnoli, F., Dahlberg, S., Falker, S., Rounioja, S., Tschop, J., Morfeldt, E., Ferlenghi, I., Hilleringmann, M., Holden, D.W.,

Rappuoli, R., Normark, S., Barocchi, M.A. and Henriques-Normark, B. (2007) RrgA is a pilus-associated adhesin in *Streptococcus pneumoniae*. *Molecular Microbiology* 66, 329–340.

Nuccitelli, A., Cozzi, R., Gourlay, L.J., Donnarumma, D., Necchi, F., Norais, N., Telford, J.L., Rappuoli, R., Bolognesi, M., Maione, D., Grandi, G. and Rinaudo, C.D. (2011) Structure-based approach to rationally design a chimeric protein for an effective vaccine against Group B *Streptococcus* infections. *Proceedings of the National Academy of Sciences of the USA* 108, 10278–10283.

Okura, M., Osaki, M., Fittipaldi, N., Gottschalk, M., Sekizaki, T. and Takamatsu, D. (2011) The minor pilin subunit Sgp2 is necessary for assembly of the pilus encoded by the srtG cluster of *Streptococcus suis*. *Journal of Bacteriology* 193, 822–831.

Papasergi, S., Brega, S., Mistou, M.Y., Firon, A., Oxaran, V., Dover, R., Teti, G., Shai, Y., Trieu-Cuot, P. and Dramsi, S. (2011) The GBS PI-2a pilus is required for virulence in mice neonates. *PLoS ONE* 6, e18747.

Paterson, N.G. and Baker, E.N. (2011) Structure of the full-length major pilin from *Streptococcus pneumoniae*: implications for isopeptide bond formation in gram-positive bacterial pili. *PLoS ONE* 6, e22095.

Pelton, S.I. and Leibovitz, E. (2009) Recent advances in otitis media. *Pediatric Infectious Disease Journal* 28, S133–S137.

Pezzicoli, A., Santi, I., Lauer, P., Rosini, R., Rinaudo, D., Grandi, G., Telford, J.L. and Soriani, M. (2008) Pilus backbone contributes to group B *Streptococcus* paracellular translocation through epithelial cells. *Journal of Infectious Diseases* 198, 890–898.

Rappuoli, R., Black, S. and Lambert, P.H. (2011a) Vaccine discovery and translation of new vaccine technology. *Lancet* 378, 360–368.

Rappuoli, R., Mandl, C.W., Black, S. and De Gregorio, E. (2011b) Vaccines for the twenty-first century society. *Nature Reviews Immunology* 11, 865–872.

Regev-Yochay, G., Hanage, W.P., Trzcinski, K., Rifas-Shiman, S.L., Lee, G., Bessolo, A., Huang, S.S., Pelton, S.I., McAdam, A.J., Finkelstein, J.A., Lipsitch, M. and Malley, R. (2010) Re-emergence of the type 1 pilus among *Streptococcus pneumoniae* isolates in Massachusetts, USA. *Vaccine* 28, 4842–4846.

Rinaudo, C.D., Rosini, R., Galeotti, C.L., Berti, F., Necchi, F., Reguzzi, V., Ghezzo, C., Telford, J.L., Grandi, G. and Maione, D. (2010) Specific involvement of pilus type 2a in biofilm formation in group B *Streptococcus*. *PLoS ONE* 5, e9216.

Rosini, R., Rinaudo, C.D., Soriani, M., Lauer, P., Mora, M., Maione, D., Taddei, A., Santi, I., Ghezzo, C., Brettoni, C., Buccato, S., Margarit, I., Grandi, G. and Telford, J.L. (2006) Identification of novel genomic islands coding for antigenic pilus-like structures in *Streptococcus agalactiae*. *Molecular Microbiology* 61, 126–141.

Sadowy, E., Kuch, A., Gniadkowski, M. and Hryniewicz, W. (2010) Expansion and evolution of the *Streptococcus pneumoniae* Spain9V-ST156 clonal complex in Poland. *Antimicrobial Agents and Chemotherapy* 54, 1720–1727.

Shulman, S.T., Tanz, R.R., Kabat, W., Kabat, K., Cederlund, E., Patel, D., Li, Z., Sakota, V., Dale, J.B. and Beall, B. (2004) Group A streptococcal pharyngitis serotype surveillance in North America, 2000–2002. *Clinical Infectious Diseases* 39, 325–332.

Siira, L., Jalava, J., Tissari, P., Vaara, M., Kaijalainen, T. and Virolainen, A. (2011) Clonality behind the increase of multidrug-resistance among non-invasive pneumococci in Southern Finland. *European Journal of Clinical Microbiology & Infectious Diseases* 31, 867–871.

Sjostrom, K., Blomberg, C., Fernebro, J., Dagerhamn, J., Morfeldt, E., Barocchi, M.A., Browall, S., Moschioni, M., Andersson, M., Henriques, F., Albiger, B., Rappuoli, R., Normark, S. and Henriques-Normark, B. (2007) Clonal success of piliated penicillin non-susceptible pneumococci. *Proceedings of the National Academy of Sciences of the USA* 104, 12907–12912.

Sorensen, U.B., Poulsen, K., Ghezzo, C., Margarit, I. and Kilian, M. (2010) Emergence and global dissemination of host-specific *Streptococcus agalactiae* clones. *MBio* 1.

Soriani, M. and Telford, J.L. (2010) Relevance of pili in pathogenic streptococci pathogenesis and vaccine development. *Future Microbiology* 5, 735–747.

Spraggon, G., Koesema, E., Scarselli, M., Malito, E., Biagini, M., Norais, N., Emolo, C., Barocchi, M.A., Giusti, F., Hilleringmann, M., Rappuoli, R., Lesley, S., Covacci, A., Masignani, V. and Ferlenghi, I. (2010) Supramolecular organization of the repetitive backbone unit of the *Streptococcus pneumoniae* pilus. *PLoS ONE* 5, e10919.

Staats, J.J., Feder, I., Okwumabua, O. and Chengappa, M.M. (1997) *Streptococcus suis*: past and present. *Veterinary Research Communications* 21, 381–407.

Takamatsu, D., Wongsawan, K., Osaki, M., Nishino, H., Ishiji, T., Tharavichitkul, P., Khantawa, B., Fongcom, A., Takai, S. and Sekizaki, T. (2008)

Streptococcus suis in humans, Thailand. *Emerging Infectious Diseases* 14, 181–183.

Takamatsu, D., Nishino, H., Ishiji, T., Ishii, J., Osaki, M., Fittipaldi, N., Gottschalk, M., Tharavichitkul, P., Takai, S. and Sekizaki, T. (2009) Genetic organization and preferential distribution of putative pilus gene clusters in *Streptococcus suis*. *Veterinary Microbiology* 138, 132–139.

Tanaka, D., Gyobu, Y., Kodama, H., Isobe, J., Hosorogi, S., Hiramoto, Y., Karasawa, T. and Nakamura, S. (2002) *emm* typing of group A streptococcus clinical isolates: identification of dominant types for throat and skin isolates. *Microbiology and Immunology* 46, 419–423.

Telford, J.L., Barocchi, M.A., Margarit, I., Rappuoli, R. and Grandi, G. (2006) Pili in gram-positive pathogens. *Nature Reviews Microbiology* 4, 509–519.

Ton-That, H. and Schneewind, O. (2003) Assembly of pili on the surface of *Corynebacterium diphtheriae*. *Molecular Microbiology* 50, 1429–1438.

Ton-That, H. and Schneewind, O. (2004) Assembly of pili in Gram-positive bacteria. *Trends in Microbiology* 12, 228–234.

Turner, P., Melchiorre, S., Moschioni, M., Barocchi, M.A., Turner, C., Watthanaworawit, W., Kaewcharernnet, N., Nosten, F. and Goldblatt, D. (2011) Assessment of *Streptococcus pneumoniae* pilus islet-1 prevalence in carried and transmitted isolates from mother–infant pairs on the Thailand–Burma border. *Clinical Microbiology and Infection* 18, 970–975.

Vainio, A., Kaijalainen, T., Hakanen, A.J. and Virolainen, A. (2011) Prevalence of pilus-encoding islets and clonality of pneumococcal isolates from children with acute otitis media. *European Journal of Clinical Microbiology & Infectious Diseases* 30, 515–519.

van der Poll, T. and Opal, S.M. (2009) Pathogenesis, treatment, and prevention of pneumococcal pneumonia. *Lancet* 374, 1543–1556.

Zahner, D., Gudlavalleti, A. and Stephens, D.S. (2010) Increase in pilus islet 2-encoded pili among *Streptococcus pneumoniae* isolates, Atlanta, Georgia, USA. *Emerging Infectious Diseases* 16, 955–962.

Zahner, D., Gandhi, A.R., Yi, H. and Stephens, D.S. (2011) Mitis group streptococci express variable pilus islet 2 pili. *PLoS ONE* 6, e25124.

Index

Printed and bound by CPI Group (UK) Ltd, Croydon, CR0 4YY

11/01/2026

14804844-0003